# Hot Cracking Phenomena in Welds II

Thomas Böllinghaus · Horst Herold ·
Carl E. Cross · John C. Lippold (Eds.)

# Hot Cracking Phenomena in Welds II

Springer

Prof. Dr.-Ing. Thomas Böllinghaus
Bundesanstalt für
Materialforschung und–prüfung
Unter den Eichen 87
12205 Berlin
Germany
Thomas.Boellinghaus@bam.de

Prof. Dr. Carl E. Cross
Bundesanstalt für
Materialforschung und–prüfung
Unter den Eichen 87
12205 Berlin
Germany
Carl-Edward.Cross@bam.de

Prof. Dr.-Ing. habil. Dr. E. h. Horst Herold
Otto-von-Guericke Universität
Postfach 4120
39016 Magdeburg
Germany
Horst.Herold@mb.uni-magdeburg.de

Prof. John C. Lippold
The Ohio State University
1248 Arthur E. Adams Drive
Columbus, Ohio 43221-3560
USA
Lippold.1@osu.edu

ISBN: 978-3-540-78627-6 e-ISBN: 978-3-540-78628-3

Library of Congress Control Number: 2005921916



*Cover design:* deblik, Berlin

Printed on acid-free paper

9 8 7 6 5 4 3 2 1

springer.com

# Preface

Failure of welded components can occur during service as well as during fabrication. Most common, analyses of the resistance of welded components against failure are targeted at crack avoidance. Such evaluations are increasingly carried out by modern weldability studies, i.e. considering interactions between the selected base and filler materials, structural design and welding process. Such weldability investigations are particularly targeted to prevent hot cracking, as one of the most common cracking phenomena occurring during weld fabrication.

To provide an international information and discussion platform to combat hot cracking, an international workshop on Hot Cracking Phenomena in Welds has been created, based on an initiative of the Institute for Materials and Joining Technology at the Otto-von-Guericke University in Magdeburg and the Division V.5 – Safety of Joined Components at the Federal Institute for Materials Research and Testing (BAM) in Berlin, Germany. The first workshop was organized in Berlin under the topics mechanisms and phenomena, metallurgy and materials, modelling and simulations as well as testing and standardization. It consisted of 20 individual contributions from eight countries, which were compiled in a book that found a very ready market, not only in the welding community. As a consequence of increasing interest, it has been decided to establish the Workshop on Hot Cracking Phenomena in Welds as a regular event every three years embedded in the International Institute of Welding (IIW). Attached to the IIW Commission IX and II Spring intermediate meetings, the second workshop was organized in March 2007.

The present book assembles 22 papers from 10 different countries, which have again been published without any length restrictions on individual contributions. The authors have attached great importance to highlight the very recent developments and thus, the present book particularly compiles the major worldwide hot cracking research advancements of the latest years. In recent years, the research interest is increasingly attracted to elucidation of the mechanisms, which differ widely with the type of hot cracking. The book has thus been divided correspondingly to the workshop sessions into chapters referring to the various types of hot cracking, i.e. solidification cracking, liquation cracking and ductility dip cracking.

It can only be emphasized that in combination with the previous book, this edition represents a helpful tool for metallurgical, materials and mechanical engineering students of higher semesters. For all welding scientists and experts who want to get acquainted to the subject, the state of knowledge can very easily be drawn from the various chapters instead of gathering the necessary information piecewise from elsewhere.

The editors convey their sincere gratitude to all authors, session chairs and participants of the second workshop for their engaged contributions and for establishing a regular forum for exchanging the major research advancements in hot cracking phenomena of welds. Special thanks go to Ms. A. Cichon and Ms. I. Schülke for the tremendous work in organizing the workshop.

Berlin, Magdeburg and Columbus
May 2008

Thomas Böllinghaus
Horst Herold
Carl Cross
John Lippold

# Contents

## Part IV: Liquation Cracking



## Part V: Ductility-Dip Cracking

# Part I
# Solidification Cracking Theory

# In Search of the Prediction of Hot Cracking in Aluminium Alloys

L. Katgerman, D.G. Eskin

Delft University of Technology, Netherlands Institute for Metals Research, Delft, The Netherlands

## Abstract

Hot tearing remains a major problem of casting technology despite decades-long efforts to develop a working hot tearing criterion and to implement it into casting process computer simulation.

Existing models allow one to calculate the stress–strain situation in a casting (ingot, billet) and to compare it with the chosen hot tearing criterion. Two kinds of hot tearing criteria are available in literature: mechanical and non-mechanical one. The mechanical criteria of hot tearing are derived based on mechanical behaviour semi-solid, and the non-mechanical one is based on other properties of semi-solid. In most successful cases, the simulation shows a relative probability of hot tearing and the sensitivity of this probability to such process parameters as casting speed, casting dimensions, and casting practice.

None of the existing criteria, however, can give the quantitative answer on whether the hot crack will appear or not and what will be the extent of hot cracking (position, length, shape). This chapter outlines the requirements for a modern hot tearing criterion as well as the future development of hot tearing research in terms of mechanisms of hot crack nucleation and propagation.

## Introduction – Mechanisms of Hot Tearing

Various defects of as-cast product are still frequently encountered in casting practice. One of the main defects is hot tearing or hot cracking, or hot shortness. Irrespective of the name, this phenomenon represents the formation of an irreversible failure (crack) in the still semi-solid casting.

From many studies [1, 2, 3, 4, 5, 6, 7, 8] started already in the 1950’s, and reviewed by Novikov [9] and Sigworth [10], it appears that hot tears initiate above the solidus temperature and propagate in the interdendritic liquid film. In the course of solidification, the liquid flow through the mushy zone decreases until it becomes insufficient to fill initiated cavities so that they can grow further. The fracture has a bumpy surface covered with a smooth layer and sometimes with solid bridges that connect or have connected both sides of the crack [7, 8, 11, 12, 13, 14, 15, 16].

Research studies show that hot tearing occurs in the late stages of solidification when the volume fraction of solid is above 85–95% and the solid phase is organized in a continuous network of grains. It is also known that fine grain structures and controlled casting (without large temperature and stress gradients) help to avoid hot cracking.

During direct-chill (DC) casting of aluminium alloys, primary and secondary cooling cause strong thermal gradients in the billet/ingot, resulting in uneven thermal contraction in different sections of the billet/ingot. As a result, macroscopic stresses cause distortion of the billet/ingot shape (e.g. butt curl and swell, rolling face pull-in) and/or may trigger hot tearing and cold cracking in the weak sections. The terms “hot” or “cold” refer to the temperature range where the cracking occurs – in the semi-solid mushy zone or below the solidus, respectively. In DC casting, the name “mushy zone” is frequently applied to the entire transition region between liquidus and solidus, which is misleading, as the semi-solid mixture in the top part of the transition region is actually a slurry. Only after the temperature has dropped below the coherency temperature, a real mush is formed. On the microscopic level solidification shrinkage and thermal contraction impose strains and stresses on the solid network in the mushy zone. The deformation behaviour of the mush is very critical for the formation of hot tears. The link between the appearance of hot tears and the mechanical properties in the semi-solid state is obvious and has been explored for decades; see for example reviews [9, 17].

Another important correlation – between the hot cracking susceptibility and the composition of an alloy – has been established on many occasions. A large freezing range of an alloy promotes hot tearing since such an alloy spends a longer time in the vulnerable state in which thin liquid films exist.

A lot of efforts have been devoted to the understanding of the hot tearing phenomenon. Compilations of research in this field have been done by Novikov [9], Sigworth [10], and Eskin et al. [17]. Several mechanisms of hot tearing are already suggested in literature. Some of those are outlined in Table 1.

**Table 1.** Summary of hot tearing mechanisms

| Mechanism | Suggested and developed by* | Ref.** |
|---|---|---|
| ***Cause of hot tearing*** | | |
| Thermal contraction | Heine (1935); Pellini (1952); Dobatkin (1948) | [18, 2, 19] |
| Liquid film distribution | Verö (1936) | [20] |
| Liquid pressure drop | Prokhorov (1962); Niyama (1977) | [37, 39] |
| Vacancy supersaturation | Fredriksson et al. (2005) | [21] |
| ***Nucleation*** | | |
| Liquid film or pore as stress concentrator | Patterson et al. (1953, 1967); Niyama (1977); Rappaz et al. (1999); Braccini et al. (2000); Suyitno et al. (2002) | [44, 22, 39, 38, 23, 48] |
| Oxide bi-film entrained in the mush | Campbell (1991) | [8] |
| Vacancy clusters at a grain boundary or solid/liquid interface | Fredriksson et al. (2005) | [21] |
| ***Propagation*** | | |
| Through liquid film by sliding | Patterson (1953); Williams & Singer (1960, 1966); Novikov & Novik (1963) | [44, 24, 25] |
| By liquid film rupture | Pellini (1952); Patterson (1953); Saveiko (1961); Dickhaus (1994) | [2, 44, 26, 27] |
| By liquid metal embrittlement | Novikov (1966); Sigworth (1996) | [9, 10] |
| Through liquid film or solid phase depending on the temperature range | Guven & Hunt (1988) | [28] |
| Diffusion of vacancies from the solid to the crack | Fredriksson et al. (2005) | [21] |
| ***Conditions*** | | |
| Thermal strain cannot be accommodated by liquid flow and mush ductility | Pellini (1952); Prokhorov (1962); Novikov (1966); Magnin et al. (1996) | [2, 37, 9, 29] |
| Pressure drop over the mush reaches a critical value for cavity nucleation | Niyama (1977); Guven & Hunt (1988); Rappaz et al. (1999); Farup & Mo (2000) | [39, 28, 38, 41] |
| Strain rate reaches a critical value that cannot be compensated by liquid feeding and mush ductility | Pellini (1952); Prokhorov (1962); Rappaz et al. (1999); Braccini et al. (2000) | [2, 37, 38, 23] |
| Thermal stress exceed rupture or local critical stress | Lees (1946); Langlais & Grizleski, (2000); Lahaie & Bauchard (2001); Suyitno et al. (2002) | [30, 31, 32, 48] |
| Stresses and insufficient feeding in the vulnerable temperature range | Bochvar (1942); Lees (1946); Pumphrey & Lyons (1948); Clyne & Davies (1975); Feurer (1977); Katgerman (1982) | [33, 30, 34, 35, 40, 36] |
| Thermal stress exceeds rupture stress of the liquid film | Saveiko (1961) | [26] |

* The list of the authors is by no means complete. The references have been chosen to represent the development of ideas.

** References are given in the same order as the authors in the second column.

One can see that, over the years, much more effort has been put on the conditions required for hot tearing occurrence rather than on the mechanisms of crack initiation and propagation. And when it comes to the nucleation and propagation of hot tears, an educated guess frequently replaces experimental proof.

Over the years, different macroscopic parameters, such as stress and strain, were considered as critical for the development of hot tearing. Today, the strain rate is believed to be the most important factor and some modern models are based on it. The physical explanation of this approach is that semi-solid material during solidification can accommodate the imposed thermal strain by plastic deformation, diffusion-aided creep, structure re-arrangement, and filling of the gaps and pores with the liquid. All these processes require some time, and the lack of time will result in fracture. Therefore, there exists the maximum strain rate that the semi-solid material can endure without fracture during solidification. Prokhorov [37] was the first to suggest a criterion based on this approach. More recently, an elaborate, strain-rate based hot tearing criterion was proposed by Rappaz et al. [38].

On the microscopic level, the important factor is believed to be the feeding of the solid phase with the liquid. Within this approach, the hot tear will not occur as long as there is no lack of feeding during solidification. Niyama [39] and Feurer [40] use hindered feeding as a base for their porosity and hot tearing criteria. The feeding depends on the permeability of the mush, which is largely determined by the structure. Later, a two-phase model of the semi-solid dendritic network [41], which focuses on the pressure depression in the mushy zone, has been suggested to describe the hot tear formation. This approach treats the semi-solid material as solid and liquid phases with different levels of stress and strain. The pressure drop of the liquid phase in the mush is considered as a cause of a hot tear. An extension of the two-phase model that includes plasticity of the porous network is also reported [42]. Logically, bridging and grain coalescence, which determine the transfer of stress and limit the permeability of the mush, are the other important microscopic factors for the development of hot tearing [43].

Yet the hot tearing theories that operate with macroscopic mechanical behaviour and microscopic phenomena like feeding and porosity formation do not take into account the mechanism of crack nucleation and propagation and, in this, are intrinsically weak. The other approach to the description of the hot tearing phenomenon is the application of fracture mechanics that describes initiation and propagation of cracks. The liquid film surrounding the grain at late stages of solidification is considered as a stress concentrator of the semi-solid body [44, 45]. In this theory, a liquid-filled cavity acts as a crack initiator. The propagation of the crack is determined by the critical stress [45]. The critical stress can be estimated using the Griffith energy balance approach modified by taking into account the plasticity [46, 47]. Recently, a formulation of hot tearing as a phenomenon related to micro-porosity has been proposed [48]. In this approach the porosity and the hot tearing are considered as sequential events. As a result, there is a possibility to predict simultaneously the occurrence of micro-porosity and hot tearing. The model uses the feeding difficulties at the last stage of solidification as a starting point of cavity nucleation. The nucleus then grows and becomes at the end of solidification either a micro-pore or a hot tear as determined by the Griffith model for brittle crack growth. The fracture-mechanical concept of hot tearing can be enriched with the application of a liquid-metal-embrittlement mechanism [10].

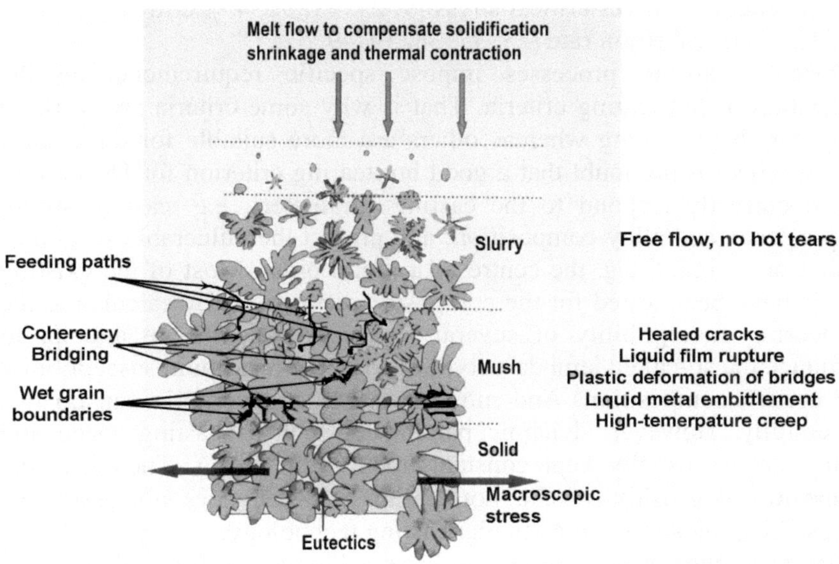


**Fig. 1**. Different length scales of equiaxed dendritic solidification along with suggested hot tearing mechanisms (the initial sketch is adopted from [49])

It is obvious that actual hot tearing mechanism includes phenomena occurring on two scales: microscopic (crack nucleation and propagation, stress concentration, structure coherency, wet grain boundaries, feeding) and macroscopic (stress, strain, or strain rate imposed on the structure). Figure 1 illustrates these scales during equiaxed dendritic solidification.

## Current Hot Tearing Criteria and their Applicability to DC Casting

The existing hot tearing criteria as reviewed elsewhere [17, 50] can be conditionally divided into the two categories: non-mechanical and mechanical. The former type of criteria deals with vulnerable temperature range, phase diagram, and process parameters; and is represented by the criteria of Clyne & Davies [6], Feurer [40], and Katgerman [36]. The latter type of criteria involves critical stress [9, 27, 31, 32, 45], critical strain [9, 29, 51], or critical strain rate [23, 37, 38, 39, 49, 52].

Different casting processes impose specific requirements on the application of hot tearing criteria. That is why some criteria are working better for shape casting whereas others are more suitable for direct-chill casting. There is no doubt that a good hot tearing criterion for DC casting should correctly respond to the casting parameters, e.g. casting speed, ramping rate, and alloy composition; and predict the vulnerable section of a billet or an ingot, e.g. the centre of a round billet. Most of the existing criteria have been tested for the composition sensitivity by calculating the hot tearing susceptibility of several binary alloys with an attempt to reproduce the so-called lambda-curve showing the maximum susceptibility at a certain composition. And most of the existing criteria can do this successfully. However, dynamic parameters such as casting speed and strain rate are usually kept constant upon such testing. Therefore, the compositional sensitivity of a hot tearing criterion does not assure its successful application to a particular casting technology.

The basic phenomena that lead to hot cracking are well established and understood, but a generic criterion that will predict hot cracking under varying process conditions is still not available. Although the earlier simple criteria based on the thermal history of the casting have been extended and improved to include shrinkage and deformation, they are still unable to give reliable predictions under all process conditions. Most of the existing hot tearing criteria do not incorporate the nucleation and propagation of a hot tear, focusing more on the macroscopic and microscopic conditions that may result in rupture.

The ultimate hot cracking criterion needs to combine aspects of thermal history, shrinkage and porosity formation, and constitutive behaviour in combination with the evolution of the semi-solid microstructure. Current research efforts are aimed, in particular, at the quantitative description of structure evolution and its correlation to cracking.

Recently, several mechanical and non-mechanical hot tearing criteria were evaluated by implementing them into a thermo-mechanical model of

DC casting [50]. The criteria show different results in predicting the hot tearing susceptibility as shown in Table 2.

**Table 2**. Sensitivity of hot tearing criteria to the casting parameters and practice upon direct-chill casting [50, 53]

| Criterion | Hot tearing increases with casting speed | More hot tears in the billet centre | Ramping casting speed during start-up of the casting reduces hot tearing | Correlation with actual cracking observed in practice |
|---|---|---|---|---|
| Clyne & Davies [6] | No | No | No | N/A |
| Katgerman [36] | Yes | Yes | No | N/A |
| Feurer [40] | Yes | Yes | No | N/A |
| Novikov [9] | No | No | No | N/A |
| Magnin et al. [29] | Yes | No | No | No |
| Prokhorov [37] | Yes | Yes | No | No |
| Rappaz et al. [38] | Yes | Yes | Yes | No |
| Braccini et al. [23] | Yes | Yes | No | N/A |
| Suyitno [48] | Yes | Yes | Yes | Yes |

The criteria of Clyne & Davies and Novikov give results that are inconsistent with casting practice, not showing any sensitivity to the casting speed and position within the billet volume. It is noteworthy to mention that these criteria are very successful in predicting the compositional dependence of hot tearing and are frequently used for shape casting. The criteria of Feurer, Katgerman, Magnin et al., Prokhorov, Rappaz et al., and Braccini et al. respond correctly to the casting parameters, demonstrating that the increasing casting speed results in an increasing hot tearing susceptibility in the centre of billet, which is in accordance with casting practice. However, most of the tested criteria, except those by Rappaz et al. [38] and Suyitno [48, 53], are not sensitive to the ramping of casting speed during the start-up phase of casting (which is a usual practice to prevent hot cracking). And, when confronted with casting practice, the criteria of Prokhorov, Magnin et al., and Rappaz et al. predict the occurrence of hot cracks whereas no cracks have been found in billets cast under given conditions. Only the criterion of Suyitno adequately responds to all tested parameters, i.e. casting speed, ramping rate, grain size, position in a billet, and casting practice. The sensitivity of this criterion is, however, a function of correctly chosen values of properties such as Young's modulus of the mush, surface tension between liquid and solid, and permeability of the mush. These parameters are scarcely available and needed to be determined experimentally, while the existing experimental techniques are not reliable.

## Outline of Hot Tearing Mechanisms as a Base for a New Hot Tearing Criterion

Up to now, hot tearing is considered as a phenomenon linked to casting and welding processes, and the existing theories, models and criteria are biased by their applicability to solidification. Hot tearing is, however, just another example of material failure. Therefore, it should be treated adequately, using the well-developed apparatus of fracture mechanics. The challenge nowadays lies not in the adequate description of macroscopic and microscopic stress–strain situations and their correspondence to the parameters and properties of the mushy zone, but rather in finding real factors causing the nucleation and propagation of a hot crack. In fact, some existing theories and models of hot tearing partially describe these factors with the crack initiator presented as a cavity filled with liquid or a pore [38, 48], or an oxide bi-film [8] and with the crack propagation path through the liquid film covering grain boundaries [10]. What is lacking is the completeness of the model. A comprehensive model and a based-on-it criterion should include nucleation and propagation of hot tears and connect these processes to the microstructure evolution during solidification of the semi-solid material; to the macroscopic and microscopic thermo-mechanical situation in the mushy zone; and to the mechanical (or fracture-mechanical) properties of the mushy zone. The last two components are well covered by a large body of publications, though many mechanical properties are still needed to be determined and the fracture mechanics potential has not been fully exploited. The correspondence between the structure evolution during solidification and the crack nucleation ad propagation is studied in much less detail. Let us consider the possible mechanisms of crack nucleation and propagation.

The mechanism of hot tear formation and propagation can be elucidated from observations of fractures. Unfortunately, the reports on hot crack fractures in metallic materials are rare. Most of such reports describe the cracking of semi-solid alloys at relatively large fractions of liquid, when grain boundaries are completely covered with liquid. In this case the mechanism of crack propagation – through liquid film by grain separation – is obvious. An example of such alloys is the classic Al–4% Cu alloy. However, alloys with high fractions of liquid in the vulnerable solidification range are in practice not susceptible to hot tearing [17, 54]. Cavities and gaps between grains that may form in the mushy zone of such alloys due to thermal contraction or external tension are easily filled with liquid due to the adequate permeability of the mushy zone and sufficient amount of available liquid that is represented in the final structure by non-equilibrium eutectics [54]. Much more important is the mechanism of crack formation and propagation in alloys containing little solute that are most susceptible to hot tearing. But the information on semi-solid fracture in such alloys is only starting to emerge. If one wanted to summarize the findings available to date, it would appear that bridging of grain boundaries is an essential feature of the fracture surface. Moreover, the closer the semi-solid material gets to the temperature range of its maximum vulnerability to hot cracking, i.e. 90–95% solid, the greater fraction of grain boundaries is connected to each other, or coalesced [55]. In this case, there is no chance for the crack propagation through a continuous liquid film as such a film does not exist. Our recent observations [56] show that a hot tear apparently propagates through the liquid film in more alloyed materials and through solid bridges in less alloyed materials as illustrated in Fig. 2. Although in both cases the fracture surface appears to be brittle, one can suggest that different cracking mechanisms are acting. In real castings, cracks appear above the solidus as well as below the solidus. In the latter case, these cracks are called "cold". There are reports, however, that these cracks, appearing at high temperatures in the fully solid material, can originate in areas of local remelting caused by intense local deformation [57]. Thus formed liquid also assists in crack propagation. These observations make such "cold" cracks in fact similar to hot tears.

The nucleation of hot cracks is an almost unexplored phenomenon. It is obvious that under any stress-strain conditions, there should be a certain, critical size of a defect (flaw, nucleus) that would enable crack growth. The problem of the crack initiation is today solved by an educated guess, as there are almost no experimental observations on the crack nucleation upon natural hot tearing. Usually, the development of a hot crack is studied on samples with a notch, hence – with the artificial crack initiator [43, 59]. Based on these observations and "post-mortem" examination of hot tear surfaces, the following crack nuclei have been suggested: (1) liquid film or liquid pool; (2) pore or series of pores; (3) grain boundary located in the place of stress concentration; (4) inclusions that can be easily separated from the surrounding liquid or solid phase, e.g. intermetallic particle or oxide film. In should be noted that pores that are frequently cited as potential hot tear nuclei can originate from gas precipitation [8], solidification shrinkage [8], or vacancy supersaturation [21]. Figure 3 summarizes some of the possible crack initiators.

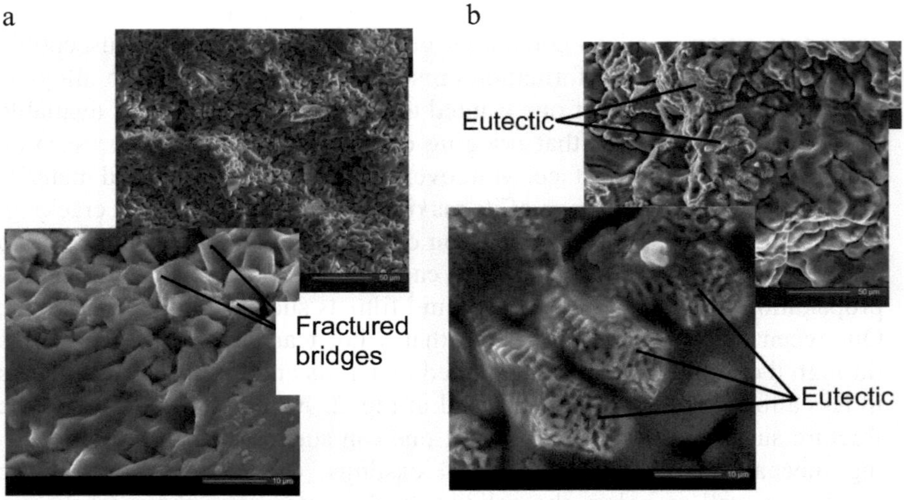


**Fig. 2**. Fracture surfaces of hot tears in 200-mm round billet produced by DC casting at a casting speed of 200 mm/min: (**a**) Al–1% Cu and (**b**) Al–3% Cu

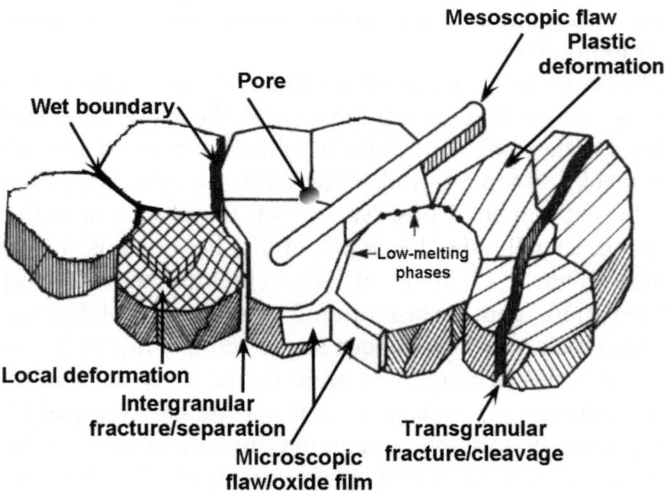


**Fig. 3**. Schematic illustration of possible hot crack initiators and some crack propagation mechanisms (the initial sketch is adopted from [58])

Despite a large body of literature on hot tearing, only few efforts have been spent on the mechanism of hot tear propagation. One can mention the application of the Griffith criterion for brittle fracture [48, 53] and vacancy-diffusion-controlled growth [21]. The apparatus for the description of crack propagation is well developed within fracture mechanics for various situations, including those resembling hot cracking, i.e. liquid film rupture, pore coalescence, high-temperature creep, and liquid-assisted fracture [58]. It is clear that the propagation of the hot crack in the potential presence of liquid (above or below the solidus) should involve the following aspects: liquid feeding (involves permeability and, inevitably structure evolution), pore coalescence, stress transfer by solid bridges, plastic deformation and creep of solid bridges in the absence of liquid, and brittle fracture of solid bridges in the presence of liquid. It is also obvious that the hot crack, like any other crack, can develop catastrophically (which is usually assumed), have sustained growth, or stop. Liquid feeding plays a dual role. Firstly, adequate feeding of the shrinking material with liquid does not eliminate the causes of hot tearing but rather "patches" the consequences, which is reflected in the term "crack healing". One can say that a semi-solid alloy containing enough liquid at the last stage of solidification, having a microstructure that enables adequate permeability of the mush, and subjected to tensile stresses is a self-healing material. On the other hand, the development of solid bridges between grains at high solid fractions in the absence of any liquid would build up enough strength and ductility to prevent any brittle rupture. Hence, the liquid feeding should be just enough to supply some liquid to solid bridges that enables their liquid embrittlement, otherwise only the mechanisms of ductile fracture, e.g. high-temperature creep and pore coalescence will be active [8]. The evidence of plastic deformation during hot tearing has been observed in direct observations [59] and upon examination of fracture surfaces [9].

We can suggest the following approach to treating the nucleation and propagation of hot cracks. Several distinct mechanisms are operational in different temperature or compositional ranges or, in other words, at different fractions of solid. Here we will consider only the decreasing temperature as the factor affecting the solid fraction, though the composition is obviously the other factor that acts in a similar manner. It is important to note that the microstructure, e.g. grain size and morphology, affects the critical temperatures and fractions of solid. At relatively low fractions of solid below the coherency temperature, the permeability of the mushy zone enables adequate feeding of the solidification shrinkage and most of the precipitating gas bubbles can float to the liquid part of the sample. In this case, the tensile stresses caused by non-uniform thermal contraction of the coherent dendrites may cause the formation of cavities and gaps that are immediately filled with liquid, or "healed". On further decreasing temperature, the tensile stress builds up to such an extend that the liquid film separating grains ruptures and the formed gap cannot be filled with liquid due to the already insufficient permeability of the mush and to the increasing capillary pressures required to fill ever narrowing openings between grains.

On further cooling, the bridging between solid grains replaces former entanglement and touching of grains, and the stress can now be transmitted over larger distances through the rigid solid skeleton, hence the semi-solid body acquires macroscopic strength. This critical temperature is called the "rigidity temperature" and can be determined experimentally [9, 60]. Note that the rigidity temperature strongly depends on the structure [60]. The permeability at such fractions of solid is not enough to completely compensate the solidification shrinkage, gas cannot escape from narrow passages and, as a result, porosity is formed at available interfaces, mainly on grain boundaries. The non-uniform thermal stress causes rather significant strains in the semi-solid material that can be or cannot be sustained by the solid bridges. Even limited access of the liquid to the solid bridge will result in its brittle fracture by the mechanism of liquid-metal embrittlement. This is partially reflected in the proposal of van Haaften et al. [61] to use the fraction of grain boundaries covered with liquid rather than the fraction of liquid in the constitutive equation for the mechanical behaviour or semi-solid aluminium alloys. The same mechanism may act at subsolidus temperatures when some amount of non-equilibrium liquid is present at grain boundaries or other stress concentrators because of non-equilibrium character of solidification [8] or local remelting [57]. There is also a possibility that the semi-solid material fails macroscopically in a brittle manner (because of film rupture and liquid-metal embrittlement) with ductile rupture of some solid bridges on the microscopic level [59]. And finally, the fraction of bridged grain boundaries becomes so overwhelmingly large and the remaining liquid is so scattered in the solid network that the semi-solid material behaves like completely solid and fails in a ductile manner by ductile pore coalescence and high-temperature creep. The crack initiator in all these cases could be represented by pore, liquid pool or film, interface with an intermetallic particle, or non-metallic inclusion. The outline of these mechanisms is given in Table 3. Figures 1 and 3 illustrate the correlation between these mechanisms and the development of the structure during solidification.

A criterion that can predict not the probability but the actual occurrence and extent of hot tearing should be based on the application of multi-phase mechanics and fracture mechanics to the failure of semi-solid materials, which is today limited by the lack of knowledge about the actual nucleation and propagation mechanisms. The mechanisms outlined in Table 3 are based on the common sense and interpretation of very few experimental observations. What is needed is a thorough and systematic study of fractures occurring in solidifying materials with the aim to single out the nature and the critical dimensions of defects or structure features that can cause the nucleation of hot cracks. It is also necessary, in our opinion, to acknowledge that different mechanisms of crack propagation are possible at different fractions of solid. Therefore, different models could be applied to the development of hot tears in ingots, billets and castings in dependence on the alloy composition, structure, and the level of stresses that are present. For example, coarse-grained material with a large solidification range and high coherency temperature is likely to fail due to liquid film rupture. In the case of alloys with a considerable amount of eutectics, e.g. foundry alloys of the Al–Si system, the healing of cracks is most probable. In contrast, a fine-grained alloy that develops coherency late in solidification and does not contain much of eutectics, e.g. a commercial wrought alloy, will undergo complex failure involving liquid-metal embrittlement and plastic deformation of solid bridges with a resultant mixed brittle/ductile fracture.

**Table 3**. Possible mechanisms acting within "hot tearing" phenomenon

| Temperature range/fraction of solid | Nucleation of crack* | Propagation of crack | Fracture mode |
|---|---|---|---|
| Between coherency and rigidity temperatures 50–80% solid | Grain boundary covered with liquid; shrinkage or gas pore. | a. Liquid film rupture<br>b. Filled gap | a. Brittle, intergranular<br>b. Healed crack |
| Below rigidity temperature 80–99% solid | Pore, surface of particle or inclusion, liquid film or pool, vacancy clusters | a. Liquid film rupture; liquid metal embrittlement of solid bridges<br>b. Plastic deformation of bridges | a. Brittle, intergranular<br>b. Plastic deformation of bridges possible |
| Close to the solidus 98–100% solid | Pore, surface of particle or inclusion, segregates at grain boundary, liquid at stress concentration point, vacancy clusters | a. Liquid metal embrittlement<br>b. Plastic deformation of bridges, creep | a. Brittle, transgranular propagation is possible<br>b. Macroscopically brittle or ductile, transgranular propagation is possible |

* The crack initiator should be located in the place of stress concentration.

## Concluding Remarks

The existing hot tearing criteria based on different principles have limited applicability to commercial casting processes, e.g. to direct-chill casting, due to their probabilistic character. The best of the available criteria can successfully predict the probability of hot tearing in its dependence on some casting parameters but fail to quantify the event, in other words cannot forecast the actual occurrence of hot cracks in ingots and billets. There are two main challenges in this endeavour. Firstly, we are lacking the knowledge of the actual causes of crack nucleation. That is to say we do not know exactly what defects or structure defects can act as crack initiators under particular temperature–stress conditions. Secondly, there is a possibility that different mechanisms of crack propagation and final failure are acting in dependence on the fraction of solid at which the fracture occurs and on the alloy structure. The application of multi-phase mechanics and, eventually, fracture mechanics to the phenomenon of hot cracking looks quite promising. The quest for a new hot tearing criterion should focus on these two research areas

## Acknowledgements

This chapter is written within the framework of the research program of the Netherlands Institute for Metals Research (www.nimr.nl), projects MP4.97014 and MC4.02134.

## References

1. H.F. Bishop, C.G. Ackerlind, and W.S. Pellini: Trans. Am. Foundrymen's Soc., 1952, vol. 60, pp. 818–833.
2. W.S. Pellini: Foundry, 1952, vol. 80, pp. 124–133, 192–199.
3. J.C. Borland: Brit. Weld. J., 1960, vol. 7, pp. 508–512.
4. S.A. Metz and M.C. Flemings: Trans. Am. Foundrymen's Soc., 1970, vol. 78, pp. 453–460.
5. U. Feurer: Giessereiforschung, 1976, vol. 28 (2), pp. 75–80.
6. T.W. Clyne and G.J. Davies: in Solidification and Casting of Metals, Metals Society, London, United Kingdom, 1979, pp. 275–278.
7. B. Rogberg: Scand. J. Met., 1983, vol. 12, pp. 51–66.
8. J. Campbell: Castings, Butterworth-Heinemann, Oxford, United Kingdom, 1st edn. 1991; 2nd edn., 2003.
9. I.I. Novikov: Goryachelomkost tsvetnykh metallov i splavov (Hot shortness of non-ferrous metals and alloys), Nauka, Moscow, USSR, 1966.
10. G.K. Sigworth: Trans. Am. Foundrymen's Soc., 1996, vol. 104, pp. 1053–1062.
11. J.A. Spittle and A.A. Cushway: Met. Technol., 1983, vol. 10, pp. 6–13.
12. L. Ohm and S. Engler: Giessereiforschung, 1990, vol. 42 (4), pp. 149–162.
13. M.L. Nedreberg: PhD Thesis, University of Oslo, Oslo, Norway, 1991.
14. J.A. Spittle, S.G.R. Brown, J.D. James, and R.W. Evans: in Proc 7th Int. Symp. on Physical Simulation of Casting, Hot Rolling and Welding, National Research Institute for Metals, Tsukuba, Japan, 1997, pp. 81–91.

15 W.-M. van Haaften, W.H. Kool, and L. Katgerman: in Continuous Casting, K. Ehrke and W. Schneider, ed., Wiley-VCH, Weinheim, Germany, 2000, pp. 239–244.
16. I. Farup, J.-M. Drezet, and M. Rappaz: Acta Mater., 2001, vol. 49, pp. 1261–1269.
17. D.G. Eskin, Suyitno, and L. Katgerman: Progr. Mater. Sci., 2004, vol. 49, 629–711.
18. R.W. Heine and P.C. Rosenthal, Principles of Metal Casting, McGraw-Hill, New York, USA, 1955.
19. V.I. Dobatkin: Nepreryvnoe lit'e I liteinye svoistva splavov (Direct-Chill Casting and Casting Properties of Alloys), Oborongiz, Moscow, Russia, 1948.
20. J. Verö: Met. Industry, 1936, vol. 48, pp. 431–494.
21. H. Fredriksson, M. Haddad-Sabzevar, K. Hansson, and J. Kron: Mater. Sci. Technol., 2005, vol. 21, pp. 521–529.
22. W. Patterson, S. Engler, and R. Kupfer: Giessereiforschung, 1967, vol. 19 (3), pp. 151–160.
23. M. Braccini, C.L. Martin, M. Suéry, and Y. Bréchet: in Modelling of Casting, Welding and Advanced Solidification processes IX, P.R. Sahm, P.N. Hansen, and J.G. Conley, ed., Shaker Verlag, Aachen, Germany, 2000, pp. 18–24.
24. J.A. Williams and A.R.E. Singer: Austral. Inst. Met., 1966, vol. 11, pp. 2–9.
25. I.I. Novikov and F.S. Novik: Dokl. Akad. Nauk SSSR, Ser. Fiz., 1963, vol. 7, pp. 1153.
26. V.N. Saveiko: Russ. Castings Production, 1961, (11), pp. 453–456.
27. C.H. Dickhaus, L. Ohm, and S. Engler: Trans. Am. Foundrymen's Soc., 1994, vol. 101, pp. 677–684.
28. Y.F. Guven and J.D. Hunt: Cast Met., 1988, vol. 1, pp. 104–111.
29. B. Magnin, L. Maenner, L. Katgerman, and S. Engler: Mater. Sci. Forum, 1996, vol. 217–222, pp. 1209–1214.
30. D.C.J. Lees: J. Inst. Met., 1946, vol. 72, pp. 343–364.
31. J. Langlais and J.E. Gruzleski: Mater. Sci. Forum, 2000, vol. 167, pp. 31–337.
32. D.J. Lahaie and M. Bouchard: Metall. Mater. Tras. B., 2001, vol. 32B, pp. 697–705.
33. A.A. Bochvar: Izv. Akad. Nauk SSSR, Otdel. Tekhn. Nauk, 1942, (9), p. 31.
34. W.I. Pumphrey and J.V. Lyons: J. Inst. Met., 1948, vol. 74, pp. 439–455.
35. T.W. Clyne and G.J. Davies: British Foundrymen, 1975, vol. 68, pp. 238–244.
36. L. Katgerman: J. Met., 1982, vol. 34 (2), pp. 46–49.
37. N.N. Prokhorov: Russ. Castings Production, 1962, (2), pp. 172–175.
38. M. Rappaz, J.-M. Drezet, and M. Gremaud: Metall. Mater. Trans. A, 1999, vol. 30A, pp. 449–455.
39. E. Niyama: in Japan–US Joint Seminar on Solidification of Metals and Alloys, Japan Society for Promotion of Science, Tokyo, Japan, 1977, pp. 271–282.
40. U. Feurer: in Quality Control of Engineering Alloys and the Role of Metals Science, H. Nieswaag and J.W. Schut, ed., Delft University of Technology, Delft, The Netherlands, 1977, pp. 131–145.
41. I. Farup and A. Mo: Metall. Mater. Trans. A, 2000, vol. 31A, pp. 1461–1472.

42. M. M'Hamdi and A. Mo: in Light Metals 2002, W. Schneider, ed., The Minerals, Metals, and Materials Society, Warrendale, USA, 2002, pp. 709–716.
43. M. Rappaz, P.-D. Grasso, V. Mathier, J.-M. Drezet, and A. Jacot: in Solidification of Aluminum Alloys, M.G. Chu, D.A. Granger, and Q. Han, ed., The Minerals, Metals, and Materials Society, Warrendale, USA, 2005, pp. 179–190.
44. K. Patterson: Giesserei, 1953, vol. 40 (12), pp. 597–605.
45. J.A. Williams and A.R.E. Singer: J. Inst. Met., 1968, vol. 96. pp. 5–12.
46. J.J. Gilman: in Proc of 2nd Symp. on Naval Structural Mechanics, E.H. Lee and P.S. Symonds, ed., Pergamon, London, United Kingdom, 1960, pp. 43–99.
47. E. Orowan: in Fatigue and Fracture of Metals, W.M. Murray, ed., Massachusetts Institute of Technology, MIT Press, Cambridge, USA, 1950, p. 139.
48. Suyitno, W.H. Kool, and L. Katgerman: Mater. Sci. Forum, 2002, vol. 396–402, pp. 179–184.
49. J.F. Grandfield, D.J. Cameron, and J.A. Taylor: in Light Metals 2001, J.L. Anjier, ed., The Minerals, Metals, and Materials Society, Warrendale, USA, 2001, pp. 895–901.
50. Suyitno, W.H. Kool, and L. Katgerman: Metall. Mater. Trans. A, 2005, vol. 36A, pp. 1537–1546.
51. L. Zhao, Baoyin, N. Wang, V. Sahajwalla, and R.D. Pehlke: Int. J. Cast Metals Res., 2000, vol. 13 (3), pp. 167–174.
52. J.-M. Drezet and M. Rappaz: in Light Metals 2001, J.L. Anjier, ed., The Minerals, Metals, and Materials Society, Warrendale, USA, 2001, pp. 887–893.
53. Suyitno: PhD Thesis, Delft University of Technology, Delft, The Netherlands, 2005.
54. Suyitno, D.G. Eskin, V.I. Savran and L. Katgerman: Metall. Mater. Trans. A, 2004, vol. 35A, pp. 3551–3561.
55. Y. Ju. PhD Thesis, Norwegian University of Science and Technology, Trondheim, Norway, 2004.
56. D. Eskin, Suyitno, and L. Katgerman: in Aluminium Cast House Technology 2005, Proc. 9th Australasian Conference, J. Taylor, I. Bainbridge, and J. Granfield, ed., Collingwood: CSIRO Publishing, 2005, pp. 77–84.
57. A. Deschamps, S. Péron, Y. Bréchet, J.-C. Ehrström, and L. Poizat: Mater. Sci. Technol., 2002, vol. 18, pp. 1085–1091.
58. M. Janssen, J. Zuidema, and R.J.H. Wanhill: Fracture Mechanics. Delft University Press Blue Print, Delft, The Netherlands, 2002.
59. W.-M. van Haaften, W.H. Kool, and L. Katgerman. J. Mater. Eng. Perform., 2002, vol. 11, pp. 537–543.
60. D.G. Eskin, Suyitno, J.F. Mooney, and L. Katgerman: Metall. Mater. Trans. A, 2004, vol. 35A, pp. 1325–1335.
61. W.-M. van Haaften, W.H. Kool, and L. Katgerman: Mater. Sci. Eng. A, 2002, vol. 336, pp. 1–6.

# Application of the Rappaz-Drezet-Gremaud Hot Tearing Criterion to Welding of Aluminium Alloys

J.-M. Drezet[1], D. Allehaux[2]

[1] Federal Polytechnic School of Lausanne, Materials Simulation Laboratory, Lausanne, Switzerland
[2] EADS Corporate Research Centre-Paris, Suresnes, France

## Abstract

Hot tearing, a severe defect occurring during solidification is the conjunction of tensile stresses which are transmitted to the mushy zone by the coherent solid underneath and of insufficient liquid feeding to compensate for the volumetric change. The RDG (Rappaz Drezet Gremaud) criterion for the appearance of hot tears in metallic alloys [1] is based upon a mass balance performed over the liquid and solid phases and accounts for the tensile deformation of the solid skeleton perpendicular to the growing dendrites and for the induced interdendritic liquid feeding. When tackling the problem of hot tearing in welding of aluminium alloys, the RDG criterion can be used at three levels of increasing complexity by:

- ranking the alloy with regards to their sensitivity to hot cracking
- studying the risk of hot tearing in the process using only the thermal field (thermal criterion),
- and studying the influence of the mechanical behaviour of the mushy alloy on the risk of hot cracking (thermo-mechanical criterion).

Each level is illustrated by an example dealing with laser beam welding. Nevertheless, one of the critical issues in the RDG approach is the definition of a coherency point which, in low-concentration alloys, corresponds to the bridging or coalescence of the primary phase. To tackle this aspect, a 2D granular model is presented together with preliminary results.

## Introduction

Hot tearing together with microporosity have always been recognized as major defects in castings of aluminium alloys, the first one being typical of semi-continuous casting whereas the second one is dominant in shape castings. These two defects are also present in welds. As mentioned by Campbell [2], they are interconnected, as both result from a lack of feeding and nucleation of a pore/void in the remaining liquid. However, if porosity is associated with solidification shrinkage and can occur within the grains or at grain boundaries, hot tearing is clearly linked with tensile stresses in the solid and is confined to grain boundaries.

An excellent review of all the hot tearing models has been published by Eskin et al. [3] in 2004. Recent developments made both at the macroscopic and microscopic levels, have pointed out the difficulties in connecting these two scales. Macroscopic approaches are based nowadays on two-phase approaches [4, 5, 6]: they are the natural extension of the so-called RDG criterion [1], but using a more rigorous rheological approach for the compressible mushy solid phase. The main advantage of such methods is that they can take into account the whole scale of the parts to be welded, but their weakness is that averages can hardly account for the localization of strains and feeding at grain boundaries. On the microscopic scale, coalescence and percolation of grains can be accounted for at the scale of a small volume element of the mushy zone (typically a few cubic centimetres) using granular approaches [7, 8, 9]. Although such techniques cannot yet consider a whole solidification process, they provide a detailed and interesting view of the phenomena occurring during hot tearing, in particular localization of strains and feeding, gradual transition from a continuous liquid film network to a fully coherent solid, etc. These methods are briefly presented in the last section.

## The RDG Hot Tearing Criterion

In the RDG criterion [1], a deformation perpendicular to the thermal gradient is applied to the solid phase, regardless whether it corresponds to columnar or equiaxed dendrites (see Fig. 1). The component of the strains parallel to the thermal gradient is neglected on the basis that it is not a component susceptible of inducing hot tearing. Defining then the average density of the solid-liquid mixture as $\rho = \rho_s g_s + \rho_\ell g_\ell$, where $g_\nu$ and $\rho_\nu$ are the volume fraction and the density of phase $\nu$ ($\nu = s, \ell$), and using

Darcy's equation describing the interdendritic flow in a mushy zone, the following expression is derived for unsteady conditions [1, 10]:

$$\rho\frac{\partial g_s}{\partial t}+(1+\beta)g_s(\dot{\varepsilon}_{yy}+\dot{\varepsilon}_{zz})-div\left(\frac{K}{\mu}(gradP_l-\rho_l g)\right)=0. \qquad (1)$$

$\beta = (\rho_s/\rho_\ell - 1)$ is the solidification shrinkage, $\dot{\varepsilon}_{yy}$ and $\dot{\varepsilon}_{zz}$ are the two components of the strain rate of the solid perpendicular to the thermal gradient (see Fig. 1), K is the permeability of the mush, $\mu$ is the viscosity and $p_\ell$ the local pressure in the interdendritic liquid, **g** being the gravity vector. Note that $\rho_s$ and $\rho_\ell$ have been assumed constant, while $g_s$ is constant in the directions perpendicular to the thermal gradient (directional solidification). It is interesting to notice that although a deformation is applied, it is the strain rate that appears in Eq. (1). This expression is fairly general and can be interpreted as follows: solidification shrinkage (1st term) and/or deformation of the solid (2nd term) have to be compensated by liquid flow (3rd term) if pores or hot tears are to be avoided. In the case a third phase (pores or hot tears) is considered, the right hand term of Eq. (1) is simply replaced by $-\partial g_p/\partial t$, where $g_p$ is the fraction of pores (or hot tears) [11].

Under steady directional solidification conditions at a velocity $v_T$, the maximum pressure drop, $\Delta p_{max}$, across the mushy zone and associated with deformation and solidification shrinkage is given by:

$$\Delta p_{\max}=(1+\beta)\mu\int_0^L\frac{E}{K}dx+v_T\beta\mu\int_0^L\frac{g_l}{K}dx \qquad (2)$$

where E is the cumulated strain rate $E(x)=\int g_s(\dot{\varepsilon}_{yy}+\dot{\varepsilon}_{zz})dx$. These two integrals can be transformed into integrals over temperature, thus introducing a "competition" between strain rate and thermal gradient, G, for the first contribution, and the standard $v_T/G$ ratio for the shrinkage term as already derived by Niyama in his porosity criterion [12].

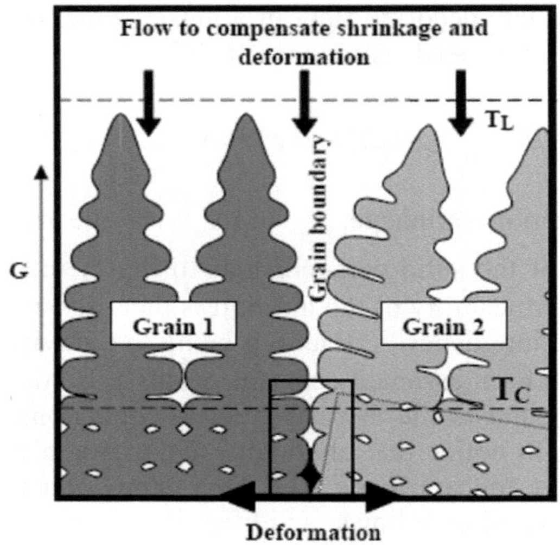


**Fig. 1.** Schematics of hot tearing formation and of the two-phase problem considered in the RDG approach [11]

The RDG criterion simply states that a hot tear forms if the local pressure in the liquid, i.e., the metallostatic pressure minus the pressure drop, falls below a given cavitation pressure. This criterion has nevertheless a few limitations:

- using only the perpendicular component of the plastic strains is not strictly valid, the longitudinal component also inducing some suction (or expulsion) of the liquid;
- the lower bound of the integrals of Eq. (2) is ill-defined. As $g_s$ tends towards unity, the permeability goes to zero and the calculation diverges. In practical situations, this bound is set up to a value of $g_s$ at which the solid is considered as coherent, i.e., the liquid remains only as liquid pockets (no continuous liquid films).
- in relation to grain boundaries, the method does not consider any localization of the strains and feeding.

Stating the local pressure in the liquid films must not fall below a cavitation pressure if no hot tears should form is equivalent to state that the strain rate must remain lower than a maximum value. In other words, the mush can sustain some deformation but its rate should remain low enough in order to permit liquid feeding. The deformation is indeed limited by the ductility of the solid + liquid mixture. Ductility curves obtained by tensile tests exhibit the typical shape of U when liquid and solid phases are present as shown in Fig. 2 [13, 14]. Some ductility is present at temperatures around the coherency temperature (beginning of mechanical resistance in tension), then decreases when approaching the solidus temperature before increasing again in the solid state. The brittle temperature range, BTR, corresponds to the temperature interval where the ductility is really low [15]. In order to avoid hot cracking, the thermomechanical path $\varepsilon(T)$ of the alloy must not cross the U ductility curve. In other words, the slope of $\varepsilon(T)$ is limited. Noticing that $\partial\varepsilon/\partial T = \partial\varepsilon/\partial t * \partial t/\partial T = \dot{\varepsilon}/\dot{T}$, the strain rate is limited too, as found in the RDG approach. Note that the quantity $\dot{\varepsilon}/\dot{T}$ has also been used as a hot tearing indicator [3].

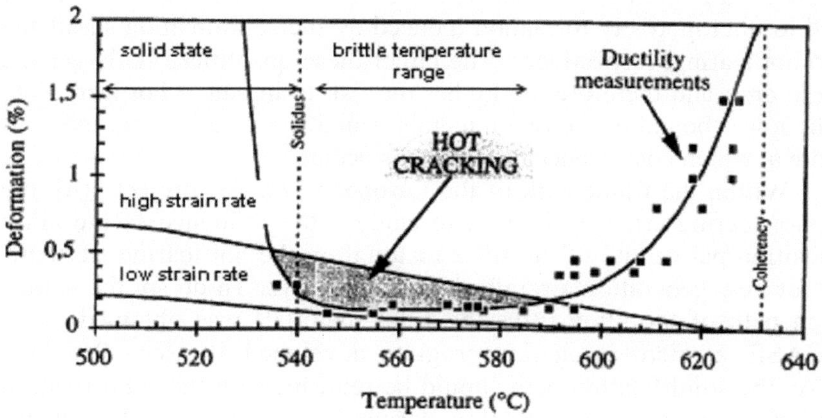


**Fig. 2.** U ductility curve for the Al-Cu 4.5 wt. pct. alloy together with two thermomechanical paths (adapted from [14])

Finally, the RDG criterion, originally derived for aluminium alloys, has been extended to steel by Drezet et al. [16] by taking into account the peritectic reaction that transforms ferrite into austenite. Moreover, Rindler et al. [17] have further worked out the criterion for steels by releasing the assumption of a uniform strain rate acting over the mush.

## Alloy Hot Tearing Susceptibility

Under the assumption that the strain rate applied to the mushy zone is uniform $\dot{\varepsilon} = \dot{\varepsilon}_{yy} + \dot{\varepsilon}_{zz}$ and that steady state is reached, the liquid pressure drop through the mush is made out of two contributions, $\Delta p_{sh}$ due to the solidification shrinkage and $\Delta p_{mec}$ due to deformation [10]. The two contributions can be written as:

$$\Delta p_{sh} + \Delta p_{mec} = \frac{180\mu}{G\lambda_2^2}\left[ v_T \beta A + \frac{(1+\beta)B\dot{\varepsilon}}{G} \right] \tag{3}$$

$$\text{with} \quad A = \int_{T_{cg}}^{T_{liq}} \frac{f_s^2 dT}{(1-f_s)^2} \quad \text{and} \quad B = \int_{T_{cg}}^{T_{liq}} \frac{f_s^2 \int_{T_{cg}}^{T} f_s dT}{(1-f_s)^3} dT \tag{4}$$

$\lambda_2$, the secondary dendrite arm spacing, is the typical length of the microstructure used to define the permeability K. The two integrals A and B are related to microporosity formation induced by the solidification shrinkage and to hot tearing, respectively. The larger these quantities, the larger the pressure drop and therefore the higher the risk to initiate a hot tear. Note that the lower bound of the two integrals A and B are set to Tcg, the temperature at which coalescence of the grains occurs.

Within the framework of the European Wel-Air project [18], the RDG hot tearing criterion is used to analyse the influence of the alloy composition per se and of the filler material on the hot tearing susceptibility for new generation aircraft aluminium alloys. To do so, the solidification path of the alloy (plus the filler material) was obtained using ProPHASE, a microsegregation program developed at Alcan Péchiney [19]. As the solidification path should be relevant with the solidification conditions undergone by the alloy during laser welding, a Scheil approach (no solid-state diffusion) is used owing to the very high cooling rate experienced by the alloy. The grain coalescence temperature is not easy to determine; the temperature corresponding to a solid fraction of 98% or the eutectic temperature if more than 2% eutectic has formed, was adopted as it correctly predicted the hot cracking susceptibility for binary alloys, the so-called Λ curve [1]. Figure 3 shows the solidification paths computed with ProPHASE for five aluminium alloys. Their composition is detailed in Table 1.

**Table 1.** Composition for five selected aluminium alloys

| Alloy | Si | Fe | Cu | Mn | Mg | Cr | Zn | Ti | Zr | Sc | Li |
|---|---|---|---|---|---|---|---|---|---|---|---|
| AA2139 | 0.04 | 0 | 4.9 | 0.3 | 0.45 | 0 | 0 | 0 | 0 | 0 | 0 |
| AA6056 | 0.85 | 0 | 0.95 | 0.45 | 0.73 | 0 | 0 | 0 | 0 | 0 | 0 |
| AA6013 | 0.8 | 0 | 0.85 | 0.5 | 1 | 0 | 0 | 0 | 0 | 0 | 0 |
| AA2098 | 0 | 0 | 3.66 | 0.01 | 0.33 | 0 | 0 | 0 | 0 | 0 | 0.99 |
| AA2022 | 0 | 0 | 5.2 | 0.4 | 0.25 | 0 | 0 | 0 | 0 | 0 | 0 |

As the end of solidification is particularly important, a zoom of the solidification paths is presented in Fig. 4. The ranking of the alloys with respect to their hot cracking susceptibility HCS is presented in Fig. 5 where the integral A is represented, knowing that B varies in a very similar way. It appears that the alloys AA2139 and AA2098 exhibit the lowest hot tearing susceptibility. Albeit, this trend must be balanced by the fact that these alloys present other shortcomings such as equiaxed grain zone (EQZ) and possible liquation cracks.

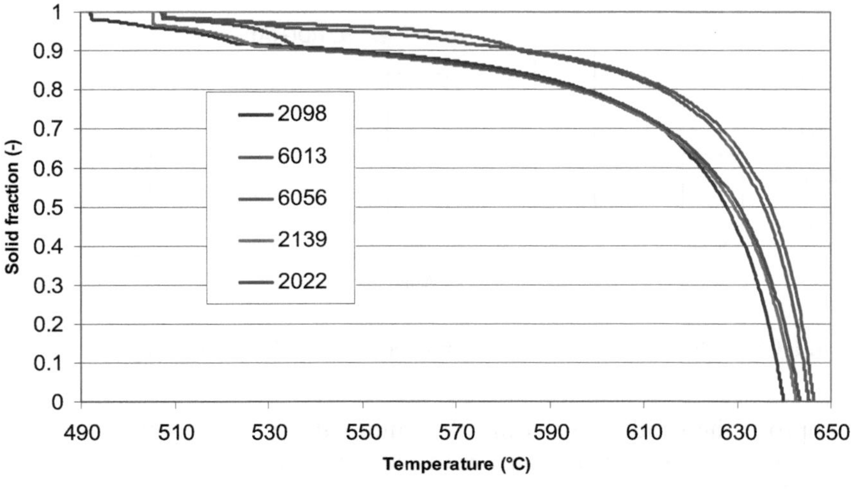


**Fig. 3.** Solidification paths of the Wel-Air alloys as computed by ProPhase

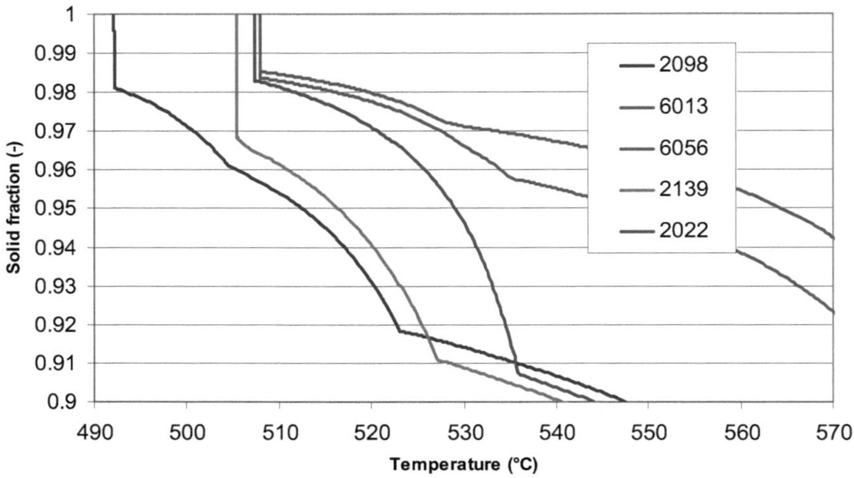


**Fig. 4.** Zoom at solid fractions higher than 90% of the solidification paths

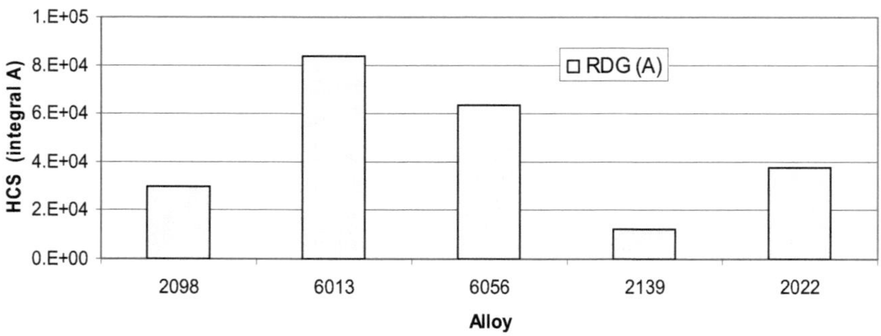


**Fig. 5.** HCS of the WEL-AIR alloys computed with the RDG criterion

To improve the weldability of the aluminium alloys, it is common to bring a filler wire in different proportion to the weld bath, the idea being to increase the Si content in order to get more eutectic. Albeit, the welded material has lower mechanical properties and the use of filler material presents some technological constraints. The influence of the amount of filler material on hot tearing susceptibility is computed for the alloy AA2098. Figure 6 shows the computed solidification paths of the 2098 alloy together with 15%, 25% and 33% of 4047 filler material and Fig. 7, the calculated hot cracking susceptibilities. It is interesting to notice that adding some silicon completely changes the nature of the phases; in particular the

$Al_2Li_3Si_2$ intermetallic phase appears when Si is present and the solidification ends at 501°C with the deposition of $Al_2Cu$. This has a large influence on hot cracking as it decreases the HCS by almost a factor 6. Nevertheless, this one remains almost unchanged when increasing the amount of 4047.

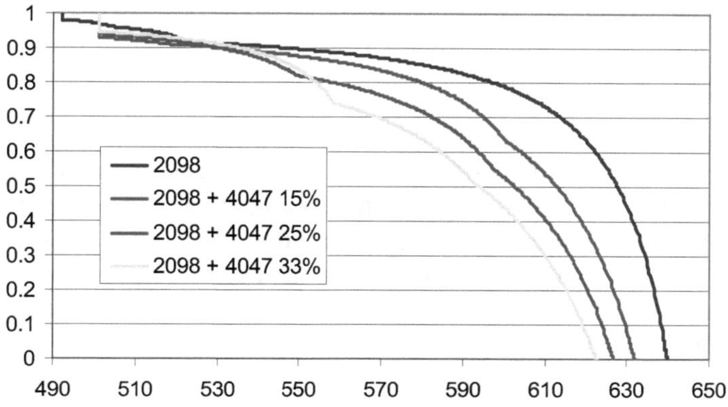


**Fig. 6.** Solidification paths of AA2098 alloy plus AA4047 filler

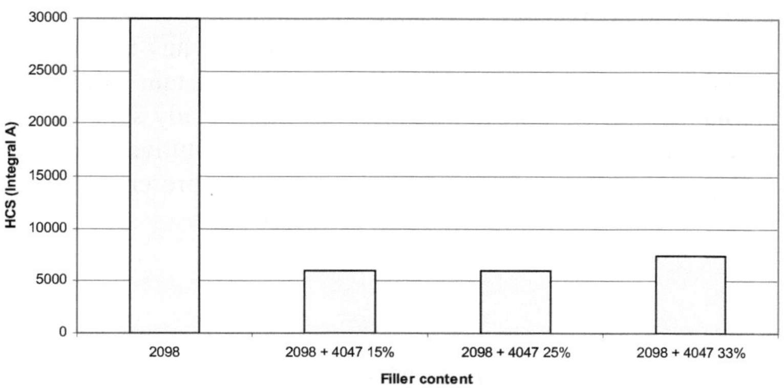


**Fig. 7.** Influence of the filler content on HCS

## Thermal Hot Tearing Criterion

The thermal criterion allows us to compute the hot tearing susceptibility of a given alloy solidified under the conditions imposed by the process, i.e. in a given thermal gradient and with a given solidification speed. The hot cracking susceptibility, HCS, is therefore defined as the inverse of the maximum strain rate sustainable by the mushy alloy. Under steady state, this quantity is a function of the two integrals A and B but also of the local thermal gradient G and solidification speed $V_T$:

$$\Delta p_{max} = \Delta p_{sh} + \Delta p_{mec} - \rho g h \leq \Delta p_{cav}$$
$$\dot{\varepsilon}_p \leq \dot{\varepsilon}_p^{max}(G, V_T, A, B, \Delta p_{cav}, \lambda_2, ...) \tag{5}$$
$$HCS = \frac{1}{\dot{\varepsilon}_p^{max}}$$

The computation of HCS for the AA6056 alloy (cf. Table 1) welded at a speed of 1 m/min with a 3 kW laser is presented as an example. The steady state temperature field in the butt joint is represented in Fig. 8. An eulerian approach is used, i.e. the material is transported at the laser speed under the fixed laser beam. The characteristics of the heat source can be found in [20]. Using the local solidification conditions along the coalescence isotherm, HCS is computed with the help of CalcoSOFT3D [21]. The results are shown in Fig. 9. As steady state is reached, HCS can be represented in a plane perpendicular to the laser speed. A maximum appears close to the bottom of the weld pool. Influence of the laser speed, power and heat source parameters can be assessed with such an approach.

Finally, note that the present approach does not require any mechanical computation, as the maximum strain rate the mush can sustain simply depends on thermal quantities (cf. Eq. 5). However, only steady state can be treated with the thermal criterion. The inclusion of a constitutive model for the mushy alloy and the investigation of transients are presented in the next section.

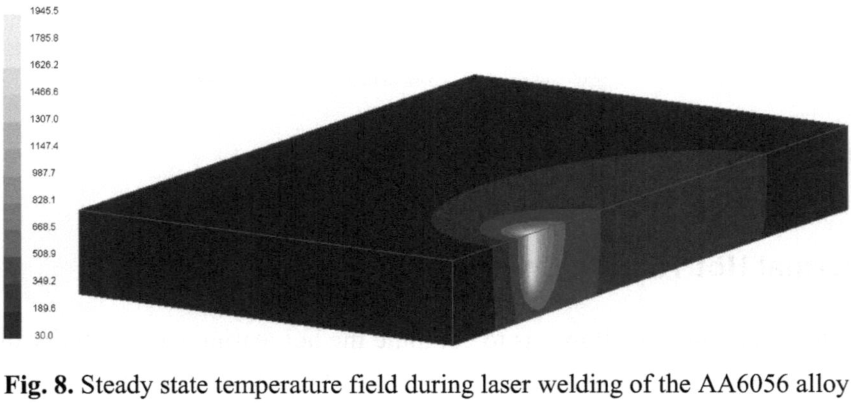


**Fig. 8.** Steady state temperature field during laser welding of the AA6056 alloy

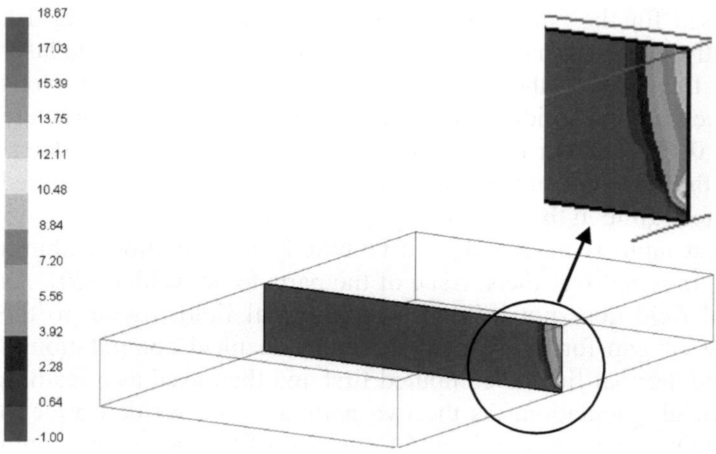


**Fig. 9.** Distribution of HCS

## Thermo-Mechanical Approach

By including the mechanical behaviour of the mush, the strain rate tensor undergone by the dendrites during solidification can be computed and therefore the induced pressure drop; the larger this value, the higher the risk to initiate a hot tear [22]. The main challenge is to establish a reliable constitutive model for the mechanical behaviour of the solidifying material. Although the mush should be treated as a compressible medium with the help of internal variable models [23], a simple incompressible model is adopted as a first step and for sake of simplicity. Indeed, many mechanical tests in the solidification interval are required to determine the numerous parameters describing the compressibility of the mushy alloy [23].

In this section, the mechanical strain rates undergone by the mushy zone at the rear of the weld pool are assessed in a simple configuration, the butt joint, in order to predict how process parameters can decrease the risk of hot tearing in steady state as well as transient regimes such as run-in and run-out. The results presented here deal again with the AA6056 alloy. Its BTR is considered to be 510°C–550°C and corresponds to solid fractions higher than 95% (cf. Fig. 4).

Thermal contraction arises as soon as the dendritic network is well developed and interconnected, that is at a solid fraction of 85% i.e. at a temperature of 600°C (coherency temperature). The rheology of the alloy is given by the classic viscoplastic Ludwik's model and is detailed in [20].

The mesh used for the butt joint configuration is presented in Fig. 10. Due to symmetry, only one half of the whole domain is meshed. The laser is supposed to travel along the z-axis and mesh is refined in the regions that undergo melting and solidification. The dimensions of the domain are given in Fig. 10. The length of the domain (dimension in z) was determined so that the transient run-in and run-out regimes are well separated by a steady state regime in the so called on-going zone.

The laser heat input is modelled by a volume heat generation within a cylinder that is moving over the surface of the parts to be welded [20]. As the mechanical field does not influence the thermal field (owing to the absence of any air gap formation), the thermomechanical computation is un-coupled. The thermal field is computed first and then used as a loading for the mechanical calculation. As the two parts are tack welded prior to laser beam welding, they are considered to be part of the same continuum and therefore no contact elements are used. All plastic deformations are reset to zero above 600°C, using the *anneal temperature feature in Abaqus 6.5 [24].

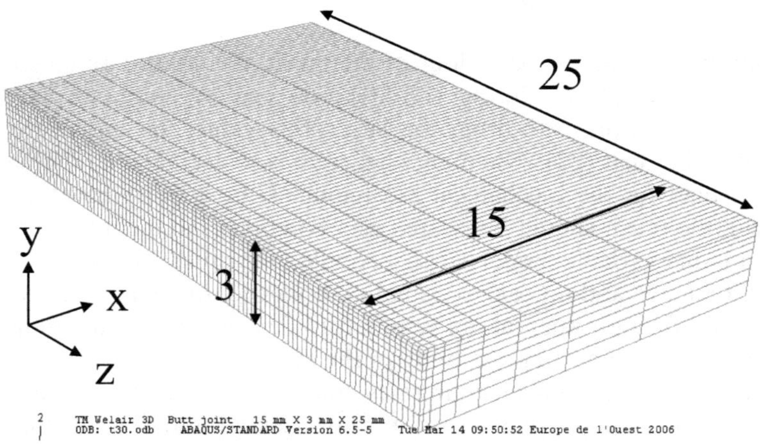


**Fig. 10**. FE mesh for the butt joint configuration (dimensions in mm)

Following the Rappaz-Drezet-Gremaud [1] hot tearing criterion and the approach of Monroe and Beckerman [25], the larger the strain rate is, the higher the risk to initiate a crack. The mean and maximum values of the trace of the total (elastic + thermal + plastic) strain rate over the BTR are taken as hot tearing (HT) indicators:

$$\begin{aligned} \text{HTI1} &= \underbrace{\text{Mean}}_{510^\circ\text{C} \leq T \leq 550^\circ\text{C}} \left(\text{tr}\dot{\varepsilon}\right) \\ \text{HTI2} &= \underbrace{\text{Max}}_{510^\circ\text{C} \leq T \leq 550^\circ\text{C}} \left(\text{tr}\dot{\varepsilon}\right) \end{aligned} \tag{6}$$

These values are saved during cooling only (i.e. during solidification) and at each integration point using the user-subroutine UVARM of Abaqus 6.5. The larger this value, the higher the risk to initiate a hot crack.

The case presented here corresponds to a laser speed of 50 mm/s, i.e. 3 m/min, with a constant power all over the specimen length. The thermal field when the laser reaches the location z = 18 mm is presented in Fig. 11. The weld pool corresponding to temperatures higher than 650°C is represented in grey.

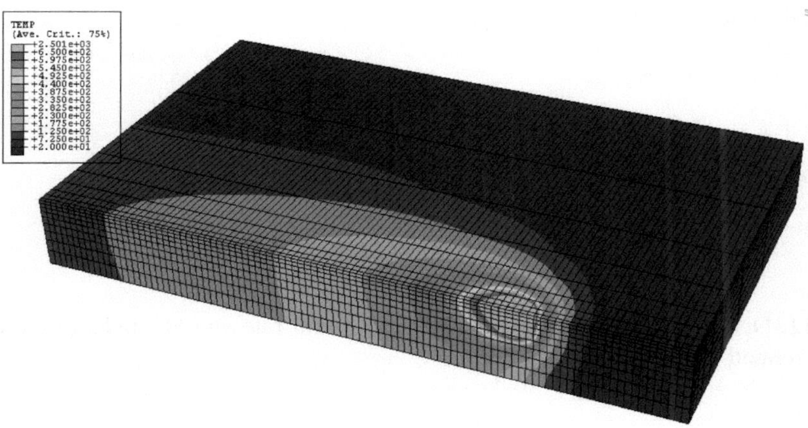


**Fig. 11.** Temperature distribution when the laser is located at z = 18 mm. The liquid pool appears in grey

The distribution of the Von Mises equivalent plastic strain rate (positive quantity) and the two hot tearing indexes, HTI1 and HTI2, is presented in Fig. 12 as a function of the position, z, along the specimen slightly below the top surface (y = 2.5 mm, x = 0). As expected, the three regimes of welding are evidenced: in the run-in, the strain rate and the hot tearing indexes exhibit high values; then they decrease and plateau in the on-going zone. Finally, in the transient run-out regime, the three quantities increase again. Both HTI1 and HTI2 present the same trends. One can notice that the equivalent plastic strain rate cannot be considered as an indicator for hot tearing since it is always positive by definition, whereas HTI1 and

HTI2 can have negative values (i.e. no risk of hot cracking). This is the case when the strain rate reaches its maximum in the run-out whereas the two HT indexes get negative. In the steady state regime, HTI1 is 0.05 /s whereas HTI2 is close to 0.16 /s. HTI1 and HTI2 both exhibit a maximum in the run-in and run-out regimes, which means that hot cracking is prone to occur at those locations. This is in accordance with industrial observations, where no hot tears are observed in the steady state regime but appear in the run-in and run-out. Finally, the HT indexes are larger in the run-in than the run-out, which seems to be less prone to hot cracking.

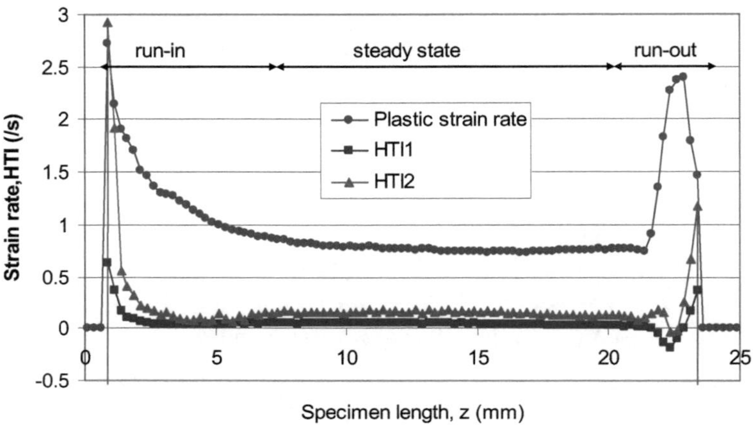


**Fig. 12.** Distribution of the equivalent plastic strain rate and HT indexes along the specimen

## Granular Model for Hot Tearing Criterion

As mentioned previously, one difficulty in the RDG approach is to estimate the temperature at coalescence, Tcg that appears in Eq. (4). Indeed, average methods are unable to account for the localization of hot tears at grain boundaries. This localization is essentially due to the fact that liquid films remain to lower temperature as compared to those located in between dendrites of the same grains (Fig. 1). In other words, the formation of a coherent solid network by coalescence or bridging of dendrites arms occurs earlier within the grains as compared with grain boundaries. Rappaz et al. [26] have introduced for that purpose a *coalescence or bridging undecooling* which, for a pure metal, is given by:

$$\Delta T_b = \frac{1}{\Delta s_f} \frac{\gamma_{gb} - 2\gamma_{sl}}{\delta} \tag{7}$$

where $\gamma_{gb}$ is the grain boundary energy, $\gamma_{s\ell}$ is the solid-liquid interfacial energy, $\Delta s_f$ is the volumetric entropy of fusion and $\delta$ is the thickness of the diffuse interfaces. For an alloy, coalescence is reached when a coalescence line (or surface) parallel to the liquidus, but $\Delta T_b$ below, is reached. Within a grain, there is no grain boundary energy and interfaces become attractive as soon as they get within interaction distance, i.e., distance $\delta$. At "repulsive" grain boundaries, $\gamma_{gb} > 2\gamma_{s\ell}$, bridging is reached at some $\Delta T_b > 0$.

This concept of bridging undercooling has been tested by experiment [27] where two crystals of a nickel-base superalloy are laser welded together under well defined conditions with increasing misorientations. At small misorientation (typically less than 15°), no hot crack forms along the weld centreline, whereas at larger values, a crack is initiated under the same conditions, thus showing the influence of the grain boundary energy. The simulations done by molecular dynamics or by phase field are interesting but correspond to very small regions near the incoming interfaces, typically a few tens of nanometers or micrometers, respectively. In practical situations, hot tears are indeed located at grain boundaries, but the *configuration* of these boundaries associated with nucleation and growth of grains is essential. For this reason, a simplified approach of coalescence for a large population of equiaxed grains was undertaken first by Mathier et al. [7] and then by Vernède et al. [8, 9]. In this *granular approach*, a random set of nucleation centres with random orientations is first generated in a given volume. Considering that the grains nucleate at the same time and that the temperature difference across each grain is small with respect to the growth undercooling, the grain boundaries correspond to the Voronoï tessellation of the nucleation centres (Fig. 13a), i.e., the grain boundary between grains I and K is the median line. Assuming globular grains, the smooth solid-liquid interfaces is first approximated by linear segment in each triangle linking a nucleation centre (open circle) and two vortices of the tessellation (open squares) (Fig. 13b).

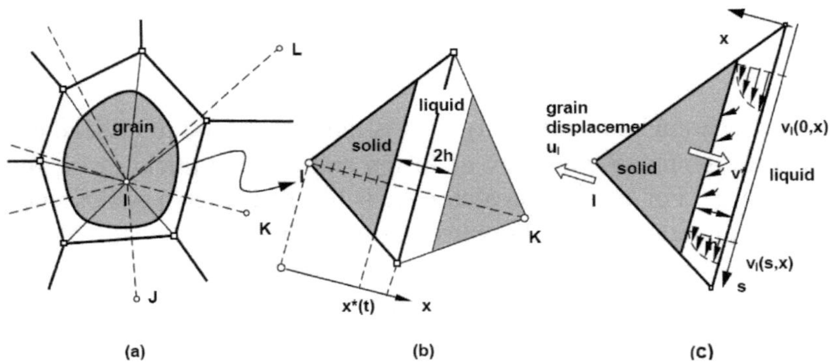


**Fig. 13.** Schematics of the granular model used for the simulation of solidification and feeding in a network of equiaxed globular grains: Voronoï tessellation (**a**), microsegregation model (**b**), feeding KPL model (**c**)

Solidification is then calculated within each triangle using a microsegregation model [9]. Therefore, at any time, the remaining width of the liquid channel in between two grains is known providing the thermal field is known. When the two solid-liquid interfaces get within interaction distances, coalescence is accounted for, using Eq. (7) and a Read-Schockley grain boundary energy model. The next step is to calculate feeding within the network of liquid films. For that purpose, a Poiseuille flow was first assumed within the channels (Fig. 13c) [8, 9]. This flow is not constant along a given channel as it has to feed solidification shrinkage and the relative movement of grains, i.e., the flow has some losses along each channel. Finally, at each vortex of the Voronoï tessellation, the sum of the (signed) incoming flows must be zero according to Kirchhoff's law (so-called Kirchoff Poiseuille with Losses, KPL, model).

The granular model has been applied to the directional solidification of an Al-1%Cu alloy (Fig. 14). The thermal gradient was 60 K/cm and the cooling rate -1 K/s (i.e., velocity of the isotherms equal to 170 μm/s). The average grain density was set to $10^8$ m$^{-2}$, i.e., average grain size of 100 μm, and the computation domain which spans across the whole mushy zone contains 14'000 grains. The central figure in Fig. 14 shows the grains with various grey levels, grains in solid contact (clusters) being represented with the same grey level while the liquid films are in black. The small figures on the left are magnifications of 4 typical regions of the mushy zone which are discussed hereafter. On the right of Fig. 14, the evolution of the volume fraction of solid as calculated in horizontal sections of the grain structure is represented together with the imposed temperature profile.

In region (a), typically for $0 < g_s < 0.89$, most the grains are isolated and surrounded by liquid films. For $0.89 < g_s < 0.97$ (region (b)), clusters of a few grains are formed but the liquid films remain continuous and interconnected. In region (c) characterised by $0.97 < g_s < 0.99$, larger clusters are visible, with a few isolated liquid films remaining inside. Finally, in region (d) ($0.99 < g_s < 1$), the solid network is continuous and liquid only remains as isolated regions. As can be seen, this granular model is able to predict the gradual transition from a continuous intergranular film network to a continuous fully coherent solid. It should be emphasised that cluster formation is directly induced by the stochastic nature of the nucleation centre locations, a feature that has not been considered in past simulation works related to hot tearing. Further analysis of the transition regions is given in [11].

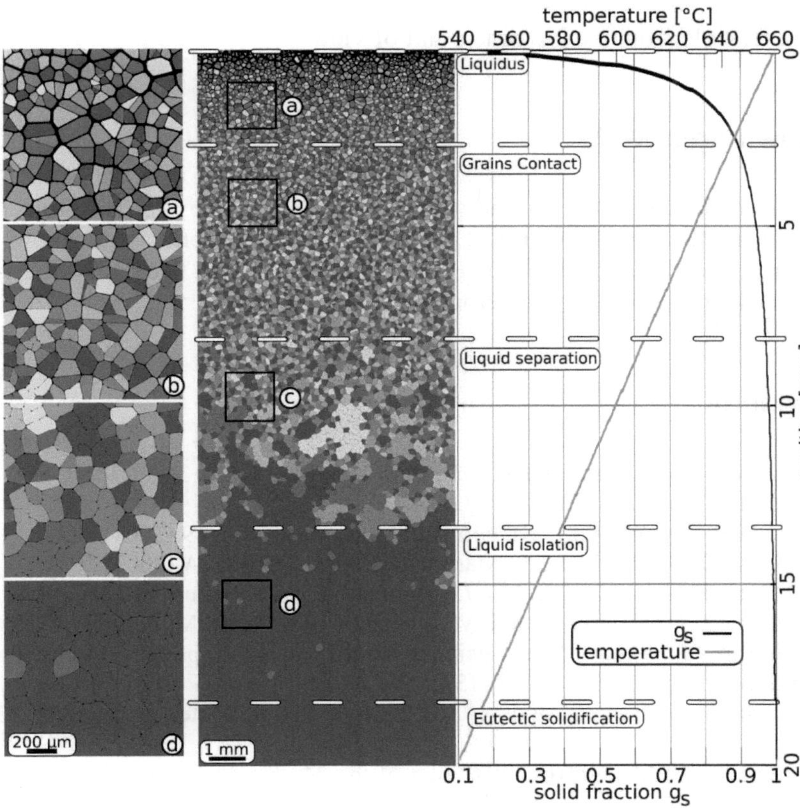


**Fig. 14.** Calculated mushy zone for an Al-1wt%Cu alloy cooled down at –1 K/s in a gradient of 6000 K/m. Grains in solid contact are shaded with the same grey level [11]

## Conclusion

Hot tearing is a complex defect that involves many phenomena, in particular thermal and solidification aspects, stress-strain in an increasingly coherent solid, feeding in a gradually disappearing liquid film network. The RDG criterion provided the first two-phase approach, which was further improved using a more rigorous formalism and the complex rheology of porous media. Nevertheless, these approaches are still based on averages and do not consider any localization of strains and feeding at grain boundaries. Granular models, while still limited to small portions of a solidification process, have certainly the potential to answer some of these questions, once mechanical aspects will be fully built in and the model will be extended to 3 dimensions. The numerical simplicity of such approaches makes it feasible from a CPU time point of view.

## Acknowledgements

The European Commission and the whole consortium of the Wel-Air project, Specific Targeted Research Project from the 6th framework, are acknowledged for financial support. Isabelle Bordesoules from Alcan Péchiney CRV is greatly thanked for providing the Prophase computations.

## References

1. M. Rappaz, J.-M. Drezet, and M. Gremaud: Met. Mat. Trans. 30A (1999), 449.
2. J. Campbell: *Castings* (Elsevier, 2003).
3. D.G. Eskin, Suyitno and L. Katgerman: Progr. Mater. Sci. 49 (2004) 629.
4. M. M'Hamdi, A. Mo, and C.L. Martin: Met. Mat. Trans. 33A (2002) 2081.
5. M. M'Hamdi, H.G. Fjaer, A. Mo, D. Mortensen, and S. Benum in *TMS* 2004.
6. V. Mathier, J.-M. Drezet, and M. Rappaz: Two-Phase Modelling of Hot Tearing in Aluminium Alloys using a Semi-Coupled Approach, Modelling Simul. Sci. Eng. 15 (2007) 121–134.
7. V. Mathier, M. Rappaz, and A. Jacot: Mod. Simul. Mater. Sc. Engng 12 (2004) 479.
8. S. Vernède and M. Rappaz: Transition of the Mushy Zone from Continuous Liquid Films to a Coherent Solid, Phil. Mag. 86 (2006) 3779–3794.
9. S. Vernède, P. Jarry, and M. Rappaz: A Granular Model of Mushy Zones-Formation of a Coherent Solid and Localization of Feeding, Acta Materialia 54 (2006) 4023-4034.

10. J.-M. Drezet and M. Rappaz: In the proceedings of the First EsaForm Conference on Material Forming, Ecole des Mines de Paris, CEMEF, Sophia Antipolis, France, (Mar. 1998) pp. 49–52.
11. M. Rappaz, J.-M. Drezet, V. Mathier, and S. Vernède: In Materials Science Forum vols. 519–521 (July 2006) pp. 1665–1674, Trans Tech Publications, Switzerland.
12. E. Niyama, T. Uchida, M. Morikawa, and S. Saito: AFS Int. Cast Metals J. (Sept. 1982) 52.
13. I.I. Novikov and O.E. Grushko: In Materials Science and Technology, (Sept. 1995) vol. 11, p. 926.
14. B. Magnin et al.: In Materials Science Forum vols. 217–22 (1996) pp. 1209–1214, Trans Tech Publications, Switzerland.
15. W. Rindler, E. Kozeschnik, and B. Buchmayr: Computer simulation of the Brittle Temperature Range (BTR) for hot cracking in steels, in Steel Res. 2000, vol. 71 (11), pp. 460–465.
16. J.-M. Drezet, M. Gremaud, R. Graf, and M. Gaümann: A new hot tearing criterion for steel, proceedings of the 4th European Continuous Casting Conference, IOM communications, Birmingham, UK, (Oct. 2002) pp. 755–763.
17. W. Rindler, E. Kozeschnik, N. Enzinger, and B. Buchmayr: A modified hot tearing criterion for steels, in Math. Modelling of Welding Phenomena 6, (2002) pp. 819–835.
18. EU- FP6 Project: WEL-AIR.
19. C. Sigli, R. Dif, R.B. Commet, and T. Warner: In Materials Science Forum vol. 426–432, Issue 1, (2003) pp. 351–356.
20. J.-M. Drezet et al.: In Mathematical Modelling of Weld Phenomena 8, Ed. H. Cerjak (2007) pp. 137–152.
21. Calcosoft3D user manual, distributed by Calcom ESI, http://www.calcom.ch.
22. J.-M. Drezet and M. Rappaz: "Prediction of hot tears in DC cast billets" in Light Metals, Ed. J.-L. Anjier, TMS, (2001) New Orleans, pp. 887–893.
23. O. Ludwig, C.-L. Martin, J.-M. Drezet, and M. Suéry in Met. Mat. Trans, vol 36A, (June 2005) pp. 1525–1535.
24. Abaqus user manual, http://www.abaqus.com.
25. C. Monroe and C. Beckerman: Development of a hot tear indicator for steel castings, in Mat. Sc. and Eng. A, vol. 413–414 (2005), pp. 30–36.
26. M. Rappaz, A. Jacot, and W. J. Boettinger: Last-stage solidification of alloys: theoretical model of dendrite-arm and grain coalescence, in Met. and Mat. Trans. vol. 34A, (Mar. 2003) pp. 467–479.
27. N. Wang, S. Mokadem, M. Rappaz and W. Kurz: Acta Mater. 52 (2004) 3173.

# Weld Solidification Cracking: Critical Conditions for Crack Initiation and Growth

C.E. Cross, N. Coniglio

Federal Institute for Materials Research and Testing, Berlin, Germany

## Abstract

A perspective will be given that outlines important considerations in evaluating and predicting weldability. An examination will be made of the local conditions necessary for solidification crack initiation and growth in a weld. This will be done in light of two prominent thermo-mechanical approaches involving critical strain and critical strain rate. Critical conditions will be identified based upon values available in the literature. Methods used to measure strain and strain rate will be compared. The interpretation of crack length measurements commonly used to quantify weldability will be questioned, based upon our current understanding of the problem. Complications and problem areas needing better definition will be identified and discussed, including strain distribution in the mushy zone, segregation at grain boundaries, effect of impurities, and effect of cooling rate on solidification path. Finally, a suggestion will be made for a new approach to weld development using in-situ strain rate measurement and new composition-strain rate maps that define the boundary between crack and no-crack conditions.

## Introduction

The topic of solidification cracking as it applies to weldability has for many years been a highly contentious subject, particularly with regard to defining its cause. This is due in large part to its complex nature, involving interplay between thermal, mechanical, and metallurgical components. Welding research has concentrated on specific aspects of this subject (e.g. weldability testing and crack length measurement), while ignoring the underlying cause. Weldability tests normally permit only a relative ranking of crack susceptibility. No one mechanism appears to explain the behavior of all alloys. The ability to predict, a priori, the cracking behavior of a particular alloy has remained to date an unattainable goal. However, with advancing ideas regarding cracking mechanisms, together with improved methods of testing and simulation, welding science is moving towards a point where cracking may be predictable for a given alloy and welding application.

The goal of this chapter is to provide a perspective that both identifies and quantifies the critical conditions important to solidification crack initiation and growth. A new way of looking at critical cracking conditions will be presented that is critical of traditional 50 year old ductility curve approach. Of practical interest is an examination of how critical conditions are represented in conventional weldability testing. Also, a new methodology will be proposed for predicting weldability using critical stain rate – dilution maps.

## Thermo-Mechanical Analysis

Assuming that all thermo-metallurgical conditions are held constant (e.g. constant alloy composition and cooling rate), it is useful to concentrate on what conditions are required to initiate and propagate cracking from a purely thermo-mechanical aspect. This serves to decouple the problem from variable metallurgical influences and allows a focus on mechanical concerns. In this regard, cracking is caused from the strains that arise from the weld thermal experience, including both thermal and solidification shrinkage, affected by weld heat input and conditions of restraint.

### Crack Initiation

#### *Critical Strain*

This topic has traditionally been approached assuming that the mushy zone, and its associated liquid films at grain boundaries, can withstand a limited strain before failure. This follows from early work of Pellini [1], and has been further developed by Prokovorov [2] and Senda et al. [3] who have established characteristic ductility curves for specific alloys. These curves define the critical strain that can be tolerated over the solidification range (brittle temperature range- BTR), bounded by liquidus and non-equilibrium solidus temperatures. Examples of several different ductility curves are given in Fig. 1 for different alloy systems. It is observed that minimum ductilities vary between 0.1–0.5% for aluminum alloys [4], 2–5% for plain carbon steels [5], and 1–9% for stainless steels [5].

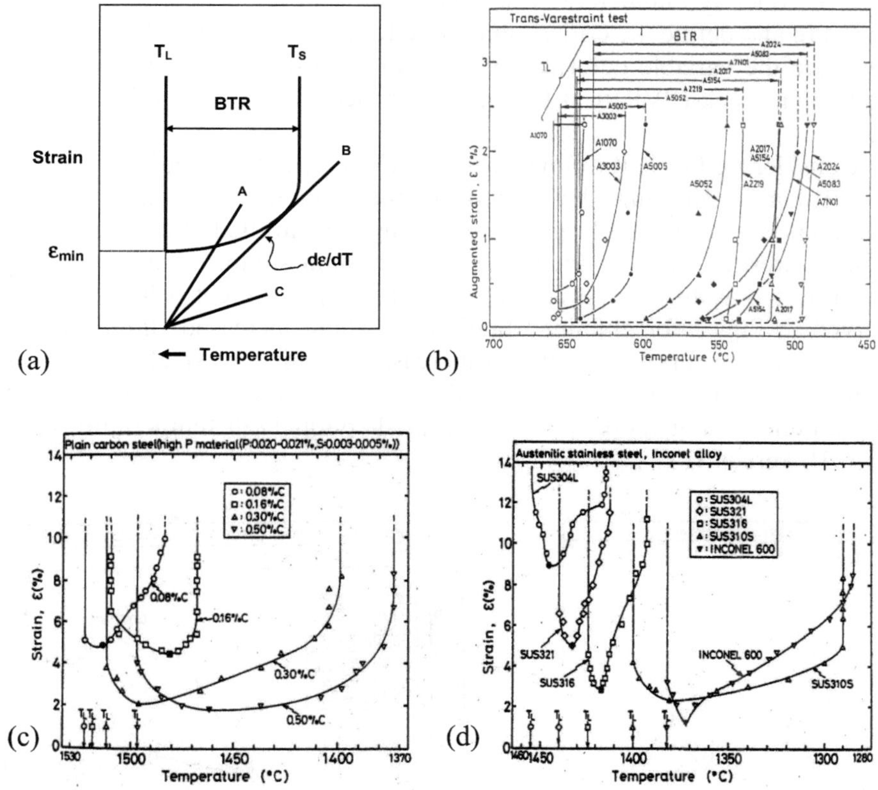


**Fig. 1.** Ductility curve comparison showing (**a**) general schematic, (**b**) aluminum alloys [4], (**c**) plain carbon steel alloys [5], and (**d**) austenitic stainless steel alloys [5]

Developers of the ductility curve concept believed that strain accumulation is an important factor in determining whether cracks will form. It was argued that strain accumulates from the beginning of solidification and, if accumulated in sufficiently high amounts during solidification (as represented by line A in Fig. 1a), the ductility limit will be exceeded and cracking will result. This rate of strain accumulation with temperature drop ($d\varepsilon/dT$) can be related to the strain rate ($d\varepsilon/dt$) accordingly:

$$\frac{d\varepsilon}{dT} = \frac{dt}{dT} \cdot \frac{d\varepsilon}{dt} = \frac{\dot{\varepsilon}}{\dot{T}} \tag{1}$$

Consequently, numerous tests have been developed specifically to measure and define the critical strain rate for different alloys. Among these tests are the variable deformation rate (VDR) test [6], programmable deformation crack (PVR) test [7], and controlled tensile weldability (CTW) test [8].

Considering Eq. 1, it follows from the U-shape of the ductility curves in Fig. 1 that ductility must be dependent upon strain rate. In fact, this is how these curves are typically generated, using a controlled strain rate (slow bend) varestraint test to determine the minimum strain to cause cracking when tested at different strain rates (see Fig. 2). Different behavior is observed in Fig. 2 for aluminum and steel, where increased strain rate reduces ductility for aluminum, but increases ductility for steel [4, 9]. This behavior is reflected in the shape of the respective ductility curves, with aluminum having more of a truncated shape at high temperature (e.g. Fig. 1b versus 1c).

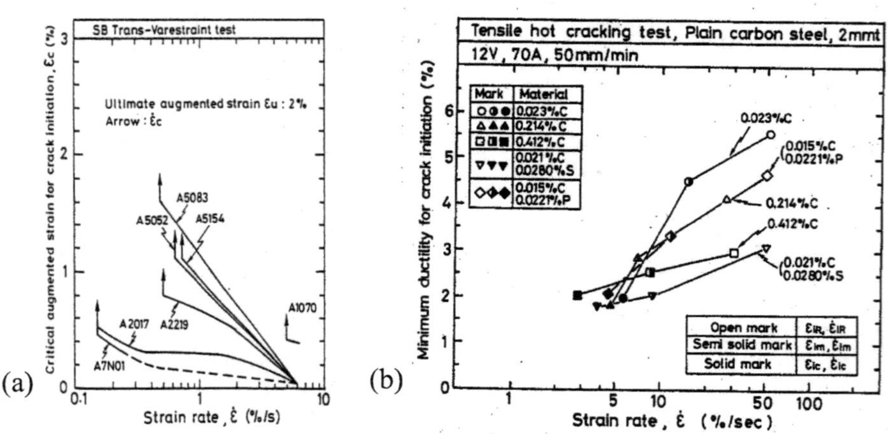


**Fig. 2.** Strain rate dependence of hot crack ductility for (**a**) aluminum alloys [4] and (**b**) plain carbon steel alloys [9]

In addition to strain rate effects, Pumphrey and Jennings argued that a large solidification range permits the build-up of more strain (assuming a constant strain rate) and a greater likelihood that a critical strain will be reached [10]. While alloys with a large solidification range are often found more susceptible to cracking, this is not always the case. The most notable exception is the Al-Mg alloy system, which demonstrates both good weldability and a large solidification range. Although this exception brings into serious question the validity of the strain build-up model, other arguments have been proposed to explain this inconsistent behavior, among them including liquid wettability conditions that promote solid-solid bridging between dendrites [11].

In more recent analyses, numerical simulations have been used to model cracking behavior based upon strain in the mushy zone, building upon the limited ductility concept. Feng simulated weld centerline strain for welds made on aluminum 2024 plate (911–775 K solidification range) as shown in Fig. 3 [12]. Here it is observed that welds started at the plate edge (x = 0 mm) accumulate significantly more strain than welds started 12.5 mm from the edge (1/4 of plate length). Such behavior is commonly observed in practice, particularly for low restraint conditions [13]. Note that strain values are around 2% at the coherent temperature of 865 K for the edge-started weld. Ploshikhin et al. have also simulated weld metal strain and predict cracking will occur when grain boundary deformation exceeds 15 μm [14].

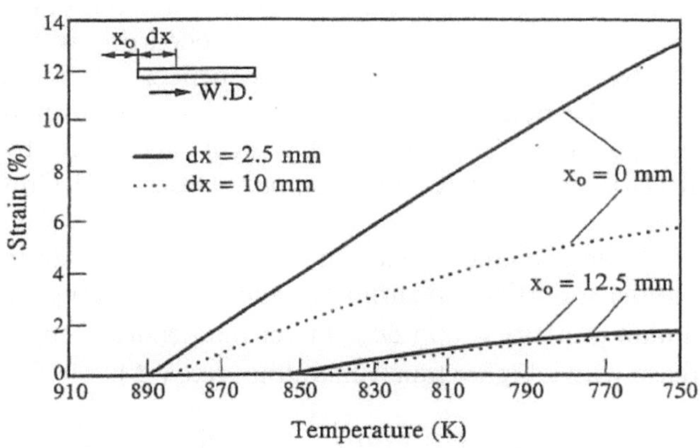


**Fig. 3.** Predicted strain along centerline of aluminum 2024 weld for two different weld starting positions: at plate edge and 12.5 mm from plate edge [12]

## *Critical Strain Rate*

Following from the ductility curve-strain buildup concept outlined above, there appears to be a critical strain rate that defines the boundary between crack and no-crack conditions. This critical condition is represented by line B in Fig.1a, where the deformation curve becomes tangent to the ductility curve. Although critical strain rate characterization is limited, critical strain rate for temperature drop (CST) have been reported [15, 16], expressed in terms of strain per temperature (dε/dT). Measured values for critical strain rate are compared in Table 1, where it is observed that values range between 0.1 and 5.0%/s. These values can be taken as a direct comparison of weldability, with low values representing poor weldability.

**Table 1.** Critical strain rate to initiate weld solidification cracking derived from [4]

| Test | Aluminum Alloy | Critical Strain Rate (%/s) |
|---|---|---|
| Slow Bend Trans-Varestraint Test | 2017 | 0.15 |
| | 5083 | 0.47 |
| | 2219 | 0.50 |
| | 5052 | 0.64 |
| | 5154 | 0.70 |
| | 1070 | 5.00 |

Meanwhile, with significant advancements in solidification theory applied to castings, there is now a better understanding of a mechanism for solidification cracking, which points to a direct connection with strain rate. In particular, attention has been paid to the pressure drop occurring in the liquid film between dendrites, and its resulting consequences on liquid fracture. Fracture itself may be tied to decohesion of oxides, pore formation, or possibly even cavitation. This subject has been addressed by Feurer [17], Campbell [18], and most recently Rappaz, Drezet and Gremaud (RDG) [19].

In the RDG theory [19], an account is made of the local thermal strain conditions in addition to the solidification shrinkage and liquid feeding. In this analysis, the maximum pressure drop ($\Delta p_{max}$) over the dendrite length ($L$) is predicted to increase with a higher transverse strain rate ($\dot{\varepsilon}$):

$$\Delta p_{\max} = (1+\beta)\mu \int_0^L \frac{E}{K} dx + v\beta\mu \int_0^L \frac{(1-f_S)}{K} dx \qquad (2)$$

where $E(x) = \int f_S \dot{\varepsilon} dx$, $\beta$ = shrinkage factor $\left( \rho_S / \rho_L - 1 \right)$, $K$ = permeability, $v$ = growth velocity, $f_S$ = fraction solid, and $\mu$ = viscosity.

The significance of applying RDG theory to welding is profound, as it gives elevated importance to strain rate and it brings into serious question the relevance of ductility curves as well as numerous existing crack prediction simulations based upon strain. Although ductility concepts have become firmly entrenched and are widely accepted in the welding community, they have no foundation in any known liquid fracture mechanism. Furthermore, the existence of ductility curves can be explained using RDG concepts. The U-shaped ductility curves of Fig. 1 can be taken to represent time-temperature nucleation curves (see Fig. 4), replacing the strain axis with a time axis. The application of higher strains means longer test duration, and hence more time for crack initiation and growth. The C-curve kinetics for crack nucleation follows from a low driving force at high temperatures (i.e. low pressure drop Δp), and an increased resistance to deformation at low temperature (i.e. increased coherency).

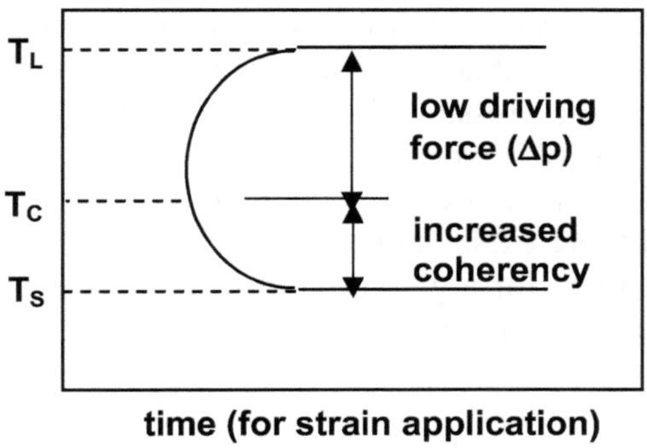


**Fig. 4.** Schematic showing C-curve kinetics for hot crack nucleation and growth

## Crack Growth

Only a limited amount of work has been devoted to understanding the fundamentals of crack growth in a liquid film. This curious lack of attention may come from a presumption that conditions appropriate for crack initiation must also result in crack growth. This assumption finds support in many weldability tests (e.g. circular patch test), where it is observed that once a crack is initiated, the crack will follow behind the weld pool until the torch is extinguished. Another common problem is the failure to recognize crack initiation and growth as two distinguishable events, involving two different mechanisms.

Two approaches for modeling liquid crack growth can be found: one involving breaking of bonds [20]; and one involving the flow of liquid [21]. Considering the relatively slow speed of crack propagation in welding (i.e. at torch velocities around 4 mm/s), it may be difficult to justify a mechanism based upon bond breakage. For the liquid flow model, within a differential period of time, the transverse strain will be compensated by both advancement of the crack and the back-flow of liquid, as defined by a simple mass balance [21]:

$$\dot{\varepsilon}(\lambda - x) = \dot{x}\,h + v_L h\,, \tag{3}$$

where $\dot{\varepsilon}$ is the rate of strain applied transverse to the growth direction, $x$ is the position of the crack relative to the solidus temperature, $\lambda$ is the size of the mushy zone, $\dot{x}$ is the rate of crack growth, $h$ is the liquid film thickness, and $v_L$ is the flow velocity of liquid to feed displacement.

Based upon direct observation of weld crack growth, $x$ appears to remain at a fixed position within the mushy zone (i.e. at a constant solid fraction, perhaps defined by the coherent temperature) and must therefore grow at the same velocity as the weld torch [22]. It follows that if the crack were to grow faster than the torch, it would advance toward the weld pool and encounter conditions of hydrostatic pressure (i.e. more conducive for porosity rather than a planar defect). If it were to grow slower than the torch and drop behind the solidus temperature, the crack would immediately terminate. This suggests that there exists a critical strain rate, below which the growth cannot be maintained for a given liquid flow rate. Data given in Table 2 from Matsuda et al. supports this idea [6].

**Table 2.** Critical strain rate to stop cracking as a function of the aluminium base metal, filler, and welding speed, as measured by Matsuda et al. [6]

| Test | Welding Speed (mm/min) | Aluminium Alloy (Base Metal/Filler) | Critical Deformation Rate (mm/s) |
|---|---|---|---|
| VDR Variable Deformation Test | 240 | 1100/1070 | 0.15 |
| | 400 | 1100/1070 | 0.25 |
| | 600 | 1100/1070 | 0.40 |
| | 800 | 1100/1070 | 0.50 |
| | 400 | 5052/1070 | 0.05 |
| | 400 | 5083A/5183B | 0.18 |

## Measuring Local Strain

Key to the establishment of critical cracking conditions, verification of numerical simulations, and eventual prediction of cracking behavior is the ability to measure transverse strain rate in the mushy zone behind a moving weld pool. This poses some unique challenges primarily associated with the high temperatures encountered in welding. Different methods have been used over time, including real-time observation of scribe marks [23] and moiré-fringe analysis of grid patterns [24].

Use has also been made of extensometers spanned across the weld, attached either above or below the plate surface [25]. When applied to the top surface, small diameter pins are attached on opposite sides of the joint, and extension arms connect these pins to a remote extensometer (i.e. removed from the torch path and possible exposure to heat). This allows measurement of strain as the welding torch passes between the affixed pins. When applied below the surface, the extensometer can be placed directly at the point of measurement, but this requires use of through-thickness welds and a thin plate (i.e. plane strain) to provide any useful information. An example of one such measurement for an autogenous Al 6060 weld is given in Fig. 5a, where an extensometer (10.5 mm gage) has been placed at the mid-span of a 100 mm long joint and the torch passes over the extensometer at approximately 15 s. A compression cell is observed ahead of the weld pool (negative strain), followed by a tensile cell behind the weld pool, with a steadily increasing strain rate (approximately 0.5%/s in mushy zone).

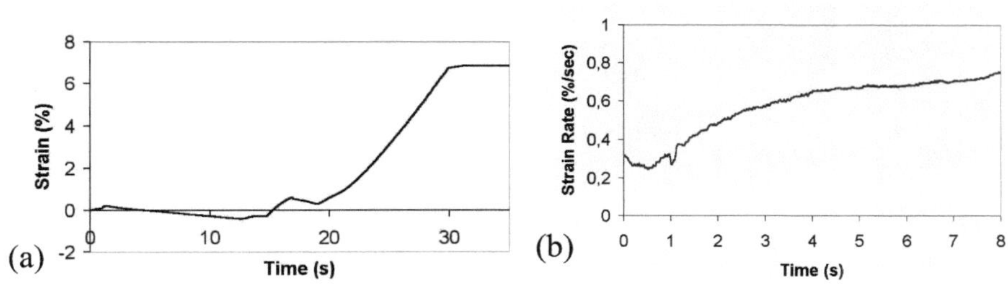


**Fig. 5.** (**a**) Output from extensometer placed across the root of an autogenous, full-penetration GTA weld (10.5 mm gage) made on 4 mm aluminum 6060-T6 plate, (**b**) first derivative of strain-time curve in (a) showing transverse strain rate

Extensometer measurements are limited to a fixed location and gage length. A much more thorough characterization can be made using a digital image correlation (DIC) technique, which employs the computer aided tracking of a random speckle pattern painted onto the plate surface prior to welding [26]. This allows transverse strain to be monitored during welding over a large area as demonstrated in Fig. 6a, and permits strain across the mushy zone to be examined as shown in Fig. 6b. When using light optics, this method is best suited for use on the bottom surface of the weld, where arc light and gas fumes cannot interfere with speckle resolution. Use of a laser based DIC system, however, would not suffer from this limitation.

One technique has been developed to measure strain directly in the weld mushy zone making use of small discontinuities found on the weld surface [27]. The relative movement of these discontinuities is tracked using high speed photography, typically providing an effective gage length of around 1 mm. Measurements made using this technique, referred to as measurement by means of in-sito observation (MISO), are given in Table 3 for values representing the minimum strain to initiate cracking (at high strain rate) in a tensile test. When compared against corresponding values based upon applied strain in a trans-varestraint test (TVT), it is observed that MISO values are significantly higher. This discrepancy may be due in part to a hinge effect, discussed below in the weldability section, and reflects a difficulty to control applied strain in the TVT.

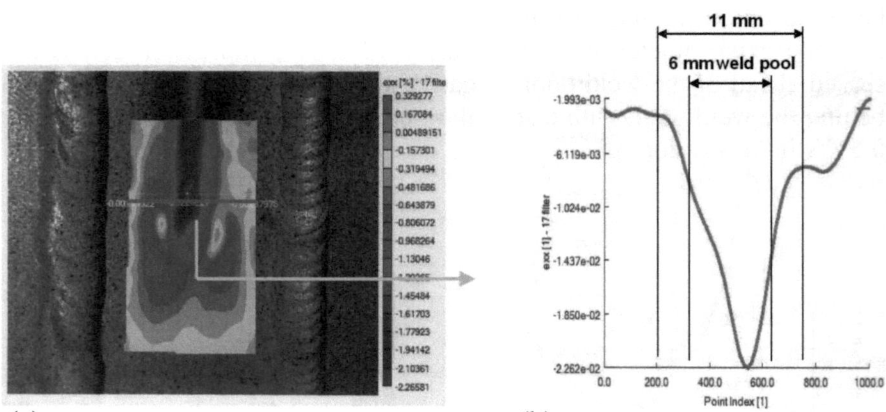


(a) (b)

**Fig. 6.** (**a**) Light optical DIC measurements made on bottom side of plate during autogenous, full-penetration, bead-on-plate GTA welding of aluminum 6060 plate. Note black paint speckle pattern sprayed onto plate prior to welding. (**b**) Strain across mushy zone of weld pool showing a constant outward displacement with distance from weld centerline

**Table 3.** Minimum strain for cracking (in %) comparing MISO measurements with applied strain in TVT test [27]

| Alloy | MISO | Trans-Varestraint |
|---|---|---|
| plain carbon steel (30C-P) | 2.0 | <0.5 |
| 304L | 9.0 | 0.5–0.75 |
| 310S | 2.3 | <0.15 |
| Al 5083 | 2.1 | <0.15 |

## Weldability Testing

A broad variety of weldability tests have been developed over the years that rely upon special conditions of joint configuration or applied strain to generate cracks as outlined in Table 4. In most cases, weldability is determined based upon an evaluation of crack length measurements (i.e. crack length parametrics). In other cases specifically involving application of controlled strain rate, weldability can be related directly to a critical strain rate (crack versus no-crack) for a given alloy and welding condition. As argued above, this latter case holds the greatest promise for predicting cracking because it relates to an actual cracking mechanism. Crack length measurements, on the other hand, can only be used to rank the cracking susceptibility of different alloys (i.e. provide a relative comparison), and even this has come under increased scrutiny as discussed below.

**Table 4.** Comparison of Weldability Tests and Method of Evaluation

| weldability tests | | | | |
|---|---|---|---|---|
| intrinsic | | | | |
| | circular patch test | (CPT) | crack length parametrics | |
| | houldcroft test | | | |
| extrinsic | | | | |
| | longitudinal varestraint test | (LVT) | | |
| | transverse varestraint test | (TVT) | | |
| | modified varestraint test | (MVT) | | |
| | sigma-jig test | | | |
| | variable tensile strain test | | critical strain rate | |
| | variable speed varestraint | | | |
| | variable deformation rate test | (VDR) | | |
| | program. verformungsrisstest | (PVR) | | |
| | cont. tensile weldability test | (CTW) | | |

### Crack Length Parametrics

Crack length measurements are most often reported in terms of total accumulated crack length (TLC), as it is generally assumed that for fixed testing conditions, an alloy with higher susceptibility must somehow result in more extensive cracking. In longitudinal varestraint specimens, for example, the TCL constitutes the sum of several smaller crack lengths, where the number of cracks is bounded by the number of grain boundaries and the length of each crack is bounded by size of the mushy zone. It is found that cracking increases with applied strain up to some saturation limit, as each grain boundary is fractured and opened to its maximum length. Ductility theory proponents would argue that higher applied strain results in more grains exceeding the critical strain value. However, with regard to critical strain rate theory, this ductility argument can be countered by noting that the application of higher strain also means longer test duration, and hence more time for crack initiation and growth.

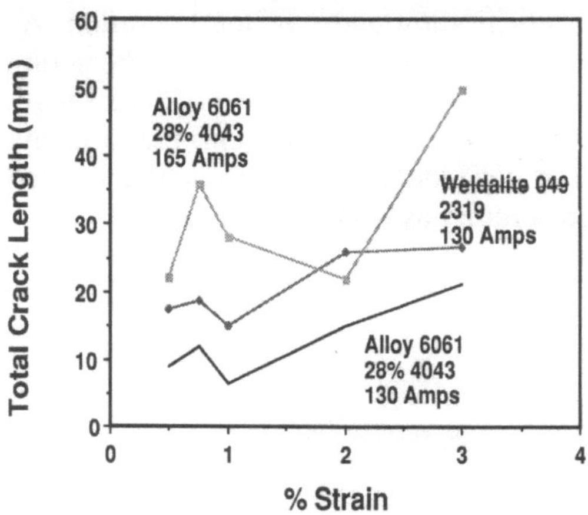


**Fig. 7.** Weldability comparison of aluminum alloys 6061 (4043 filler) and Weldalite 049 (2319 filler) using the trans-varestraint test run at high strain rate (50%/s) [28]

Additional problems are encountered when attempting to rank weldability in terms of crack length measurement, particularly when comparing alloys with different thermal properties. For example when comparing aluminium alloys 6061 against 2094 (Weldalite 049) in Fig. 7, conditions of constant heat input (i.e. constant current) suggest that alloy 6061 is more weldable [28]. However, the exact opposite conclusion is reached for conditions of constant penetration. To achieve the same penetration, Alloy 6061 requires higher heat input (i.e. higher current) which results in a larger mushy zone and corresponding higher TCL values.

To avoid the above noted problems associated with TCL comparisons, it has been found useful to relate the maximum crack distance (MCD) in a varestraint test (i.e. obtained at high applied strain, measured normal to isotherms) to a corresponding temperature range. This requires the aid of a thermocouple inserted into the weld pool. The resulting value, or solidification cracking temperature range (SCTR) [29], is believed to be independent of welding parameters and appears to represent a characteristic property reflecting relative weldability. As noted earlier, a large temperature range suggests the possibility for more strain build-up, but it also represents a more difficult condition for liquid feeding and a subsequent higher liquid pressure drop.

In normal varestraint testing, both TVT and LVT, strain is applied at relatively high rates, with different values reported as presented in Table 5. It should be noted that the local weld metal strain rate, measured with the MISO technique, is more than twice the applied strain rate. Under normal circumstances, a fixed strain is applied over a duration of approximately100 ms, which has led to a general assumption that these tests involve instantaneous strain application. Also, these strain rates far exceed critical values for crack initiation by several orders of magnitude (recall Table 1), which brings into question how results from this type of test can ever be used to simulate real-world behavior.

**Table 5.** Comparison of weldability test strain rates

| Applied Strain Rate | MISO Strain Rate | Weldability Test | Reference |
|---|---|---|---|
| 50%/s | – | TVT | [28] |
| 40%/s | 100%/s | LVT | [22] |
| – | 130%/s | Tensile | [5] |

Work by Robino et al. [22] has recently demonstrated a problem with assuming strain to be instantaneous in a LVT test. Using high speed photography, a crack was observed to initiate in the mushy zone (at $f_s$=0.78) at a local strain of 1% during the application of a 4% augmented strain. This crack then proceeded to grew simultaneously in opposite directions, both towards the weld pool and away from it. The advancing crack tip grew at the velocity of the weld torch (3.3 mm/s), thus maintaining its relative position at $f_s$=0.78 and terminating as the applied strain becomes dissipated. The retreating crack tip grew at a higher velocity and terminated when it intersected the advancing solidus temperature. During the period of strain application (approx. 120 ms), the weld pool advanced 0.4 mm, which is the same order of magnitude as the measured crack length of 0.7 mm.

What this means is that MCD measurements coming from varestraint tests should be expected to vary with torch travel speed and applied strain rate and cannot be related to a characteristic temperature range (e.g. SCTR) in any straight forward manner. It also suggests the importance of standardizing strain rates and welding speeds for varestraint testing for purposes of crack length and SCTR comparisons. Varestraint tests run at sub-normal speeds will likely result in MCD values that are even more unrepresentative of a characteristic temperature range.

One additional problem with varestraint testing is worth pointing out here, as it helps to explain the discrepancies observed in MISO measurements given in Table 3. When using the TVT test, it has been found that the strain experienced in the mushy zone can far exceed the applied strain. This is believed due in part to a hinging effect, whereby plastic deformation is concentrated in hot material along the weld seam as shown schematically in Fig. 8a. This is further demonstrated in strain gage measurements compared in Fig. 8b and 8c, showing behavior of wrought material versus weld material for aluminium 6061 tested at room temperature. The wrought material strain (0.6%) closely matches the applied strain (0.5%). However, the weld metal shows much higher strain values (over 2%), because it is weaker than the base metal (50% joint efficiency).

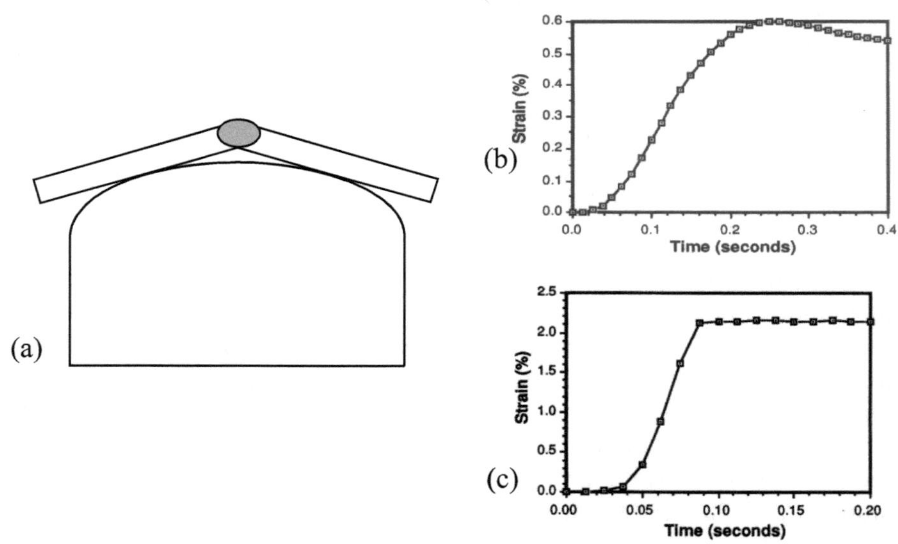


**Fig. 8.** (**a**) Schematic illustration showing hinge effect in TVT testing, (**b**) strain gage placed on Al 6061 plate for TVT test with 0.5% strain die block, (**c**) strain gage place on weld metal in pre-welded Al 6061 coupon, TVT tested with 0.5% die block [28]

## Topics of Limited Understanding

The following topics are believed to have major influences on weld solidification cracking, but are themselves not well understood and need further clarification in order to expand our understanding of cracking behavior.

### Strain Rate Partitioning

Most of the local strain rate measurements considered above, using extensometer and DIC methods, have been made in the vicinity of the mushy zone, but not in the mushy zone itself and not across a single grain boundary. In order to relate local measurements to actual strain conditions across a single grain boundary, for ideal application of theory, it is important to know how strain (and hence strain rate) is partitioned within the mushy zone. For this purpose, it becomes convenient to treat the mushy zone as a composite material consisting of parallel strips of liquid and solid phases, with behavior proportioned to the relative amount of each phase:

$$\dot{\varepsilon} = f_S\dot{\varepsilon}_S + f_L\dot{\varepsilon}_L \tag{4}$$

This becomes further complicated if it is considered that the strain partitioned to grain boundaries is different from the strain partitioned interdendritically.

### Grain Boundary Segregation

Microsegregation of alloying elements due to partitioning between dendrites during solidification has been extensively modeled using the Scheil equation and its various different forms (e.g. to account for back-diffusion) [31]. However, the concentration of segregates at weld metal grain boundaries is not well understood or quantified. It has been well documented that segregates (e.g. sulfur in steel) migrate to the weld centerline and that this behavior contributes significantly to weld cracking, particularly for tear-drop shaped weld pools [32]. While models for cracking so far have considered idealistic interdendritic conditions for reasons of simplicity, it is clear that intergranular conditions are more appropriate.

### Role of Impurities

In ferrous alloys, the impurities sulfur and phosphorous play a major role in determining weldability and this has been fairly well documented. In aluminum alloys, however, the role of impurities is less clear and this remains a little studied subject [33]. Impurities oxygen and hydrogen, for example, could conceivably affect weldability; where oxide films may serve as crack initiation sites, and hydrogen may serve to reduce these oxides. Hydrogen also results in pore formation that may help feed shrinkage. Dissolved hydrogen could also make it easier to initiate cracks through cavitation. The impurity iron has been shown to affect both porosity and cracking behavior in aluminum castings [34], but its affect in welding aluminum remains largely unexplored.

### Solidification Path

The solidification path of an alloy, and its corresponding solid fraction versus temperature curve, is an important input to the RDG model. While there is now popular use of thermodynamic based software predictions, it is of course important to also perform experimentation to confirm solidification behavior. It is of particular interest to understand the effect of cooling rate on phase formation. For example, solidification microstructures are compared in Fig. 9 for aluminum 6060 showing a significant change in constituent phase distribution with cooling rate, in addition to the finer dendrite spacing and grain structure. A single sensor-differential thermal analysis (SS-DTA) technique developed specifically for welds, has proven useful in identifying phase reactions and solidification ranges [35, 36].

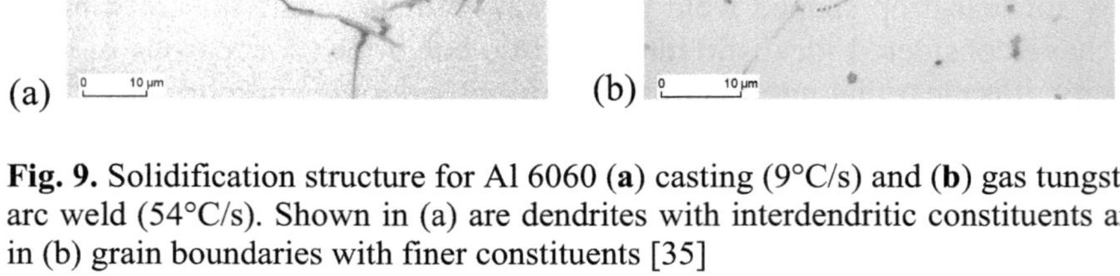


**Fig. 9.** Solidification structure for Al 6060 (**a**) casting (9°C/s) and (**b**) gas tungsten arc weld (54°C/s). Shown in (a) are dendrites with interdendritic constituents and in (b) grain boundaries with finer constituents [35]

## Crack Prediction

Future approaches for predicting the onset of cracking for a given industrial application will involve applying material information to a cracking mechanism with the aid of numerical simulation. Inputs to this simulation will include pertinent welding information (heat input, travel speed), alloy composition (base metal and filler dilution), and restraining conditions (weld geometry, fixture design). The ability to accurately predict cracking behavior will require a thorough understanding of the cracking mechanism itself, something that does not yet exist.

Meanwhile, and until a reliable cracking model is developed, a more pragmatic approach may prove useful. This could, for example, involve development of composition-strain rate maps as demonstrated in Fig. 10 [8]. Based upon Controlled Tensile Weldability (CTW) testing, a boundary has been established for the aluminum 6060/4043 alloy system that defines how much 4043 filler metal is required to avoid cracking for a given local strain rate. Using this map, a welding engineer can make informed decisions regarding weld procedure development (including filler dilution) based upon local strain rate measurements. Such measurements could conceivably be made using a laser scanner across the mushy zone. While the given data is for the Al 6060-4043 alloy system, a similar type of map could be generated for any ferrous or non-ferrous alloy.

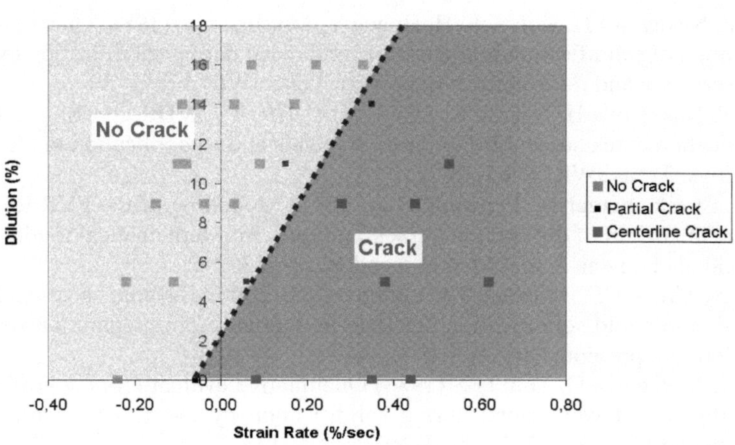


**Fig. 10.** Critical strain rate – ductility map showing conditions required for weld solidification cracking in Al 6060 + 4043 filler using gas-tungsten arc process (full penetration, 4 mm plate, 4 mm/s travel speed) [8]

## Summary

The long-held belief in the welding community that solidification cracking is a ductility limited phenomena, and that ductility curves represent a characteristic critical strain behavior, has been questioned. Now that mechanisms are being developed to explain how a liquid fails, a mechanism is now available to help guide how we look at weld cracking. Strain rate, before just a secondary effect needed to achieve a critical strain, has now become a key controlling factor for both crack initiation and growth. Common weldability tests involving the application of a fixed strain at high rates, far exceeding critical strain rate values, are likewise coming under scrutiny as to just what it is they represent.

## References

1. Pellini WS (1952) Strain theory of hot tearing. Foundry 80: 125–199
2. Prokhorov NN (1956) The problem of the strength of metals while solidifyin-during welding. Svar Proiz 6: 5–11
3. Senda T, Matsuda F, Takano G (1973) Studies on solidification crack susceptbility for weld metals with trans-varestraint test. J Japan Weld Soc 42: 48–56
4. Nakata K, Matsuda F (1995) Evaluations of ductility characteristics and cracking susceptibility of Al alloys during welding. Trans JSRI 24: 83–94
5. Matsuda F, Nakagawa H, Kohmoto H, Honda Y, Matsubara Y (1983) Quantitative evaluation of solidification brittleness of weld metal during solidification by in-situ observation and measurement (report II). Trans JWRI 12: 73–80
6. Matsuda F, Nakagawa H, Nakata K, Okada, H (1979) The VDR cracking test for solidification crack susceptibility on weld metals and its application to aluminum alloys. Trans JWRI 8: 85–95
7. Herold H, Streitenberger M, Pchennikov A (2001) Modelling of the PVR test to examine the origin of different hot cracking types. In: Mathematical modelling of weld phenomena 5, Inst Metals, London, pp 783–792
8. Coniglio N, Cross CE, Michael Th, Lammers M (2006) Defining a critical weld dilution to avoid solidification cracking in aluminum. In preparation for Welding Journal, presented at AWS, Atlanta
9. Matsuda F, Nakagawa H, Tomita S (1986) Quantitative evaluation of solidification brittleness of weld metal during solidification by in-situ observation and measurement (report III). Trans JWRI 15: 125–133
10. Pumphrey WI, Jennings PH (1948) A consideration of the nature of brittleness at temperatures above the solidus in castings and welds in aluminum alloys. JIM 75: 235–256
11. Borland JC (1960) Generalized theory of super-solidus cracking in welds and castings – an initial development. Brit Weld J 7: 508–512

12. Feng Z (1994) A computational analysis of thermal and mechanical conditions for weld metal solidification cracking. Welding in the World 33: 340–347
13. Cross CE, Böllinghaus Th (2006) The effect of restraint on weld solidification cracking in aluminium. Welding in the World 50: 51–54
14. Ploshikhin V, Prikhodovsky A, Makhutin A, Ilin A, Zoch HW (2005) Integrated mechanical-metallurgical approach to modeling of solidification cracking in welds. In: Hot cracking phenomena in welds, Springer, pp 223–244
15. Arata Y, Matsuda F, Nakata K, Shinozaki K (1977) Solidification crack susceptibility of aluminum alloy weld meatls (report II). Trans JWRI 6: 22–35
16. Herold H, Pchennikov A, Streitenberger M (2004) Assessment of hot cracking initiation by laboratory test procedures and FEM simulation associated experimental measurements during welding of large weld components. J Japan Weld Soc 22: 211–217
17. Feurer U (1977) Influence of alloy composition and solidification conditions on dendrite arm spacing, feeding, and hot tear properties of aluminum alloys. In: Proc international symposium engineering alloys, Delft, pp 131–145
18. Campbell J (1991) Casings, Butterworth-Heinemann
19. Rappaz M, Drezet JM, Gremaud M (1999) A new hot-tearing model. Met Mat Trans 30A: 449–455
20. Murakawa H, Serizawa H, Shibahara M (2005) Prediction of welding hot cracking using temperature dependent interface element. In: Mathematical modelling of weld phenomena 7, Maney, pp 539–554
21. Braccini M, Martin CL, Suery M, Brechet Y (2000) Relation between mushy zone rheology and hot tearing phenomena in Al-Cu alloys. In: Modeling of casting, welding, and advanced solification processes IX, Shaker Verlag, Germany, pp 19–24
22. Robino CV, Reece M, Knorovsky GA, DuPont JN, Feng Z (2005) Prediction of maximum crack length in longitudinal varestraint testing. In: Proc. 7th int. conf. trends in welding research, ASM Int, pp 313–318
23. Chihoski RA (1972) The character of stress fields around a weld arc moving on aluminum sheet. Weld J 51: 9s–18s
24. Johnson L (1974) Moire techniques for measuring strains during welding. Exp Mech 14: 145–150
25. Kannengiesser T, Mclnerney T, Florian W, Böllinghaus T, Cross CE (2002) The influence of local weld deformation on hot cracking susceptibility. In: Mathematical modelling of weld phenomena 6, Maney, pp 803–817
26. Sutton MA, Wolters WJ, Peters WH, Ranson WF, McNeill SR (1983) Determination of displacements using an improved digital correlation method. Image and Vision Computing 1: 133–139
27. Matsuda F, Nakagawa H, Nakata K, Kohmoto H, Honda Y (1983) Quantitative evaluation of solidification brittleness of weld metal during solidification by means of in-situ observation and measurement (report I). Trans JWRI 12: 65–72
28. Cross CE, Tack WT (1993) Factors affecting the weldability evaluation of aluminum alloys. unpublished research, Martin Marietta Aerospace, Denver

29. Lippold JC (2005) Recent developments in weldability testing. In: Hot cracking phenomena in welds, Springer, pp 271–290
30. Unpublished research (2006), BAM, Germany
31. Brooks, JA (1990) Weld microsegregation: modelling and segregation effects on weld performance. In: Weldability of Materials, ASM, pp 41–47
32. Savage WF, Aronson AH (1966) Preferred orientation in the weld fusion zone. Welding J 45: 85s–89s
33. Coniglio N, Cross CE (2007) Weld parameter and minor element effects on solidification crack initiation in aluminium. In: this conference proceedings, pp. 277–310
34. Lu L, Dahle AK (2005) Iron rich intermetallic phases and their role in casting defect formation in hypoeutectic Al-Si alloys. Met Mat Trans 36A: 819–835
35. Coniglio N, Cross CE (2006) Characterization of solidification path for aluminium 6060 weld metal with variable 4043 filler addition. Welding in the World 50: 14–23
36. Alexandrov BT, Lippold JC (2006) A new methodology for studying phase transformations in high strength steel weld metal. In: Proc 7th int conf trends in welding research, ASM Int, pp 975–980

# Consideration of the Welding Process as a Thermo-Physical Mechanism to Control Cracking in Weldments

H. Herold, M. Streitenberger

Otto-von-Guericke University, Institute for Materials and Joining Technology, Magdeburg, Germany

## Abstract

Prevention of cracking in weldments is the result of understanding and evaluating extensive experience from cases of damage. The welding process-con trolled mechanics for the initiation of weldment cracking by extensive strain due to thermal expansion of the moving weld pool with its surrounding temperature field, and a local characterisation of the dynamic non-linear behaviour during solidification in, around and behind the weld pool in interaction with the weldment restraint are described in terms of thermo-physical mechanics. Different experience from solidification cracking assessment results in a concept for controlling weldment cracking by means of thermo-physical mechanics.

## Introduction

Cracking in weldments is an unwished side effect of the fusion welding process when applied to metals in manufacturing structural components and providing them with special-use properties for lifetime. Historically, in any metal-working process priority has been given to high toughness in combination with all other aspects of material resistance –just to mention the Damascus blade and all types of novel steel-makers' products to which a special metal-working treatment was applied to exclude brittle failure and any cracking. Thus, weldment cracking should be avoided, but if occurring it will indicate that in the respective operation, e.g. fusion welding, tolerances have been exceeded.

Failure analysis and assessment concepts are made during the lifetime of welded structures if cracks have formed and grown to macroscopic scale, endangering the reliability of structural components. Cracks originate at the weakest link of material behaviour in the microstructure, caused by locally positioned thermo-mechanical overstress within the welded structural component on which general thermo-mechanical stress is imposed by welding.

On the example of solidification cracking, the paper follows the philosophy: When we can describe and analyse the hot cracking phenomena in both material behaviour and welded structural components, concentrating on the crack – initiating thermo-mechanical processes due to welding and on the state of art in hot-cracking assessment, then we can develop more efficient engineering tools for their prevention in stages of manufacture.

## Weldment Cracking

The welding procedure supplies the thermal energy for local fusion with the intensity and efficiency of the heat source (arc, beam). The material of the assembled weld component absorbs the heat by its thermal capacity and mass (volume), with heat losses due to heat transfer by resistance, convection and emission. The heat dissipates thermodynamically into the structures' volume by heat conductivity, changing the physical properties of the metal by thermal expansion. Quasi-steady temperature fields and temperature gradients around the moving welding pool characterize the changing local behaviour, which results in superposition of both the microstructural material transformation and thermo-mechanical structural kinetics, balancing the local thermal, mechanical and constitutional yielding.

Thermal cycles are used to describe the constitutional and thermo-dynamic reaction of the composition of the alloy on solidification and HAZ. They also form the basis for calculation programs for temperature-time-transition, correlation of microstructure and mechanical behaviour, grain growth, and cooling time concepts in general. Quasi-steady temperature fields describe the 2D or 3D thermo-physical and thermo-dynamical heat transport around the weld pool in relation to the welded component, which is used for the calculation of temperature-induced mechanical reactions as distortion, displacement, stress and strain with the help of Finite Element Analysis (ANSYS, ABACUS).

Whether a designed component of a selected material is weldable or not cannot be measured directly. It is the result of balancing equilibrium criteria which decides whether the welding-process treated material of the component to will yield wether micro-fissuring will be initiated, or whether break at usage will occur.

## Classification of Weldment Cracking

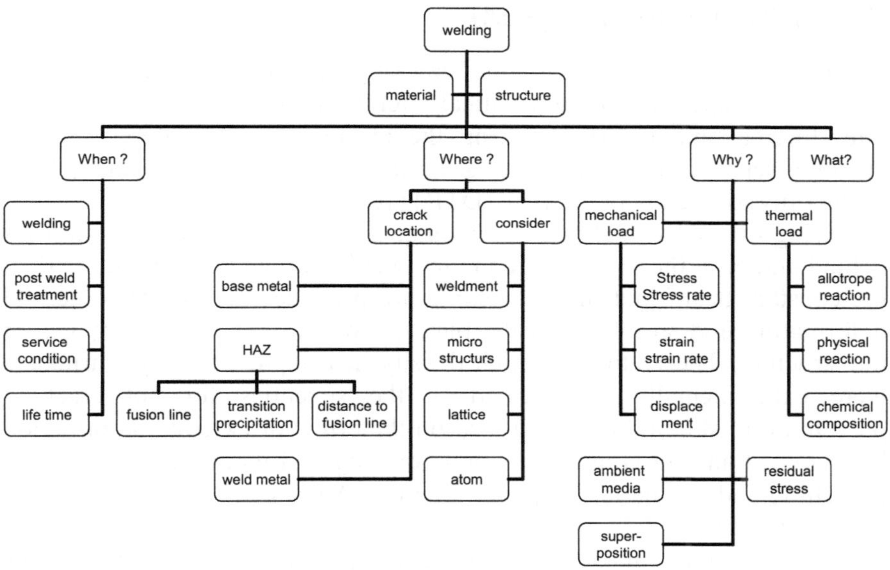


**Fig. 1.** Classification of weldment cracking

The classification of weldment cracking in Fig. 1 verifies not only welding processes in fabrication but also post-weld heat-treatment like stress relief, special cladding in automotive industry and higher temperature service conditions on lifetime elicited by manufacture. Each further heat-treatment, post-weld treatment, repair welding, service conditions at higher temperatures will result in local material reactions, superposing that of the prior welding and damaging the material locally, and will initiate weldment cracking.

When we investigate how a crack is initiated, where and why the crack occurs and what the crack's path or way is, than we consider existing cracks at weldments. The different weldment cracking types can be traced back to the weakest link of their microstructure where, on the one side, the local material failure of the welding procedure conditions occur due to metal-constitutional, metal-physical and thermo-mechanical conditions, and, on the other side, metal-physical laws of dislocation movement in heat-affected and deformed material act. The failure of a welded component is always related to local micro-structural disruptions of the weakest link of the whole structure.

## Tools for the Characterisation of Solidification Cracking

### *Finite Element Analysis (FEA)*

In the sixties, the analysis of hot cracking proved that crack initiation depends on superposition of a critical internal deformation or deformation rate (thermal and external) during the existence of a material-specific cracking-sensitive temperature range (brittleness temperature range (BTR) or ductility loss temperature rage (DTR) [1, 2, 3].

During one-sided welding processing, the liquid weld metal solidifies behind and/or at the sides of the molten weld pool, while at the same time the constant heat input of the welding arc produces non-uniform thermal deformation and shrinkage in the plates, especially at transient stages of the structure. The features of high density heat input and large shrinkage were experimentally investigated in the seventies for hot cracking prevention in shipbuilding, focussed on the effect of external deformation behaviour, transient strain, seam-end cracking and its prevention, and for modelling and simulation by Finite Element Analysis (FEA) [2, 3, 4, 5, 6, 7, 8, 9, 10, 11].

In the eighties, mostly experimental and simulative investigations followed, concentrated on the effect of root gap changing during welding of butt-weld joints [2, 3, 4, 5], with partially systematic and quantitative theoretical considerations. The results of welding shrinkage computed by FEA were compared with experimental results in the nineties, when the effects of local heating, welding sequences and constraint conditions on the transverse shrinkage of large plates [3, 4, 5, 6] as well as the effect of precision ship assembling came under discussion. All these investigations described and calculated the correlation between solidification cracking, heat input, welding processing and deformation fundamentally and in detail, but the results were not related to each other for assessing the hot crack risk during one-sided welding of real components due to transient stages.

Further FEA simulation [4] was executed to determine the local residual stress and strain distribution along the seam in a welded component with respect to the position of hot cracks. Hot cracking prevention has been carried out in this principle case by welding-procedure associated effects, minimizing the non-thermal high-temperature transverse strain to between

1400 and 1000°C by reduced heat input per unit length and increased weld speed all over the welded seam as well as by special designing and assembling for welding [5, 6].

Welding reliability in structural transient stages has become more important as refers to hot cracking prevention and hot cracking assessment.

### *Microscopy and Fractography*

Microscopy offers increasingly precise and fundamental contribution in identifying micro-structural and allotrope constituents (by analysing mixed crystals, precipitations, inter-metallic and other special phases), their size and distribution within the grain or at the grain boundaries. Furthermore, micro-fractography gave progress to classification and better understanding of grain-boundary constituents and interactions that bring about solidification cracking.

### *Weldability Tests*

Numerous special weldability test procedures were created to quantify the various effects of sensitivity on hot cracking. The different hot cracking test procedures and their comparability were discussed on international scale [7], which resulted in successful European standardisation. While the theoretical hot cracking research is focussed on local thermo-mechanical kinetics and metallurgical dynamics during solidification, the hot-cracking test methods applied in practice are based on global concepts with measurable constitutional, mechanical, thermal and time-affected test criteria [8]. Their development has changed from descriptions of hot cracking phenomena by yes/no criteria (test procedure with self-loading specimens [9]) to quantification of hot cracking sensitivity by thermo-physical simulation as well as directly in weldment during welding procedures, using external loading for the specimens [10].

First, the PVR-test [11] was applied to quantify hot cracking sensitivity of weld and base metal. In this, the material was evaluated, ranking from hot cracking sensitivity to solidification cracking, liquation cracking as well as ductility dip cracking. Crack sensitive bands in the HAZ (for steady stages) with determined distances to the fusion line in accordance with the crack sensitising temperature ranges induced by the welding process were defined [12]. Second, the PVR test was modelled with the help of FEA [13] to describe the origin of hot cracking in welding. That is the correlation of the moving temperature field and internal deformation within the BTR, derived mathematically as the changed deformation BTR within the BTR dependent on time. Third, the PVR-test was applied to qualify welding procedures by tailored-to-material heat input, suited filler metal and consumables, examining the effect of shielding gases and dilution level on hot cracking [14]. The PVR test was compared with other test procedures applied in Magdeburg, like MIS Test and DSI hot ductility test which work with externally loaded specimens [15].

The analysis of different hot cracking test procedures with externally loaded specimens proved that for the formation of all hot cracking types superposition of a critical deformation speed within a material-specific temperature range is necessary. The premises for any hot cracking assessment in welding are defined equilibrium criteria, quantifying the local processing conditions for crack initiation of metal-physical associated crack mechanisms at the moment of instability. The critical values for weldment reliability cannot be directly examined in hot cracking tests.

## Role of Fracture Mechanics in Respect to Weldment Cracking

The guide *Fracture mechanical assessment of cracks on welded components* was intended for life-time reliability assessment of locally damaged structures to exclude critical failure [16]. Measurement of weld-zone fracture toughness by $K_{IC}$, CTOD and critical J-Integral value was recommended by AWS and BS [17] not only for failure assessment.

About 60% of applied fracture-mechanical failure analyses were related to weld imperfections which had grown into macroscopic scale during lifetime, thus causing local failure at the weakest link in the microstructure in a once thermo-mechanically overstressed area within the entire mechanically loaded welded component. In classic fracture mechanics, different crack-criteria balance the equilibrium conditions relating to stress, stress-intensity factor and crack size. If we can understand the special criteria for the initiation of welding cracks or hot cracking, then we can, based on this knowledge, develop tools for its prevention.

In the early nineties, the need for higher guarantees for reliability of welded structural components resulted in the *IIW-case study collection on the assessment of significance of weld imperfections* In this, based on examples, recommendations on significance and assessment of weld imperfections with respect to structural stress-strain analysis were offered, comparisons between predicted and true failure loads were made, material and welding-process conditions were analysed, and conclusions about repair and other were drawn [18].

Figure 2 considers the question: Which aspects of the classic fracture-mechanic associated assessment can be transferred to a welding process associated assessment? On the left-hand side, a fracture-mechanic test procedure for measuring fracture toughness ($K_{IC}$, CTOD and critical

J-Integral) as a method for the assessment of failure due to measurable cracks that have grown in a welded component under load, and related Failure Assessment Diagrams FAD for calculation of the failure risk can be see. Failure assessment is a special tool for defining component reliability, derived from fracture mechanics, in order to consider failure due to special service conditions (as relate to lifetime reliability of plants at high-temperature, creep, corrosion, e.g.).

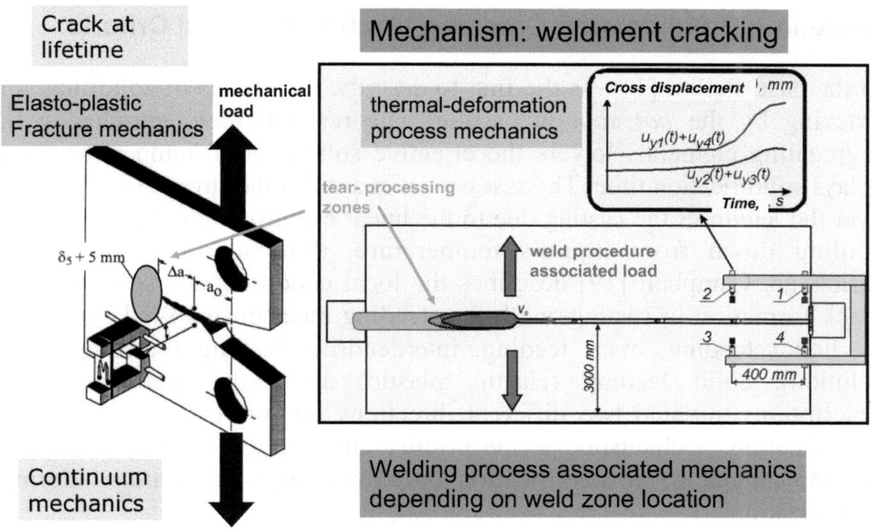


**Fig. 2.** Comparison of classic fracture mechanics and welding-process associated assessment

On the right, the welding-process associated test procedure determines the material resistance to solidification cracking in welding procedures as a method for assessing solidification-cracking initiation on the structural component due to the welding procedure applied.

Both assessment concepts differ, as a rule, in the way a load is applied (externally applied mechanical load and welding-process induced complex thermal-physical and thermo-mechanical load), and in the type the inhomogeneous processing zone (estimation of crack initiation in welding as the critical the crack stage or crack propagation due to crack growth at lifetime) is defined The concepts agree with the rule of geometrical relations between the moving processing zone and the structural component during welding.

It also follows that material properties as toughness defined in notch impact tests as well as weldment cracking sensitivity defined in weldability tests are not suitable for assessing structural failure or crack initiation at welded structural components.

## Assessment of Solidification Cracking

### Modelling Solidification Cracking Mechanisms and Criteria

In the past, Pellini [1] was the first to describe the effects of solidification cracking by the *hot spot* in casting. The residual melt, enriched with segregating elements, lowers the effective solidification temperature and delays solidification time. The assessment quantifies the stress-concentration over the length of the casting due to the linear expansion coefficient during cooling down from liquid`s temperature, thus producing strain by shrinkage. Campbell [19] describes the local conditions for solidification crack formation in castings with five feeding mechanisms fundamentally, by liquid feeding, mass feeding, interdendritic feeding (eutectic, solid solution), solid feeding (elastic, plastic) and burst feeding. These descriptions initiated two different directions for further investigation of solidification mechanisms, giving priority either to the effects of thermo-mechanical changes at solidification, or to the changing constitution during solidification.

### Criteria Determined by Constitutional Solidification

The following models of solidification (without considering solidification cracking) quantify equilibrium conditions in respect to their constitutional change. Hansen correlates solidification speed with shrinkage speed as hot-ductile crack criteria [20]. The microstructure-process-relation by Hunt [21] quantifies the secondary dendrite arm distance as a function of the time for solidification within the brittleness temperature range (BTR). Feurer [22], defines the *hot tearing susceptibility* by lacking volume due to solidification shrinkage and continuous feeding. In the crack-susceptible liquid-solid state, the *cracking susceptibility coefficient* by Clyne et al. [23, 24] correlates the time needed for liquid flow and mass flow at solidification with the time for stress relaxation and tearing between the dendrites depending on its solid fraction, using the solidification model by Scheil. The welding pool can be understood as a small and rapidly solidifying casting between components, or rather as continuous casting, if we consider different types of feeding mechanisms and microstructure-process-relations, quantifying the secondary dendrite arm distances as a function of solidification time during the BTR.

Rappaz-Drezet-Gremaud established the *hot tearing initiation model* (RDG criterion), combining local strain with mass balance between dendrites and solidifying interdendritic liquid [25]. This was applied in laser beam welding with Al-alloys, with special attention to preventing solidification cracking at the transient stages at weld start and weld end [26]. The RDG criterion was extended to steel by Drezet et al. [27], taking into account the peritectic reaction that transforms δ-ferrite into austenite. Further developments of this RDG criterion by Rindler et al. [28] release the assumption that the uniform strain-rate acts troughout the mush state (BTR).

For all models, the strain-rate acting throughout the mush state (within the BTR, during existing time of BTR) is the most important factor in the solidifying weld metal both for local constitutional and thermo-physical mechanical determination.

## Criteria Determined by Structural Mechanisms

As established by Prokhorov [2] and Matsuda [3], in opposite to castings, the whole welded structural component including weld metal absorb all thermo-mechanically induced local deformation, strain and strain-rate during solidification shrinkage within the BTR, which are associated to the welding-process, for the prevention of hot cracking. According to the theory of technological strength by Prokhorov [29], solidification cracks will always occur when a minimum deformation δ[%] and a critical deformation rate $B_{cr}$ [%/°C] is reached or exceeded in the BTR. The deformation rate $B_{BTR}$ is to be understood as modification of the deformation $d\varepsilon_i/dT$ in the BTR ($B=d\varepsilon_i/dT$ [%/°C]). The amount of deformation $\varepsilon_I$ corresponds to the sum of thermal shrinkage $\varepsilon_T$ and external deformation $\varepsilon_{ext}$ ($\varepsilon_i = \varepsilon_T + \varepsilon_{ext}$).

For easier application, Prokhorov's more theoretical criteria were transformed to centreline solidification cracking, defining the critical speed of cross displacement within the BTR $v_{cr}$ [23, 24], and modified by the amount of cross displacement changing during time of the existing BTR [30]. Both criteria are useable for solidification cracking assessment, balancing the critical value of criteria with the FEA computed welding process, material and structure induced values of local cross displacement and their speed.

## Assessment Concepts

Hot cracking occurs when as well the critical cross displacements during the existence time of the BTR *Δuy min* are exceeded by the total cross displacement *Δuy BTR*, due to component design and assembly for welding, and/or the critical speed of cross displacement is exceeded by the local speed of cross displacement [31, 32, 48].

The welding-process associated local speeds of cross displacement $v_q$ are calculated by FEA with the help of the program ANSYS at discrete points along the seam during one-side welding of the actual welding components, taking into account the effects of design and assembly by dimension of run-outs, tacks and magnetic clamping. The speed of cross displacement $v_q$ in Fig. 3 will maximize systematically when the arc comes near the plate's end, and the far side of the weld pool passes from the plate to the run-out due to the stiffness jump. It increases locally during fusion of each tack, additionally releasing assembling stresses. These short and rapid increases of cross displacement with regard to time and versus the welded seam are the most important factors to initiate centerline solidification cracking of longer butt weld seams. However, the estimation of the critical $v_{cr}$ needs an empirical assessment diagram for determination of the critical speed of cross displacement $v_{cr}$ in correlation with the applied material and welding process.

The assessment diagram in Fig. 4 offers the complexity of hot cracking by three sub-diagrams. The assessment diagram for centreline solidification cracking can be seen on the right side. The multitude of calculated curves of the weld component and process-associated speed of cross displacements $v_q$, measured with the help of the welding-process associated hot-cracking test method, are plotted here as cooling time of the weld metal versus the change of the measured and calculated cross displacement ΔU (t) in [mm]. The critical speed of cross displacement $v_{cr}$ is defined here as the threshold curve, separating the regions with occurring hot cracking from crack-free regions, established by empirical approximation. The sub-diagram in the centre shows the correlation with the hot cracking theory by Prokhorov, setting up internal deformation versus temperature. The left sub-diagram shows the thermal cycle of the applied welding procedure.

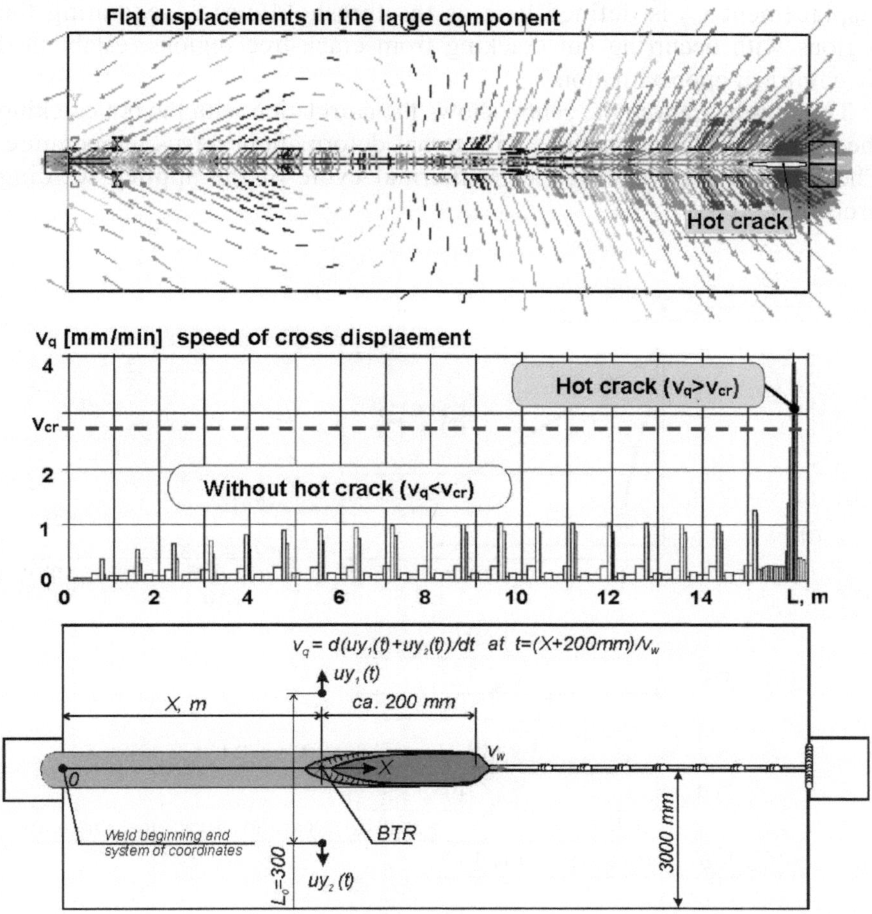


**Fig. 3.** Distribution of flat displacements (arrows value amount and direction) in large plates during one-side submerged arc welding (sub-fig. *above*) and the speed of cross displacement after passing the plate end by the welding pool $v_q$, calculated with the help of FEA simulation (sub-fig. *middle*) in correlation with the applied welding procedure with taking into account designing and assembling for welding in principle (sub-fig. *below*) [49]

The multitude of calculated curves of the weld component and process-associated speed of cross displacements $v_q$, measured with the help of the welding-process associated hot-cracking test method, are plotted here as cooling time of the weld metal versus the change of the measured and calculated cross displacement ΔU (t) in [mm]. The critical speed of cross displacement $v_{cr}$ is defined here as the threshold curve, separating the regions with occurring hot cracking from crack-free regions, established by empirical approximation.

The sub-diagram in the centre shows the correlation with the hot cracking theory by Prokhorov, setting up internal deformation versus temperature. The left sub-diagram shows the thermal cycle of the applied welding procedure.

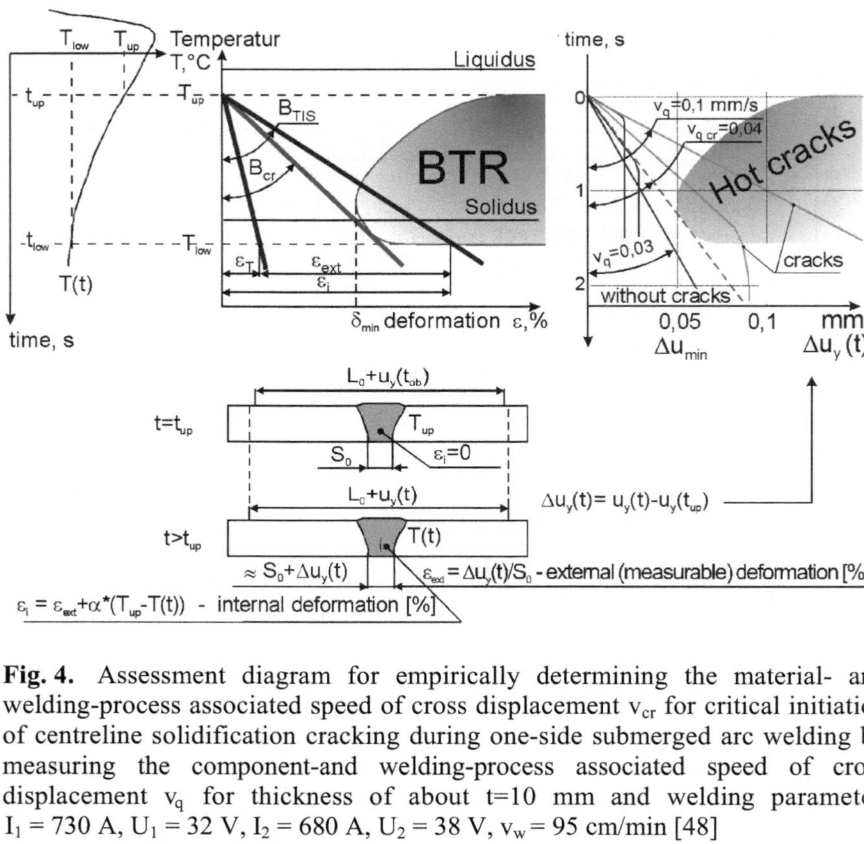


**Fig. 4.** Assessment diagram for empirically determining the material- and welding-process associated speed of cross displacement $v_{cr}$ for critical initiation of centreline solidification cracking during one-side submerged arc welding by measuring the component-and welding-process associated speed of cross displacement $v_q$ for thickness of about t=10 mm and welding parameter: $I_1 = 730$ A, $U_1 = 32$ V, $I_2 = 680$ A, $U_2 = 38$ V, $v_w = 95$ cm/min [48]

The assessment diagram in Fig. 4 is applicable in correlation with FEA simulation of large components in order to develop new methods of crack-free design and assembling for welding relevant and large components, as well as for refining metallurgical effects and modifying welding procedures. However, all variations in welding procedures will change the location of the threshold curve of the tested welding procedure, thus affecting the material and metallurgical effects as well as the component and assembling effects. This high sensitivity to welding-procedure induced effects underlines the limited expressiveness of the laboratory test procedures in view of the reliability of the weldment.

The computational procedure of solidification cracking assessment combines thermal and structural analyses, due to the applied welding procedure on the structural component of a selected material, with the implementation of local analysis of the balancing solidification cracking criteria (modified cross displacement and speed of cross displacement, existing during time of BTR). Crack initiation is proved by checking exceeded critical criteria values for each discrete FEA noodle point [33].

Solidification assessment concepts under the use of strain rates are considerably gaining in importance if the indication of different assessment concepts and criteria, applied independently of one another to similar structures (butt-joints and T-joints of the same material) agree in principle [44, 51] and follow different realization. The FEA model by Drezet et al. [44] based on the RDG criteria is applied in studying the influence of laser beam welding on the strain-rate in the BTR by FEA to prevent solidification cracking at transient stages on weld start and weld end. Successful optimization of the welding procedure was achieved by increased laser power at run-in, constant power in linear run and decreased power at run-out with effective clamping in this case. While [44] realizes compensating laser beam power along the seam, [51] uses structural design and assembling for welding in combination with thermal effects by welding direction and sequence.

The thermo-mechanical conditions for weld-metal solidification cracking of Al-alloys were modelled in the simulation model by Feng, Zaharia and David [34], the local stress-strain evolution within the BTR at the weld centreline solidification crack site having also been calculated. The combination of heat transfer model with mechanical model and solidification behaviour of the weld pool results in a concept of a computation model, which acts as key concept for quantitative assessment on thermo-mechanical FEA formulae. That means that the microscopic deformation, which relates directly to the cracking process, is controlled to a significant degree by the macroscopic deformation process, which obeys the principles of continuum solid mechanics and heat transfer. The concept was used to prove the relation between the macroscopic thermo-mechanical condition and the local effects produced by the trailing edge of the welding pool due to the welding operation (welding parameter, joint design, component restraint).

Murakawa et al. [35, 36] developed a Temperature Dependent Interface Element, as it is not possible to apply a simple thermal elastic-plastic FEA for analysing deformation and residual stress in welding in the Houldcroft test under the use of fish bone specimens. In their model, they adopted (crack) opening displacement from elastic-plastic fracture mechanics, considering the crack propagation as the formation of new surfaces, without taking into account the effects of temperature on the strength of the interface element. The influence of BTR width and the opening displacement at the crack were examined, from which followed that the opening displacement must be adjusted together with the BTR width, so that the temperature-dependent interface element represents the solidification cracking susceptibility of the material.

Ploshikhin et al. [37] developed a thermo-mechanical and metallurgical model for the solidification cracking mechanism in welds. These solidification cracking criteria are based on the tensile strain accumulated in the segregating inter-granular liquid film at the weld centreline by the solidified dendrites from the completely solid region of the base material until cracking appears. The local critical parameter $\delta_{acc}$ represents the ability of the segregating liquid film to accumulate strains without tearing, measured by the micro-structural morphology (by fraction of residual liquid, primary and secondary dendrite arm spacing). Solidification cracking will result when the maximum accumulated strain exceeds the critical value.

## Conclusion

The analysis of weldment cracking especially solidification cracking in welding proved that the crack initiation depends on the superposition of a critical deformation rate (strain rate, cross displacement, accumulated strain) within a material-specific temperature range, e.g. brittleness temperature range (BTR) with intergranular changing yielding capacity due to constitutional- time-temperature- and shrinking conditions which exceed the critical value of the defined cracking criteria.

The premises for any hot cracking assessment on welding are defined equilibrium criteria, quantifying the local processing conditions for crack initiation of metal-physical associated crack mechanisms of instability at the moment of passing the BTR. The critical values for weldment reliability cannot be directly examined in hot cracking tests.

The possibility and validity of solidification crack initiation assessment has been proved by different assessment concepts and criteria, which were created independently of one another and applied to similarly structured butt-joints, T-joints of aluminium alloys, and one-side welding of special structured components of ferritic steels, and which resulted in a similar tendency.

Solidification cracking assessment for welded structural components of austenitic steel or nickel-base alloys have not been investigated, so far.

If we can analyse the cracking processes due to welding procedure more in detail, focussing on the local microscopic conditions, then we can also develop new engineering tools, based on the state of art in actual welding-process-associated tools, by computational combination of the data of thermo-physical, mechanical, metallurgical and structural behaviour, with the aim of prevention of weldment cracking.

## References

1. Pellini WSF (1952) Foundry 80:125–133, 192, 194, 196, 199
2. Satoh K, Terasaki T et al. (1980) Changes of root gap during welding in case of butt weld joints (1st Report). J Japan Weld Soc vol 49 7:478–483 (in Japanese)
3. Ueda Y, Murakawa H et al. (1992) Simulation of welding deformation for precision ship assembling (Report I) – In-plane deformation of butt welded plate. Trans JWRI vol 21 2:125–135
4. Andersson B, Karlsson L (1981) Thermal stresses in large butt-welded plates. J Thermal Stress 4:491–500
5. Herold H, Streitenberger M, Pchennikov A, Makarov E (1998) Modelling of one sided welding to describe hot cracking at the end of longer butt weld seams. (1999) Weld in the World vol 43 2:56–64
6. Herold H, Streitenberger M, Pchennikov A (2001) Prevention of centreline solidification cracking during one-side welding. Doc IIW-Doc. IX-2000-01
7. Wilken K (1999) Investigation to compare hot cracking tests – Externally loaded specimen. (1999) Doc IIW IX-1945–99
8. PrEN ISO 17641-1 (Nov. 2003) Destructive tests on welds in metallic materials – Hot cracking tests for weldments – Arc welding processes – Part 1: General (ISO/FDIS 17641-1:1003)
9. PrEN ISO 17641-2 (Nov. 2003) Destructive tests on welds in metallic materials – Hot cracking tests for weldments – Arc welding processes – Part 2: Self-restraint tests (ISO/FDIS 17641-2:1003)
10. prCEN ISO/TR 17641-3 (Nov. 2003) Destructive tests on welds in metallic materials – Hot cracking tests for weldments – Arc welding processes – Part 3: Externally loaded tests (ISO/DTR 17641-3:1003)
11. Folkhard E (1984) Metallurgie der Schweißung nichtrostender Stähle. Springer, Wien New York, pp 153
12. Herold H, Schulze S, Streitenberger M, Spieler S (1998) Mikroschädigungen von NiMo-Werkstoffen. In: Petzow G. (eds) Praktische Metallographie 29 Fortschritte in der Metallographie. DGM Informationsgesellschaft mbH, pp 175–180

13. Herold H, Streitenberger M, Pchennikov A (2001) Hot Cracking Theory by Prokhorov and Modelling of the PVR-test. Weld World vol 45:3/4, 17–22
14. Herold H, Zinke M, Hübner A (2004) Investigation on the use of nitrogen shielding gas in welding and its influence on hot crack behaviour of high-temperature resistant fully austenitic Ni- and Fe-base Alloys. IIW-Doc IX-2110-04
15. Herold H, Pchennikov A, Streitenberger M (2005) Influence of deformation rate of different tests on hot cracking formation. In: Böllinghaus T, Herold H (eds) Hot cracking phenomena in welds. Springer, Berlin Heidelberg, pp 328–346
16. Merkblatt DVS 2401 (Aug. 2004) Fracture mechanical assessment of cracks on welded components. Fachbuchreihe Schweißtechnik Bd. 101. DVS-Verlag GmbH, Düsseldorf
17. BS 7448 Part 2: Seventh draft (revised) Draft British Standard (1996) Fracture mechanics toughness test. Part 2: method for determination of $K_{IC}$, critical CTOD and critical J valued of weld in metallic materials
18. Blauel JG (1993) IIW-case study collection on the assessment of the significance of weld imperfection. IIW-Doc. X-1280-93, XE 35-93
19. Campbell J (1991) Castings, Butterworth-Heinemann-Ltd. pp 175–240
20. Hunt JD (1979) Solidification and Casting of metals. The Metals Society London
21. Hunt JD (1979) Solidification and Casting of metals. The Metals Society London
22. Feurer U (1976) Gießerei 23:75–80
23. Clyne TW, Davies GJ (1981) Brit Foundry vol 74 4:65–73
24. Clyne TW, Kurz W (1981) Metallurgical trans vol A 12:965–971
25. Rappaz M, Drezet J-M, Gremaud M (1999) A New Hot-Tearing Criterion. Metallurgical and Materials trans vol 30A 2:449–455
26. Drezet J-M, Mathier V, Allehoux D (2006) The modeling of laser beam welding of Aluminium alloys with special attention to hot cracking in transient regimes. In: Cerjak H, Bhadeshia HK, Kozeschnik E (eds) Math. Modelling of Welding Phenomena 8, presented
27. Drezet J-M, Gremaud M, Graf R Gräumann M (2002) A new hot tearing criterion for steel. In: Proc 4th International Continuous Casting Conference, IOM communications, Birmingham, UK, pp 755–763
28. Rindler W, Kozeschnik E, Enzinger N, Buchmayr B (2002) A modifyied hot tearing criterion for steels. In: Cerjak H, Bhadeshia HK (eds) Math. Modelling of Welding Phenomena 6, IOM Communication, UK, pp 819–835
29. Prokhorov NN, Prokhorov N (1968) Schweißtechnik (Berlin) vol 19 1:8–11
30. Pchennikov A, Streitenberger M, Herold H (2004) Application and further development of Prokhorov`s solidification-cracking theory. IIW-Doc. IX-2160r1-05
31. Herold H, Pchennikov A, Streitenberger M (2004) Assessment of hot cracking initiation by laboratory test procedures and FEA simulation associated experimental measurements during welding of large weld components. Quarterly Journal of Japan Welding Society, vol 22 2:211–217

32. Pchennikov A (2005) Entwicklung von Maßnahmen zur Heißrissvermeidung beim Einseitenschweißen langer Nähte. Diss, Otto-von-Guericke Universität Magdeburg
33. Pchennikov A, Gruss H, Herold H (2006) Avoidance of hot cracking at unsteady weld start on laser beam welding of Aluminium. In: Cerjak H, Bhadeshia HK, Kozeschnik E (eds) Math. Modelling of Welding Phenomena 8, presented
34. Feng Z, Zaharia T, David SA (1997) On the thermomechanical conditions for weld metal solidification cracking. In: Cerjak H, Bhadeshia HK (eds) Math. Modelling of Welding Phenomena 3, The Institute of Materials, Cambridge UK, pp 114–150
35. Shibahara M, Serizawa H, Murakawa H (1999) Finite Element Method for hot cracking using temperature dependent interface element. Trans JWRI vol 28 1:47–53
36. Shibahara M, Serizawa H, Murakawa H (2000) Finite Element Method for hot cracking using temperature dependent interface element (Report II). Trans JWRI vol 29 1:59–64
37. Ploshikhin V, Prikhodovsky A, Makhutin M, Llin A, Zoch H-W (2005) Integrated mechanical-metallurgical approach to modelling of solidification cracking in welds. In: Böllinghaus T, Herold H (eds) Hot Cracking Phenomena in Welds 1, Springer, Berlin Heidelberg, pp 223–244

# Determination of Critical Strain Rate for Solidification Cracking by Numerical Simulation

M. Wolf, Th. Kannengießer, Th. Böllinghaus

Federal Institute for Materials Research and Testing, Berlin, Germany

## Abstract

Hot crack prevention during welding processing of metallic materials is an essential prerequisite of component safety. The causes of solidification cracking can be attributed to the occurrence of metallurgical and thermomechanical effects. Even though numerous hot cracking test procedures have been developed until now, unexpected solidification cracking during component welding cannot be avoided, since specifically in the hot cracking test the thermomechanical conditions and the local crack-critical strain rates around the weld pool may be very different from those in the component.

In the presented numerical analyses, contrary to the well-known energy distribution models (according to Goldak, for example), the experimentally determined weld pool geometry has directly been implemented as a 3D-function into the numerical simulations and thus conduced to high computational accuracy.

The investigations were carried out using nickel-base Alloy 602 CA, since its solidification resistance exhibits a significant dependence on the shielding gas, which enabled studies of hot crack-critical as well as hot crack-uncritical material behaviour. Validation was carried out with the help of the MVT-test which allowed additional variation of the specimen loading rate. Numerical calculation of the solidification crack-critical limiting temperature versus various welding parameters and the investigated shielding gases could be performed. In order to provide a transferability of hot cracking tests to components, critical specimen loading rates were determined and the local crack-critical strain rates in close vicinity of a weld pool were calculated. It is demonstrated that the local critical strains and strain rates represent a crack criterion for a transferability.

## Introduction

Hot crack prevention during fabrication and processing of metallic materials is an essential pre-requisite of component safety. The causes of hot cracking can be attributed to the occurrence of metallurgical and thermomechanical effects.

In order to define hot cracking, a solidification crack-critical limiting temperature $T_{SR}$ can be drawn upon according to which solidification cracks occur beyond 0.5 $T_M$ ($T_M$ melting point) [1]. Such solidification cracks appear at a stage where the melt is distributed like a film along grain boundaries. The precise processes involved in solidification crack formation are, however, difficult to quantify experimentally. For examining metallurgical influences on hot crack formation, hot cracking tests using externally loaded specimens are an obvious choice enabling deliberate variation of the loading conditions [1, 2].

In order to ensure transferability of hot cracking test result to real components, it is exactly the action of thermomechanical effects and of external geometric boundary conditions (restraint intensity, tacks) that needs to be quantified. [3, 4]. Specifically, local critical strains and strain rates, respectively, are difficult to determine experimentally, on account of the high temperatures arising in the immediate vicinity of a weld pool [5]. Because of the low chances of measured thermomechanical data collection, numerical simulation of hot cracking is gaining in importance. In order to take account of the fluid-mechanical behaviour of the melt in numerical calculations, this study presents a new method of directly implementing the weld pool geometry in the numerical simulations. This allows it to use the reconstructed weld pool geometry as a given function exactly considering the change of the weld pool shape versus the welding speed. In the next step, numerical MVT-test simulations will be performed in order to calculate local crack-critical strains and strain rates, respectively.

## Experimental Procedure

### Test Materials and Welding Parameters

The investigations were carried out selecting Ni-base Alloy 602 CA (Material No. 2.4633) as test material with composition given in Table 1 [6]. It was available in solution-annealed condition. In preceding MVT- and PVR-hot cracking tests, the metallurgical effect of N on hot cracking resistance was already identified [7]. It was hence possible, by the use of two different shielding gases, to investigate a hot crack-critical (shielding gas: pure Ar) as well as a hot crack-uncritical (shielding gas: 1% $N_2$ + 99% Ar) material behaviour for Alloy 602 CA.

**Table 1.** Chemical composition of Alloy 602 CA, percent by weight

| Ni | Cr | Fe | Mn | Ti | C | Zr | Cu | Mo | Si | Al | Y |
|---|---|---|---|---|---|---|---|---|---|---|---|
| 62.1 | 25.6 | 9.45 | 0.07 | 0.14 | 0.165 | 0.1 | 0.01 | 0.05 | 0.04 | 2.35 | 0.06 |

The numerical calculations were verified with the help of the MVT-Test [8], in which the specimen loading rate was varied additionally [1, 2]. For quantification of the mechanical specimen loading, strain gauges were applied to the specimens subjected to MVT-testing using various die radii. The size of MVT-specimens was 100 mm × 40 mm × 10 mm. The various welding parameters can be seen from Table 2. The tests P1 – P4 were conducted in combination with the shielding gases Ar as well as Ar + 1% $N_2$.

**Table 2.** Welding parameters for TIG-welding

| Welding parameter | Current I [A] | Voltage U [V] | Welding speed $v_s$ [mm/s] | Heat input $E_S$ [kJ/cm] |
|---|---|---|---|---|
| P1 | 250 | 18 | 5 | 9.0 |
| P2 | 182 | 17 | 3 | 10.3 |
| P3 | 205 | 17 | 1.8 | 19.4 |
| P4 | 200 | 17 | 6 | 5.7 |

## Numerical Approach

The numerical simulations for the MVT-2-Test were performed using ANSYS® FEM-software. In order to take account of the mechanical specimen deformation, special contact and target elements were employed. The material behaviour within each element was assumed to be homogeneous and isotropic. The heat conductivity and the specific enthalpy of Alloy 602 CA are largely characterized by a linear temperature dependence.

As temperature-dependent material characteristics for the thermomechanical calculations, the heat expansion coefficient, the E modulus as well as $\sigma$-$\varepsilon$-functions were available. A bilinear stress-strain behaviour was assumed. The tangential modulus for Alloy 602 CA was determined in tensile tests using 2.13 GPa.

## Test Results and Discussion

### Measurement and Simulation of Mechanical Specimen Deformation in the MVT-Facility

Figure 1 gives an example of the measured strain history depending on the cross-beam travel for a bending strain of 2% in the Varestraint as well as Transvarestraint mode. The cross-beam travel speed was 0.1 mm/s. The strain increase per millimetre of the traverse travel $\Delta\varepsilon/\Delta H$ required for quantifying the mechanical specimen deformation was also determined. The strain increases for $\varepsilon_{tot} = 2\%$ are given in Table 3.

**Table 3.** Strain increase per mm of cross-beam travel H (MVT-Test)

| Specimen thickness 10 mm | Strain increase per mm of cross-beam travel $\Delta\varepsilon/\Delta H$ [1/mm] | | | |
|---|---|---|---|---|
| Bending strain $\varepsilon_{tot}$ | Measurement | | Simulation | |
| | Varestraint | Transvarestraint | Varestraint | Transvarestraint |
| 2% | 0.040 | 0.040 | 0.041 | 0.042 |

Calibration and validation of the numerical calculations accompanying the MVT-Test was carried out using the Transvarestraint model represented in Fig. 2.

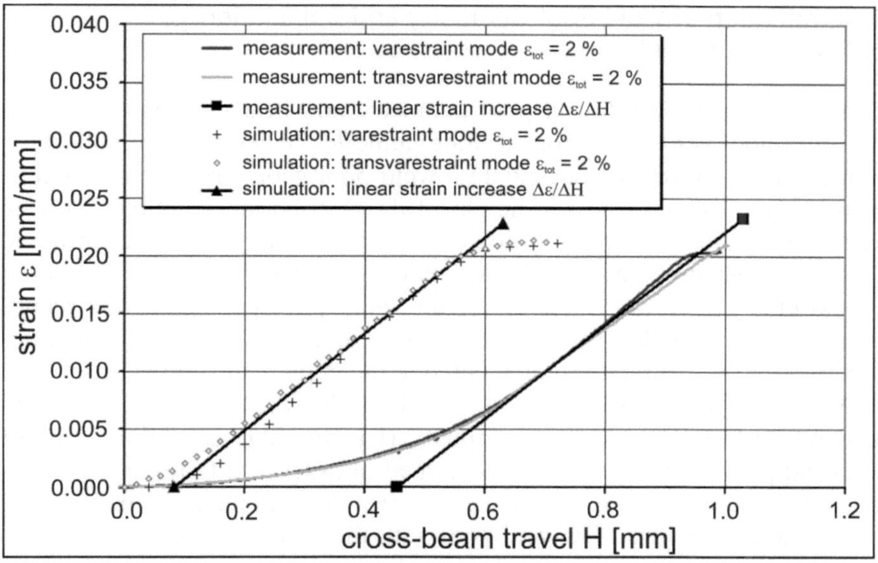


**Fig. 1.** Simulated and measured strain in the Varestraint and Transvarestraint mode (MVT), specimen thickness: 10 mm, die radius $R_M$ = 250 mm

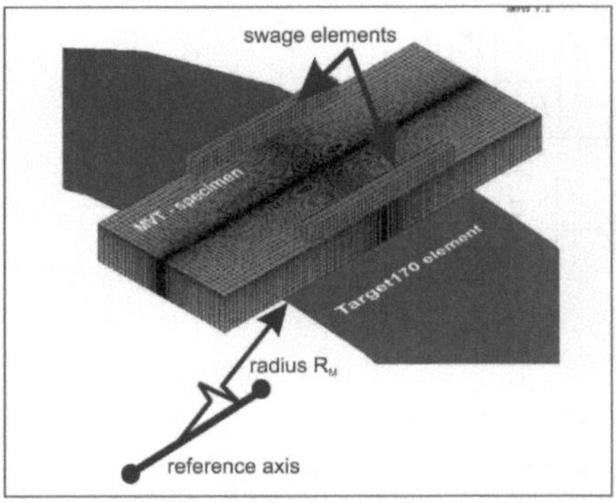


**Fig. 2.** Numerical model for the MVT-Test (Transvarestraint mode)

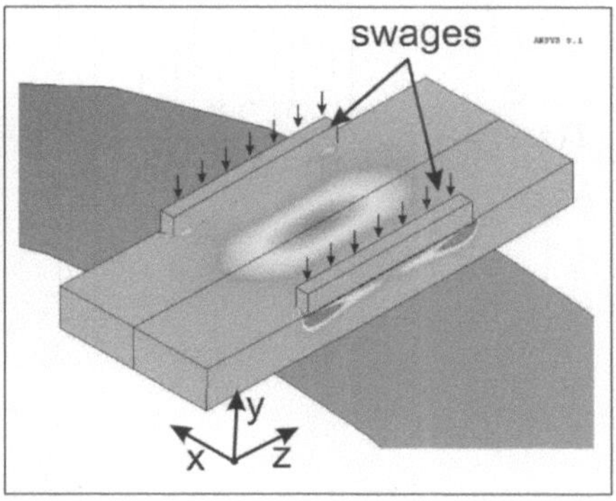


**Fig. 3.** Strain distribution in x_direction after bending (Transvarestraint mode, $R_M$ = 250 mm)

Figure 3 represents the numerically calculated strain distribution in an MVT-specimen after bending around a die with the radius $R_M$ = 250 mm.

This corresponds to a bending strain of 2%. Figure 1 and Table 3, respectively, give a comparison between the results of simulation and measurement by the example of 2% bending strain which shows a deviation of < 5%.

### Determination of Weld Pool Geometry and Calculation of Temperature Field

Exact consideration of weld pool geometry and heat distribution around the weld pool is essential for accurate numerical calculation of solidification cracking. In order to calculate a weld pool geometry as much real as possible, the exact shape and height of the distribution functions must be matched to a real arc welding process on an iterative trial and error basis. Doing this using energy distribution models as described in [9, 10] would be extremely laborious. Therefore, the weld pool geometry was reconstructed with the help of transverse sections and of the weld end crater geometries and was directly implemented in the numerical simulations.

As an analytical basic equation, Eq. (1) by Prokhorov [11] was adopted to describe the 3D-weld pool geometry (leading edge or trailing edge of the weld pool). Equation (1) was enlarged in order to differentiate geometry and shape parameters (Table 4). Whereas the geometry parameters represent the dimensions of a weld pool with the values of maximum width, maximum depth and maximum length, the shape parameters detemine the weld pool shape in the symmetry plane, the surface plane and the cross-sectional plane.

$$T(z,x) = -H_{SB} \cdot \frac{\left[1-\left(\left|\frac{z}{L_{x,SB}}\right|\right)^{\mu}\right]^{\frac{1}{\tau}}}{(1-\Delta)^{\theta}+\Delta^{\theta}} \cdot \left(\left\{\left[1-\frac{\left(\left|\frac{x}{B_{SB}}\right|\right)^{\phi}}{\left[1-\left(\left|\frac{z}{L_{x,SB}}\right|\right)^{\nu}\right]^{\frac{\phi}{\eta}}}\right]^{\frac{1}{\omega}} - \Delta\right\} + \Delta^{\theta}\right) \tag{1}$$

**Table 4.** Geometry and shape parameters for Eq. (1)

| Symbol | Definition of the geometry parameters |
|---|---|
| $H_{SB}$ | Maximum depth of the weld pool |
| $L_{x,SB}$ | Maximum length of a weld pool segment |
| $B_{SB}$ | Maximum width of the weld |
| | Definition of the shape parameters |
| $\mu$, $\tau$ | Shape of the crystallization front in the symmetry plane of the weld |
| $\eta$, $\nu$ | Shape of the crystallization front in the surface plane |
| $\phi$, $\omega$ | Shape of the crystallization front in the cross-sectional plane |
| $\Delta$, $\theta$ | Factors for the determination of turning point coordinates |

Figure 4 gives an example of an analytically calculated three-dimensional weld pool trailing edge geometry.

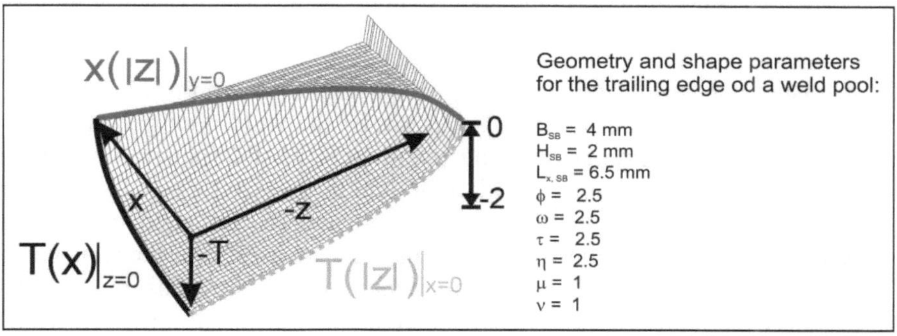


**Fig. 4.** Parameters and three-dimensional function of a weld pool geometry calculated from them

The geometry parameters $B_{SB}$, $L_{x,SB}$ and $H_{SB}$ were determined by measurement of the weld end crater length, penetration depth and weld width. The method of directly implementing the weld pool geometry in the numerical simulations is particularly suitable with regard to the defined specimen geometries and welding conditions in MVT-Tests. It was, for example, possible to generate reproducible weld pool geometries by variation of test parameters such as cross-beam travel speed $v_{trav}$ or total strain $\varepsilon_{tot}$ without the need of geometry and shape parameter matching [12, 13]. In order to determine the geometry and shape parameters, TIG-remelt welds were produced on MVT-specimens using the parameters given in Table 2.

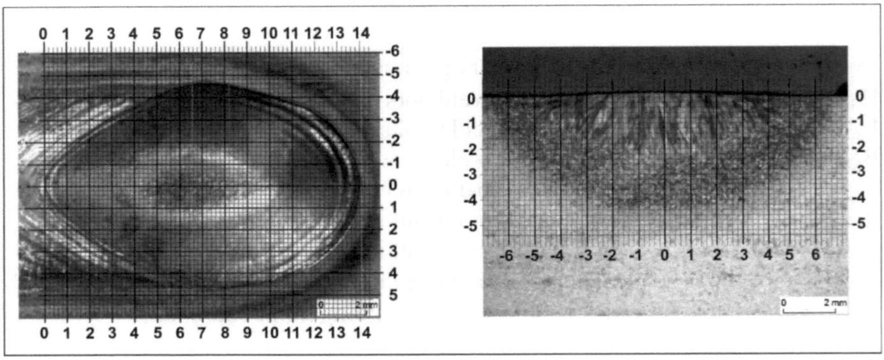


**Fig. 5.** Weld end crater (*left* hand side) and transverse section (*right* hand side) of a TIG-remelt weld of Alloy 602 CA; welding parameters P1; shielding gas: Ar + 1% $N_2$

A raster was projected with a grid spacing of 0.2 mm (Fig. 5). This raster enabled light-microscopical analyses of the weld pool geometry parameters for the individual welding parameters P1–P4 (penetration depth $H_{SB}$, weld width $B_{SB}$, total weld pool length $L_{x,SB} = L_{V,SB} + L_{R,SB}$). For shape parameter determination, the coordinates of the weld end crater functions and of the fusion line functions were measured in the transverse section. Figures 6 and 7 show a comparison between the measured coordinates and each matched curve by respective examples relating to the welding parameters P1 with Ar + 1% $N_2$, for both the crater trailing edge and the fusion line. The weld pool shape determined by this method was dependent on the shielding gas.

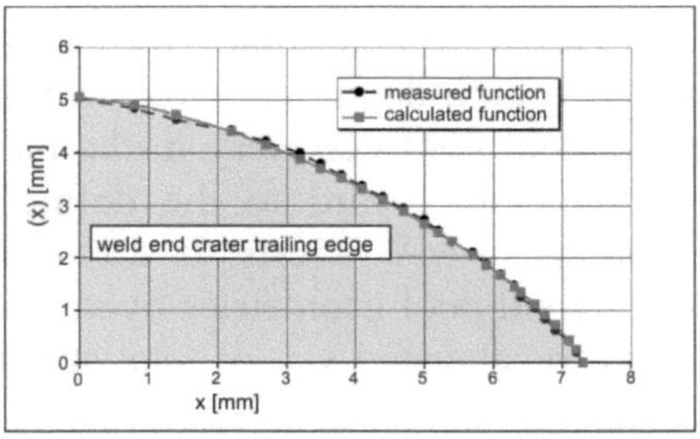


**Fig. 6.** Comparison of the crater trailing edge function for P1 with Ar + 1% $N_2$

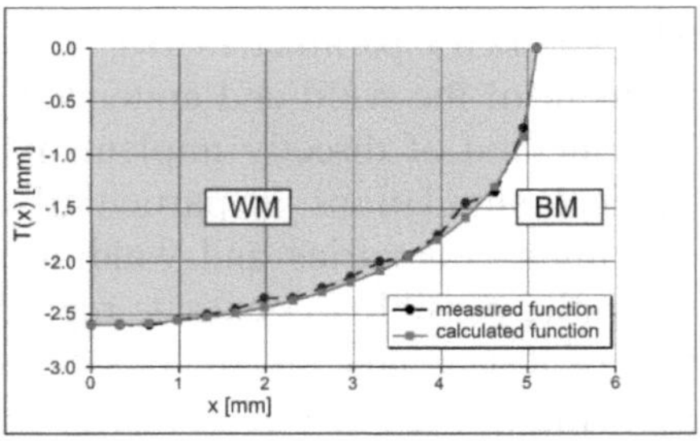


**Fig. 7.** Comparison of the fusion line function in the transverse section for P1 with Ar + 1% $N_2$

Figure 8 exemplifies a reconstructed weld pool trailing edge geometry for P1 with Ar + 1% $N_2$.

The analytical three-dimensional weld pool geometry function was implemented in the numerical model (see Fig. 9). This was accomplished for each time step by a complete selection of the nodes which are existent in the weld pool geometry at the point in time $t_i$. The movement of the heat source was subdivided into a finite number of time steps using a time increment $\Delta_t$.

The selected nodes representing the weld pool geometry were subsequently set to the liquidus temperature $T_{liq}$ ($T_{liq}$ = 1400°C).

The numerical temperature field calculations performed were validated by temperature measurement close to the fusion line. Figure 10 shows a comparison between simulation and measurement at distances of 2.7 mm and 1.0 mm, respectively, from the fusion line for P2 with Ar + 1% $N_2$. The maximum deviation in the transition area of temperature rise is approximately 15%.

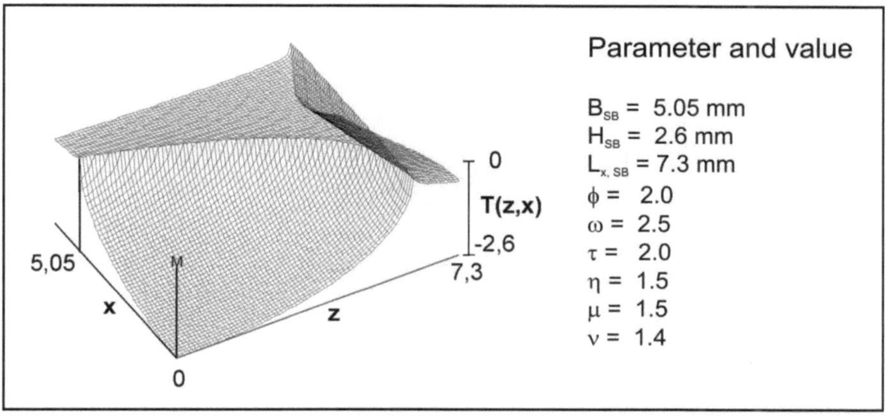


**Fig. 8.** Three-dimensional representation of the determined weld pool geometry (trailing edge) using the geometry parameters for P1 with Ar + 1% $N_2$

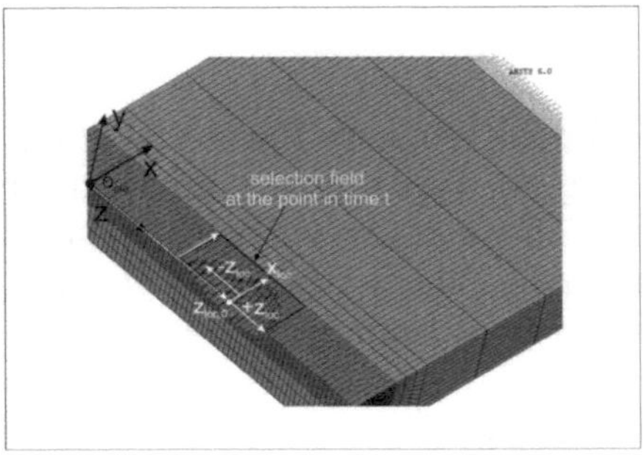


Fig. 9. Implementation of the weld pool geometry function ($T_{z,x}$) in the numerical model

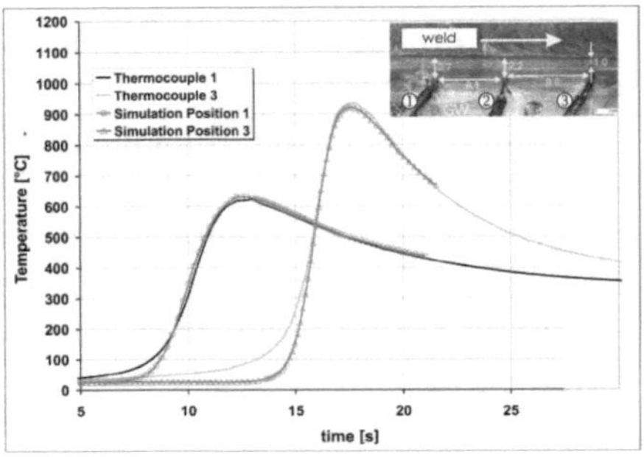


**Fig. 10.** Temperature history of simulation/measurement at distances of 2.7 mm und 1 mm, resp., from the weld for P2 with Ar + 1% N2

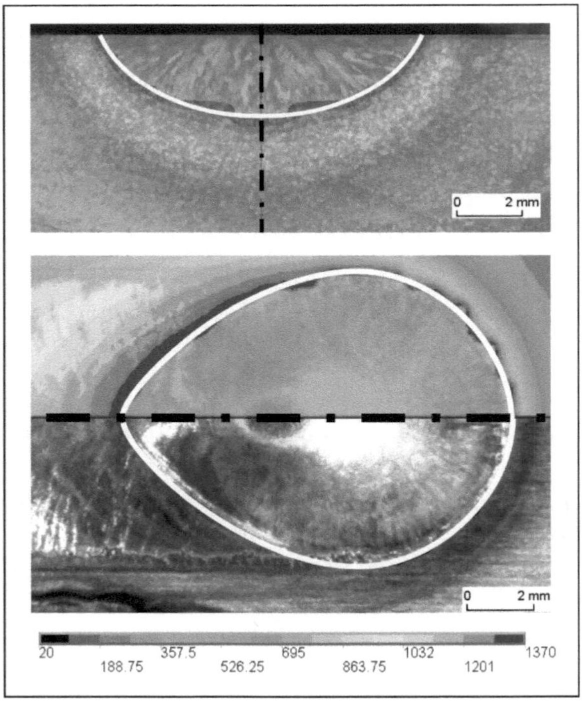


**Fig. 11.** Numerically calculated temperature distribution in the cross-section (top) and at the surface (bottom) for P2 with Ar + 1% $N_2$

In Fig. 11 shows an additional projection of the calculated temperature distribution to the transverse section and to the crater at the end of the weld. It is found that the weld pool geometry used in the simulation reproduces with adequate exactness the real weld pool geometry in the cross-section as well as at the surface.

## Numerical Calculation of Local Strain and Strain Rates

Before presenting the numerical simulation results of the MVT-Test, the numerically determined results of the thermomechanical calculations were verified.

In order to provide transferable solidification crack criteria from the MVT-Test, the local critical strain rates must be determined taking account of the critical specimen loading rates.

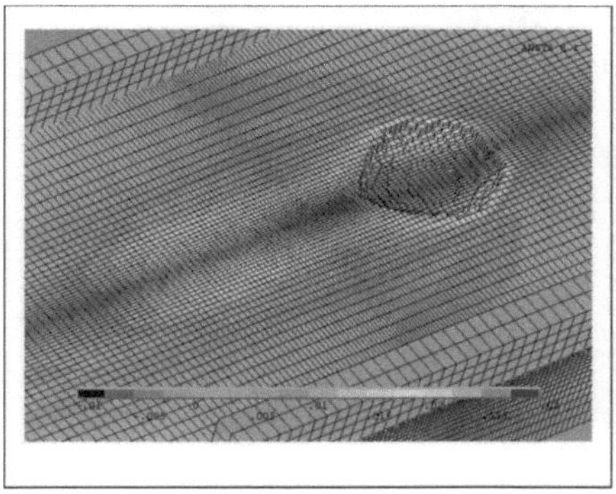

**Fig. 12.** Strain distribution for P1 with Ar, $v_{trav,crit} = 0.05$ mm/s

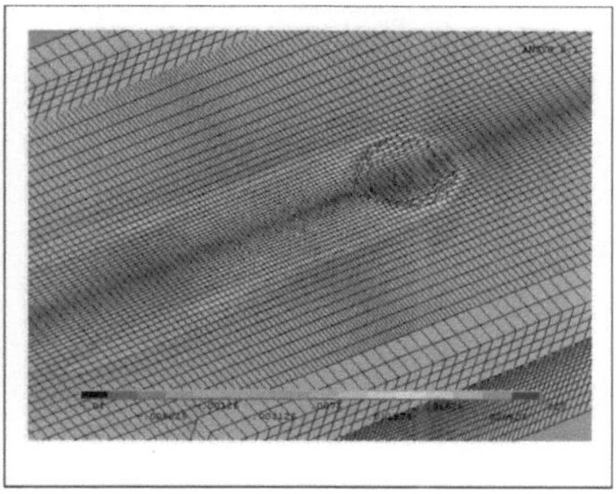

**Fig. 13.** Strain distribution for P4 with Ar, $v_{trav,crit} = 0.1$ mm/s

Figure 12 illustrates as an example the strain distribution transversely to the welding direction (x-direction) for P1, 2.3 s after the onset of mechanical loading at the point in time t = 10.3 s. The specimen loading rate for P1 (Ar) was $v_{trav,crit} = 0.05$ mm/s. It is clearly seen from Fig. 12 that the strains accumulate at the weld axis behind the weld pool due to the prevailing high temperatures. Figure 13 shows by comparison the strain distribution for P4 at the point in time t = 9.4 s. Strain accumulation is observed at the weld axis directly behind the weld pool. Contrary to P1, excessive strain concentrates on a considerably narrower zone behind the weld pool, because increasing welding speed (P4) involves a more tapered weld pool trailing edge geometry [14, 15].

This example shows the advantages of direct implementation of the weld pool geometry in the numerical simulation. Geometric influences of the weld pool geometry on the local strain rates can thus be adequately considered in the numerical calculation.

Further quantification of the relationship between the global specimen loading rate $v_{trav}$ and the local strain rate was performed by numerical simulation of four different specimen loading rates between 0.05 mm/s and 0.4 mm/s for the parameter set P1, where 0.05 mm/s represents the critical specimen loading rate $v_{trav,crit}$.

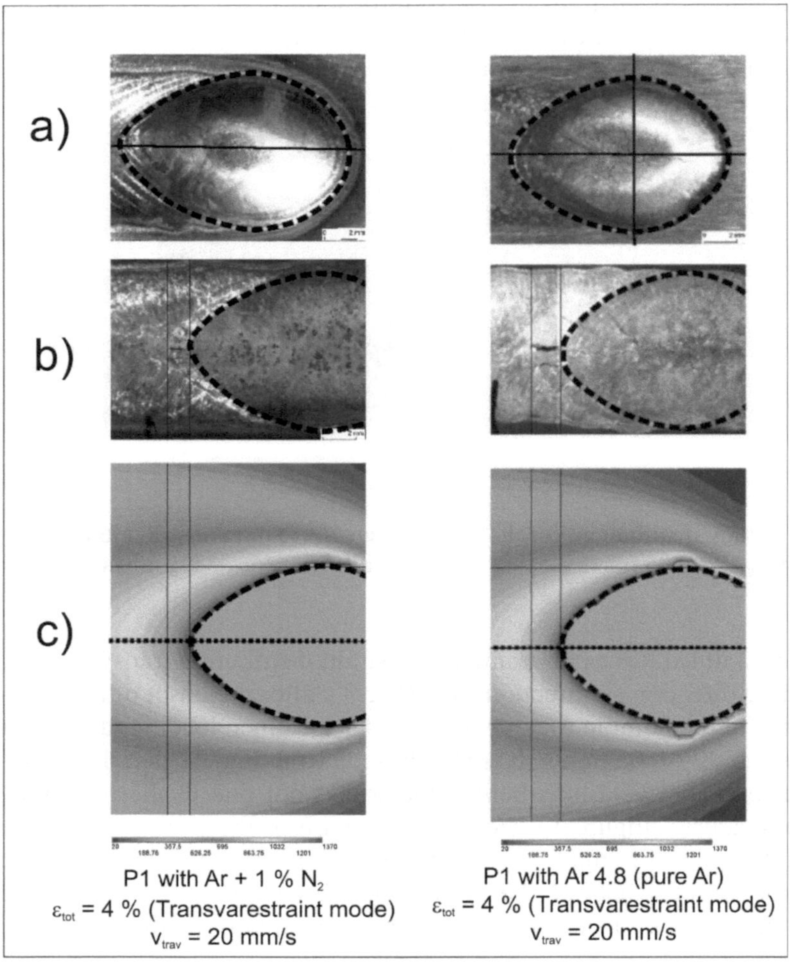


**Fig. 14.** Determination of the solidification crack-critical limiting temperature $T_{SR}$

In order to determine the solidification crack-critical limiting temperature $T_{SR}$ [1], the weld pool geometries were reconstructed from the craters at the weld end and from the transverse sections for all experiments (Fig. 14) [12]. With the help of the weld pool geometries reconstructed by this method and of numerical simulations, the temperature distribution in the immediate vicinity of the weld pool was exactly calculated. Furthermore, the maximum solidification crack lengths were experimentally ascertained with the help of the MVT-Test using rapid loading and related to the calculated temperature distributions. As a result it was found that the crack length in the temperature field calculation corresponds to the distance between the liquidus isotherm and the solidification crack-critical limiting temperature $T_{SR}$ (Fig. 14c).

Figure 15 depicts the strain history versus the specimen loading rate with additional plotting of the temperature history at the position of the crack origin. The temperature history shows the transition from the melting condition at 1400°C to the cooling phase. In investigations carried out by [12] it was found for P1 with Ar 4.8 that the solidification crack-critical limiting temperature $T_{SR}$ is 1195°C and that no solidification cracks occur below this temperature. With increasing heat input $E_S$, there is a significant rise in the solidification crack-critical limiting temperature. Accordingly, the area between the liquidus temperature and $T_{ER}$ represents the solidification crack-critical temperature interval (STI).

The local strain functions depending on the time were derived for determining the local strain rate according to the time. The functions obtained by this procedure are represented in Fig. 16. As can be seen from Fig. 16, a maximum level of local strain rate occurs in the solidification crack-critical temperature interval. This can be attributed to the fact that the material directly behind the weld pool exhibits the highest temperature in the continuum-mechanical condition. It must be presumed furthermore that there is a pronounced notch effect directly behind the weld pool due to the weld pool geometry.

The calculated values for the local strain rates can be seen from Table 5. For comparison, the calculated global strain rates are also listed in Table 5 [12].

It is found from Table 5 that the local strain rates $\dot{\varepsilon}_{loc}$ for the welding parameter set P1 with pure Ar are higher by the factor of 2.5–3.3 than the global strain rates $\dot{\varepsilon}_{glo}$. Thus, the local strain rates in the MVT-Test differ appreciably from the global strain rates. Note here the largely proportional relationship between specimen loading rate, global strain rate and local strain rate. From this it follows that exactly the critical specimen loading rate describes the solidification cracking resistance of materials linear-quantitatively. This shows that the local critical strains and strain rates can be a crack criterion for component transferability.

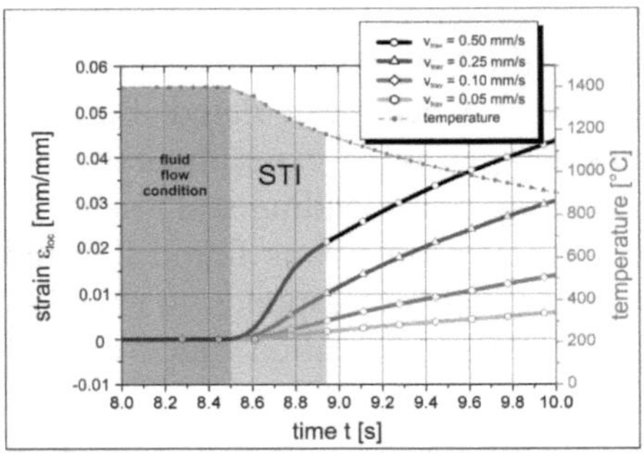


Fig. 15. Local strain versus specimen loading rate for P1 with Ar 4.8

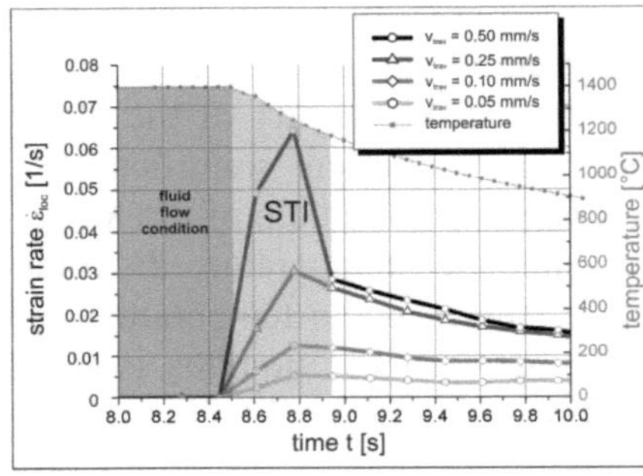


**Fig. 16.** Local strain rate versus specimen loading rate for P1 with Ar 4.8

**Table 5.** Local strain rate depending on the global strain rate for P1

| $v_{trav}$ | global strain rate $\dot{\varepsilon}_{glo}$ [1/s] | local strain rate $\dot{\varepsilon}_{loc}$ [1/s] | $\frac{\dot{\varepsilon}_{loc}}{\dot{\varepsilon}_{glo}}$ |
|---|---|---|---|
| 0.50 | 0.020 | 0.065 | 3.3 |
| 0.25 | 0.010 | 0.031 | 3.1 |
| 0.10 | 0.004 | 0.013 | 3.3 |
| 0.05 | 0.002 | 0.005 | 2.5 |

## Summary and Conclusions

Hot crack prevention during welding processing of metallic materials is an essential prerequisite of component safety. The causes of solidification cracking can be attributed to the occurrence of metallurgical and thermomechanical effects. The conclusions reached are the following:

- A new method could be developed which was used to reconstruct the weld pool geometry from transverse sections as well as from the crater of the weld end and to incorporate it as a three-dimensional analytical function into numerical simulation. The comparisons between numerical simulation and the measured temperature distributions show high correlation.

- Validation was carried out using the MVT-hot cracking test in which the specimen loading rate was varied additionally. The solidification crack-critical limiting temperature could be calculated for Alloy 602 CA depending on various welding parameters and on the investigated shielding gases.
- In view of a transferability of hot cracking tests to components, critical specimen loading rates were determined and the local crack-critical strain rates in the immediate vicinity of a weld pool were calculated.
- It could be demonstrated that the local critical strains and strain rates can be a crack criterion for a transferability.

## References


1. Hemsworth B, Boniszewski T, Eaton NF (1969) Classification and definition of high temperature welding cracks in alloys. Metal Construction and British Welding Journal 1, pp 5–15
2. Wolf M, Kannengießer T (2005) Moderne Prüfverfahren zur Bestimmung der Heißrisssicherheit beim Schweißen von hochlegierten Stählen und Nickelbasislegierungen. DVM Report 641, In: Werkstoffprüfung 2005, ISSN 1861-8154, pp 419–426
3. Kannengiesser T, McInerney T, Florian W, Böllinghaus T, Cross CE (2002) The Influence of local weld deformation on hot cracking susceptibility. In: Mathematical Modelling of Weld Phenomena 6, Book 0784 Maney Publishing, ISBN: 1-902653-56-4, pp 803–818
4. Kannengiesser T, Cross CE (2006) Effect of tack placement on local weld displacement and solidification cracking during arc welding of aluminium alloy 6083. In: Proceedings to IIW conference, 21.-22.11.2006, Bangkok, Thailand, pp 480–491
5. Matsuda F, Nakagawa H, Nakata K, Kohomoto H, Honda Y (1983) Quantitative evaluation of solidification brittleness of weld metal during solidification by means of in-situ observation and measurement. Transactions of JWRI 12 (1), pp 65–72
6. Corporate Publication (2001) ThyssenKrupp VDM, Werkstoffblatt 4137 Nicrofer 6025H/HT – Alloy 602/602 CA, Edition March
7. Zinke M, Hübner A, Herold H, Hoffmann T (2003) Erhöhung der Heißrisssicherheit von hochwarmfesten Ni-Basislegierungen durch die Anwendung stickstoffhaltiger Schutzgase beim Schweißen. ISBN: 3-87155-683-1, DVS-Report 225, pp 249–256
8. Herold H, Pshennikov A, Hübner A, Slyvinski A, Krafka H (2001) Was sagen die Heißrissprüfungen mit dem PVR- und MVT-Verfahren über die Schweißbarkeit aus. DVS-Report 216, pp 255–260
9. Goldak J, Chakravarti A, Bibby M (1984) A new finite element model for welding heat sources. Metallurgical Transaction B (15B), June, pp 299–305

10. Wei Y, Dong Z (2005) Simulation and predicting weld solidification cracks. In: Hot Cracking Phenomena in Welds, Springer-Verlag Berlin Heidelberg, ISBN: 3-540-22332-0, pp 185–222
11. Prokhorov NN, Prokhorov NN (1969) A general equation for the surface of the solidification front during welding. Svar. Proiz. 8, pp 1–4
12. Wolf M (2006) Zur Phänomenologie der Heißrissbildung beim Schweißen und Entwicklung aussagekräftiger Prüfverfahren. Dissertation, Verlag für neue Wissenschaft GmbH, ISBN-10: 3-86509-599-2
13. Wolf M, Schobbert H, Böllinghaus T (2005) Influence of the weld pool geometry on solidification crack formation. In: Hot Cracking Phenomena in Welds, Springer-Verlag Berlin Heidelberg, ISBN: 3-540-22332-0, pp 243–268
14. Katayama S (2000) Solidification phenomena of weld metals (1st report). Characteristic solidification morphologies, microstructures and solidification theory. Welding International 14 (12), S. 939–951, 2000
15. Hunziker O, Dye D, Reed RC (2000) On the formation of a centreline grain boundary during fusion welding. Acta Materialia 48, pp 4191–4202

# Part II
# Solidification Cracking of Ferrous and Nickel-Base Alloys

# Classification and Mechanisms of Cracking in Welding High-Alloy Steels and Nickel Alloys in Brittle Temperature Ranges

K.A. Yushchenko, V.S. Savchenko

E.O. Paton Electric Welding Institute, Kiev, Ukraine

Fabrication of structures from stable-austenitic high-alloy steels, including stainless steels, and nickel alloys may involve much difficulties caused by their high sensitivity to cracking during welding and during short- or long-time heating.

Cracks during the welding process may be formed in different zones of a welded joint, depending upon the chemical composition and thermal-physical properties of the materials welded, as well as upon the thermal-deformation conditions within the weld zone and welded joint as a whole (Figs. 1, 2).

The most common classification of cracks that may be formed in stable-austenitic welds and heat-affected zones is as follows:

- solidification hot cracks;
- ductility dip cracks.

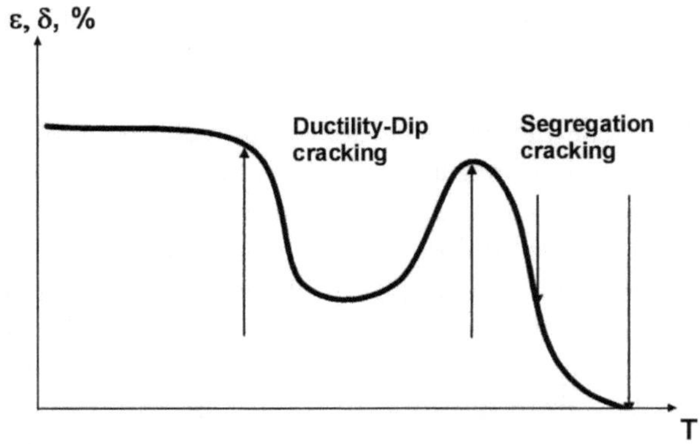


**Fig. 1.** High-temperature ductility of metal with temperature ranges of formation of characteristic types of hot cracks [1]

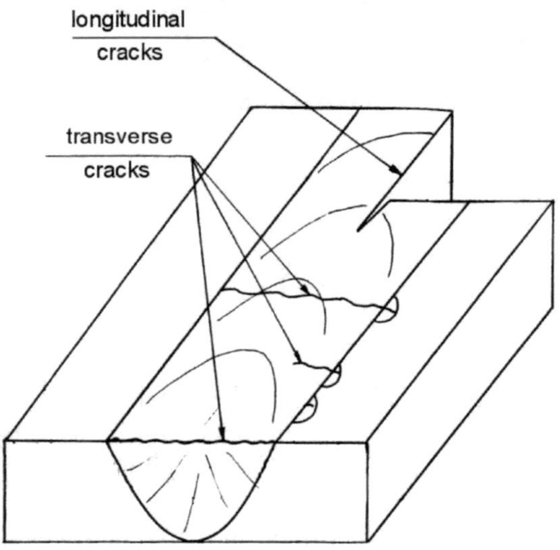


**Fig. 2.** Schematic view of the location of solidification cracks in welds

According to this classification, cracks in the HAZ metal can be of the following types:

- liquation or segregation hot cracks;
- ductility dip cracks.

Given a wide variety of types of hot cracks and, hence, the mechanisms of their formation, one of the main tasks is to identify and classify the cracks. Depending upon the location and conditions of formation of cracks, they can be classified by external attributes, as well as by using modeling, metallography and fractography examinations.

Metallography is used to reveal internal structure of the weld metal during solidification and subsequent cooling, as well as character of location of hot cracks relative to the microstructure during solidification. It is a well known fact that constitutional supercooling ahead of the solidification front, due to a difference in solubility of impurity elements in solid and liquid states of metal, results in a dendritic structure (Fig. 3) that forms interlayers in the dendrite spacings with a decreased solidification temperature. The process of cooling from the solidification temperatures ($T_s$) is accompanied by migration of high-angle grain boundaries in the solidified metal, which initially, during solidification, occupied the place of inter-dendrite boundaries (Fig. 4).

**Fig. 3.** Dendritic structure in stable-austenitic weld

In the high-temperature region of the brittle temperature range (BTR) (Fig. 1) the solidification cracks In the high-temperature region of the propagate, as a rule, along the solidification subgrain boundaries or solidification grain boundaries (Fig. 5) [2], whereas in the low-temperature region of BTR the solidification cracks propagate along the migrated grain boundaries [2].

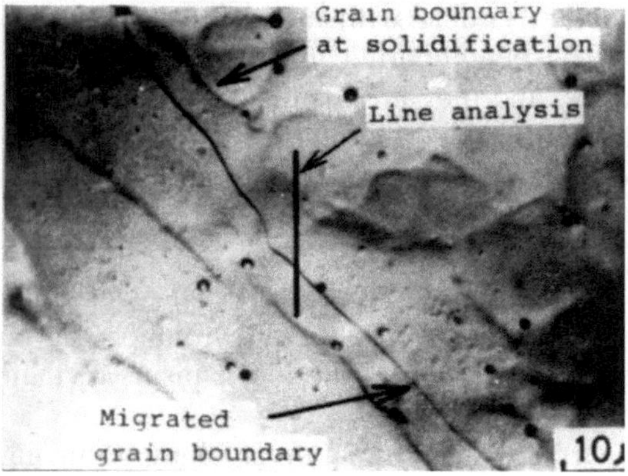


**Fig. 4.** Microstructure of weld metal with migrated boundary [2, 3]

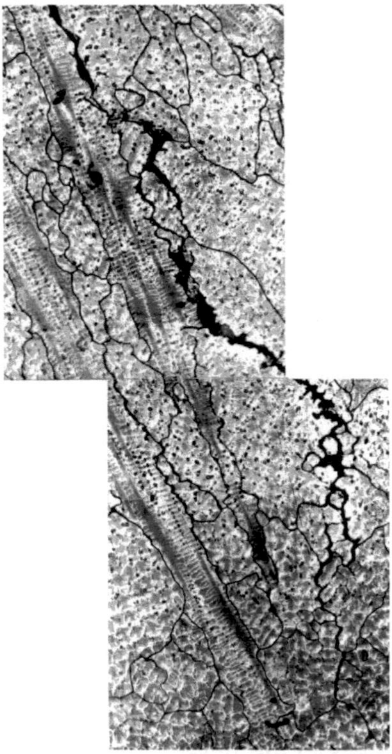


**Fig. 5.** Structure of hot solidification crack in BTR

Further cooling below $T_S$ leads to formation of a thermodynamically stable structure consisting of macro grains, with an identical crystallographic intragranular structure. The neighbouring macro grains are confined by the high-angle boundaries, and they are identical to grains in the base metal, which act as a substrate in the beginning of solidification of grains (dendrites) near the fusion line (epitaxial solidification). The ductility dip cracks initiate in welds with a stable-austenitic structure mostly at the migrated grain boundaries (Fig. 6).

Analysis of structure of the fracture surface resulting from cracking is another characteristic method, in addition to metallography, which can be used to classify the cracks and, therefore, reveal the mechanisms of their formation.

Examinations of the fracture surface in solidification cracking occurring in the solid-liquid state, according to the classification [3], show that this surface can be of the type of D, D+F or F, depending upon the temperature at the fracture location (Fig. 7).

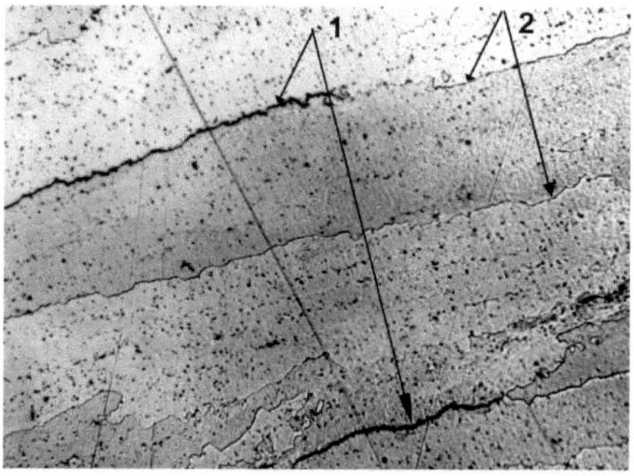


**Fig. 6.** Ductility dip crack
1 – crack;
2 – grain boundary

As found by the examinations, surface D that copies the shape of the secondary dendrite arms is formed in a region of the highest temperatures of the solidification crack range, when intergrowth of the dendrites presents a barrier to penetration of liquid metal from the weld pool to interdendritic positions.

The fracture surface becomes smooth and flat (type F) (Fig. 7c) with formation and migration of the boundaries. Surface D+F is formed at an intermediate position of the migrating boundary (Fig. 7b). It can be concluded from summarizing the fractography results obtained for the surfaces of solidification cracks [2, 3, 4], that liquid planes may exist over the entire brittle temperature range, including zones D, D+F and F, i.e. not only in the interdendritic regions (zone D) but also at the migrated boundaries (zone F). These conclusions, undoubtedly, require verification, as no coherent suggestion is available that could explain why a boundary migrated to a new position remains liquid. In our opinion, investigation of strength and ductility characteristics of metal within a range of subsolidus temperatures, including the zero strength and zero ductility characteristics, can help to clarify an extremely important issue of a real temperature at the end of complete solidification, $T_s$, and fracture of the weld metal, including in the presence of both the remaining matrix solution in solidification and other types of liquid interlayers. This will make it possible to get an insight into a very complicated multifactor mechanism of the formation of cracks in a range of solidification temperatures. Using criteria $T_{z.d.}$ (zero ductility temperature) and $T_{z.s.}$ (zero strength temperature) makes it possible to allow for the effect of thermal-deformation conditions on the formation of cracks (Fig. 8).

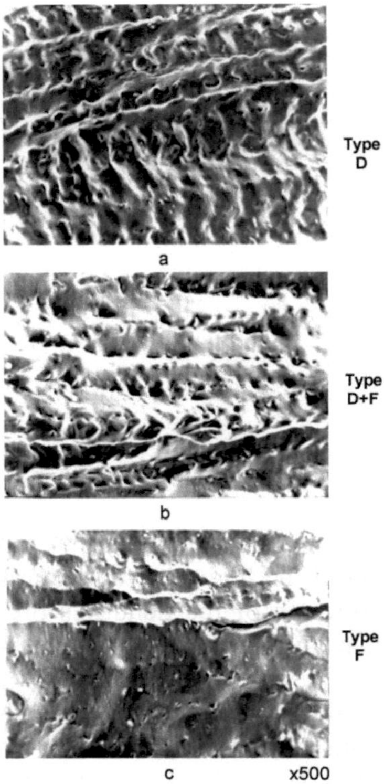


**Fig. 7.** Structure of solidification crack surface

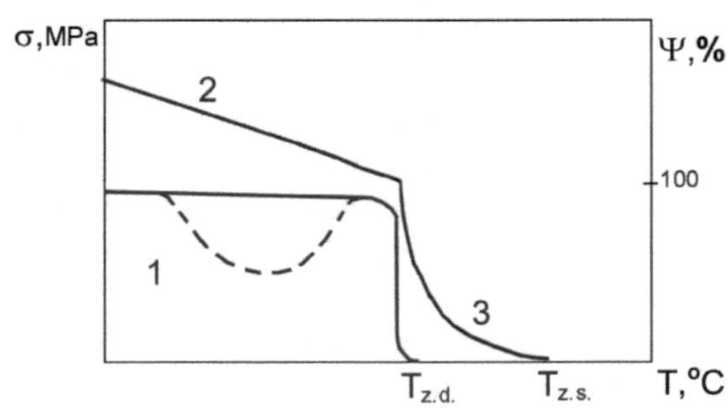


**Fig. 8.** Schematic of temperature dependence of variations in strength and ductility of metallic materials in melting temperature range
1 – ductility variation curve $\psi$=f(T);
2 – grain body strength variation curve $\sigma$=f(T);
3 – curve of variations in strength of solid body with liquid interlayer $\sigma$=f(T);
$T_{z.d.}$ – temperature of transition from intercrystalline to intracrystalline fracture
$T_{z.s.}$ – temperature of zero strength

Experimental studies of zero strength and zero ductility characteristics of high-alloy stable-austenitic cast steel (Fig. 9a) [5] and heat-resistant cast nickel alloy (Fig. 9b) [5], as well as fractograms of the fracture surfaces in tensile testing, show that the zero ductility temperature corresponds to a state before complete solidification of metal takes place. At the same time, it can be shown that there exists a temperature range in which metal, still being in the solid-liquid state during cooling, has certain strength. The zero strength temperature corresponds to the condition when the majority of metal is in the solid-liquid state, and surface tension forces are insufficient to compensate for stresses in the welded joint as a whole.

The generalizing scheme that reveals the mechanism of formation of solidification cracks in welds with a stable-austenitic structure is shown in Fig. 10.

According to this mechanism, hot cracks are formed in BTR, the upper bound of which is below $T_L$ and corresponds to the temperature of formation of the dendrite frame, which hinders penetration of liquid metal from the pool to the interdendritic regions. This is a zero strength temperature ($T_{z.s.}$), and it can be reliably determined using specialized equipment. Given that liquid metal is present at the subgrain boundaries, the crack surface in this case corresponds to type D (Fig. 7). The lower bound of BTR corresponds to the temperature of solidification of liquid boundary interlayers, and it is determined by temperature $T_{z.d.}$. It should be emphasized that the possibility of migration of liquid grain boundaries is confirmed by [6], as well as by our investigations (Fig. 11).

As the weld metal cools down and all its structural components finally solidify, ductility of the weld metal dramatically increases (in a range of 1250-1000°C), amounting to 50–70% or higher.

There are no hot cracks in this temperature range, owing to a high ductility of metal (Fig. 1). They reappear during further cooling in a range of moderate temperatures (0.4–0.7 $T_{melt}$) in a region of decreased ductility of metal. Such cracks are classified as ductility dip cracks.

The character of fracture is micro brittle and intergranular in the presence of traces of micro plastic deformation on the fracture surface. Cracks may initiate both directly during the process of cooling of the welded joint to room temperature, and (after a certain holding of the joint up to the point of its service) during service.

It should be emphasised that many structural materials having a face-centred cubic (fcc) lattice are sensitive to embrittlement in this temperature range, including:

- stable-austenitic single-phase high-alloy iron-base steels;
- nickel and nickel alloys;

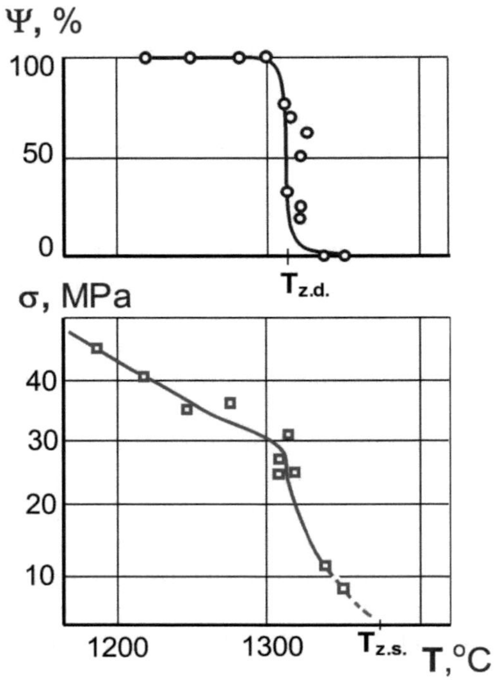


a

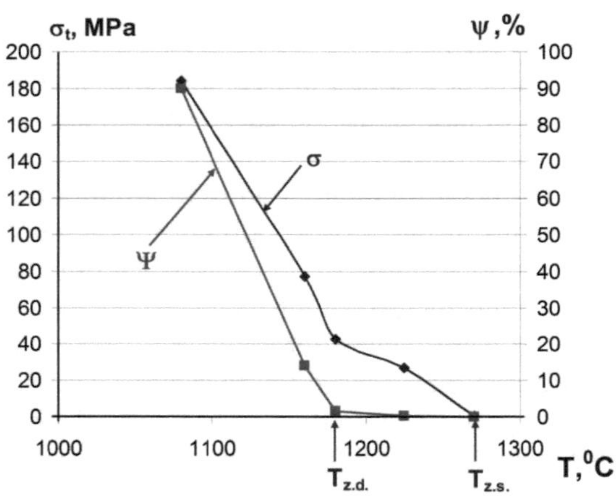


b

**Fig. 9.** Variations in strength and ductility: (**a**) – high-alloy cast metal with austenitic structure [5]; (**b**) – heat-resistant nickel alloy

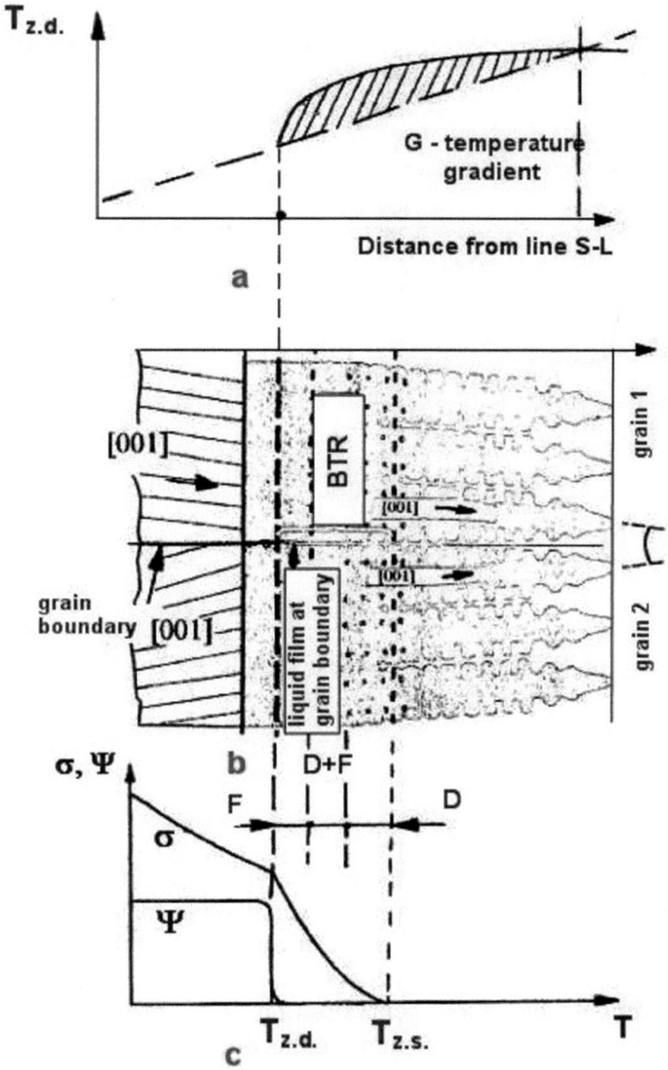


**Fig. 10.** Flow diagram of the process of solidification of metal with stable-austenitic structure in the presence of the concentration overcooling zone (**a**), variations in character of dendrite growth in bi-crystalline metal (**b**) [1], and variations in strength and ductility of metal in solidification (**c**) [5]: $T_{z.d.}$ – zero ductility temperature; $T_{z.s.}$ – zero strength temperature. Characteristics of structure of hot crack surface [6]: D – dendritic; F – flat; D+F - transient

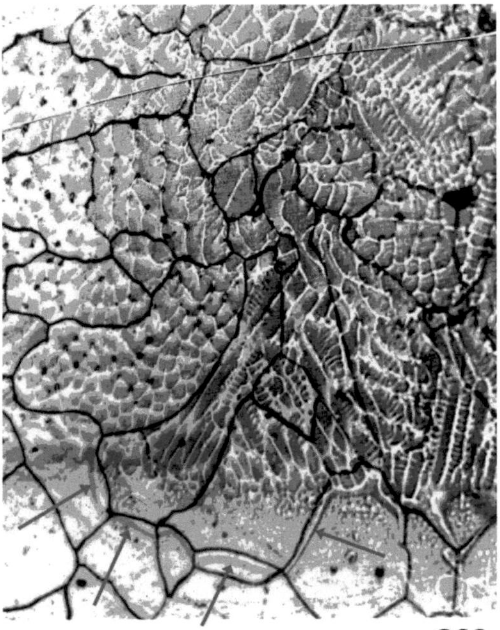


**Fig. 11.** Microstructure of welded joint in stable-austenitic steel with migrated boundary (shown by arrows)

- aluminium and aluminium alloys;
- copper and copper alloys;
- cobalt, etc. (Figs. 12, 13, 14, 15).

Therefore, the above ductility dip temperature range at (0.4–0.7) $T_{melt}$ is characteristic of a whole class of structural polycrystalline materials with an austenitic structure, which means that the causes of their embrittlement are of a common nature and fundamental character.

Results of numerous studies allow generalisation of the effect of main parameters on the processes of embrittlement in this ductility dip range.

Embrittlement is aggravated under the following conditions:

- presence of coarse-grained structure in the weld or HAZ metal;
- presence of segregation elements in the form of impurities, uncontrollable contaminants, such as [O], S, P or surface active B, Se, etc.;
- effect on metal by external energy sources, penetrating energy, including neutron radiation;
- decrease in the rate of deformation and its localisation.

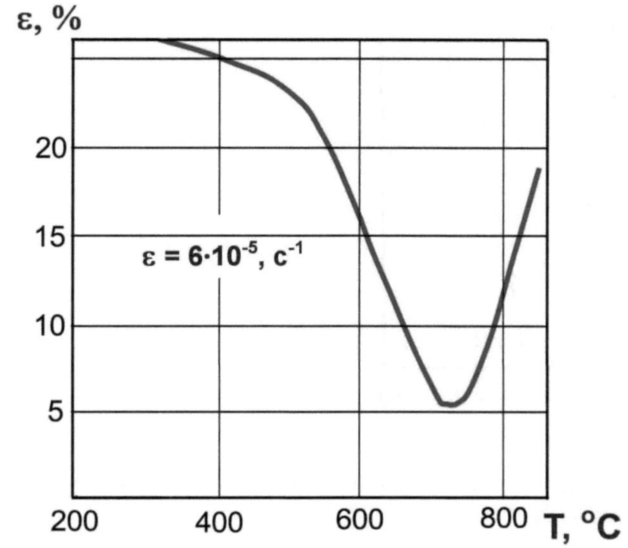


**Fig. 12.** Temperature dependence of variations in ductility of steel AISI310 with standard content of impurity elements ($S \leq 0.02\%$, $P \leq 0.035\%$)

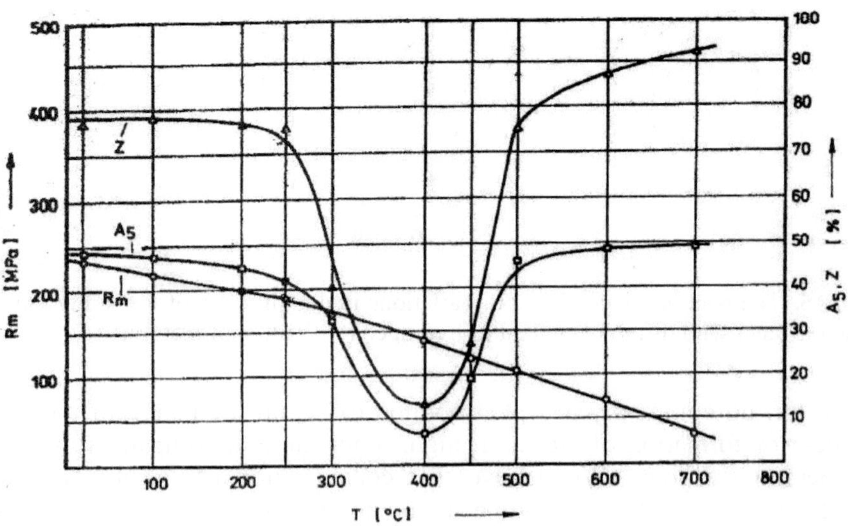


**Fig. 13.** Temperature dependence of variations in ductility ($A_5$) and strength $R_m$ of 99.99% pure copper with oxygen content at a level of 0.01% [7]

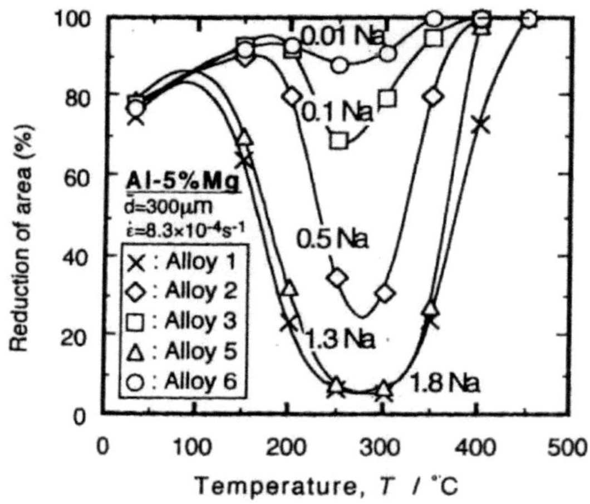


**Fig. 14.** Temperature dependence of variations in ductility of alloy Al-5Mg with different content of Na (ppm) [8]

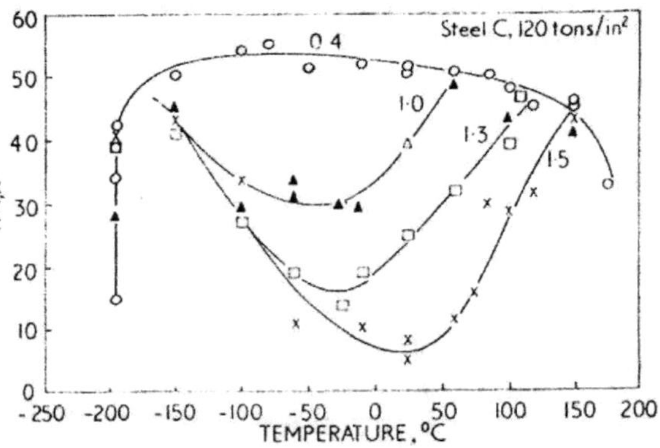


**Fig. 15.** Temperature dependence of variations in ductility RA of low-alloy high-strength steel with different content of hydrogen ($cm^3$/100 g of metal) [9]

So far, no commonly accepted opinion exists on the nature of embrittlement or formation of cracks at high, moderate and comparatively low temperatures (room). We believe that consideration of sensitivity of a welded joint to cracking with time should not be confined only to the process of cooling of the weld to room temperature during solidification. And on the whole, classification of cracks into hot and cold ones is very relative and has no clear physical explanation. At the same time, it can be stated that the presence of stresses and strain zones has a decisive effect on the processes of embrittlement and cracking of a welded joint. Besides, the decreased rate of deformation, which intensifies brittle fracture, suggests that the creep mechanism contributes to the processes under consideration.

In this case, the use of mathematical analysis for investigation of creep processes is very important in terms of specifying mechanisms of general and local plastic deformation for the class of materials considered and for formation of hot cracks. Comparing mathematical generalisation of the results obtained with known data [10] on the mechanisms of plastic deformation during high-temperature creep, and assuming similarity of the processes of creep and deformation in welds allows a conclusion on the nature of this type of the hot cracks in the welds. A large amount of experimental data on the effect of stresses and temperature on the rate of steady-state creep has been accumulated up to now [11].

At very low stresses, the creep rate changes in a linear manner from stress $\varepsilon \sim \sigma^n$, where n = 1. This dependence is characteristic of the temperatures close to solidus, and allows description of the process of diffusion creep at temperatures above 0.9 $T_{melt}$, where slip along the grain boundaries is the main form of deformation (Fig. 16). In a range of lower temperatures the main mechanism of deformation is movement of dislocation. Index "n" in this case amounts to high values [11].

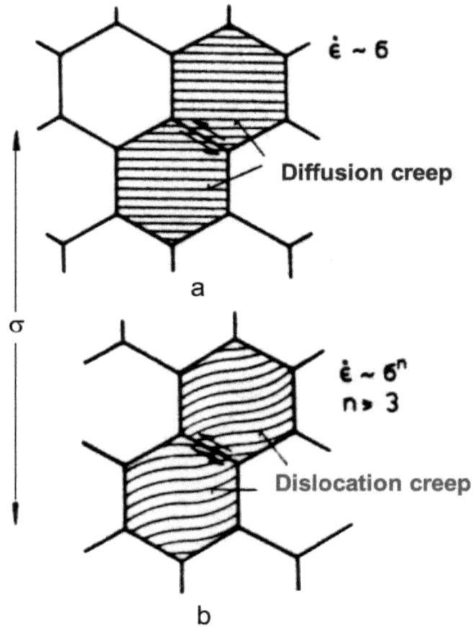


**Fig. 16.** Schematic of the main types of steady-state creep [12]:
(**a**) – diffusion; (**b**) – dislocation

Variations in the creep rate with the values of stresses at fixed values of temperature for stable-austenitic steel AISI310 were evaluated to find causes and determine the main mechanism of creep in the ductility dip temperature range. Experimental points obtained as a result of experiments are shown in Fig. 17.

Exponential dependence of variations in the creep rate for the selected temperature of 650°C is proved by the straight line describing variations of ε with σ in the selected coordinate system.

Empirical dependence of the steady-state creep rate upon the stresses was plotted on the basis of the experimental results using the following formula:

$$\dot{\varepsilon} = A \cdot \overline{e}^{\frac{Q - m\sigma}{RT}}$$

where:

$\dot{\varepsilon}$ is the variation in the deformation rate;
$A$ is the constant;
$Q$ is the creep activation energy, kJ/mole;
$\sigma$ is the stress at a given temperature, MPa;
$R$ is the gas constant, J·mole$^{-1}$·K$^{-1}$;
$m$ is the exponent, m=f(T);
$T$ is the temperature, K

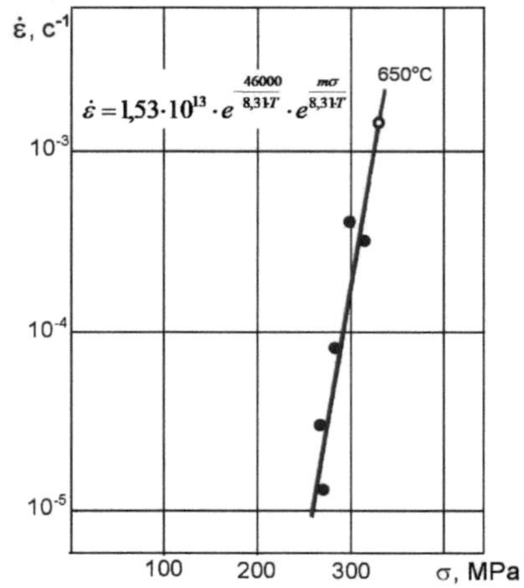


**Fig. 17.** Dependence of the rate of steady-state creep of steel AISI310 upon stresses at test temperature of 650°C. Critical strain rate, below which intergranular fracture is probable, is shown by the mark in straight line

As seen from the calculations, the creep rate can be described by the following equation:

$$\dot{\varepsilon} = 1.53 \cdot 10^{13} \cdot e^{\frac{46000}{8.31 \cdot T}} \cdot e^{\frac{m\sigma}{8.31 \cdot T}}$$

The above equation of steady-state creep describes the deformation controlled by transverse slip of the dislocations [10].

As shown experimentally, variations in the deformation rate has a strong effect on the character of fracture. The portion of a brittle intergranular component of macro fracture grows to 100% with decrease in the deformation rate.

Appearance of the fracture surface for two characteristic rates is shown in Fig. 18. Analysis of the fracture surface suggests existence of a brittle intergranular fracture with traces of micro plastic deformation developing at the decreased deformation rates (Fig. 18b), in contrast to fracture in the bulk of grains at a comparatively high deformation rate (Fig. 18a).

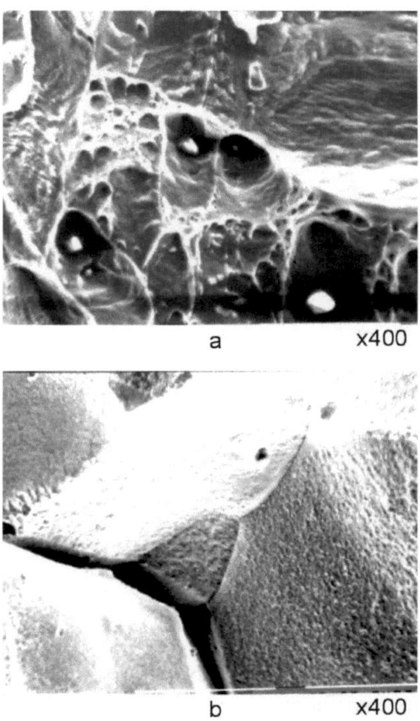


**Fig. 18.** Fractograms of fracture surfaces of steel AISI310 specimens at steady-state creep rate: (**a**) $-3.6\cdot10^{-3}$ $s^{-1}$; (**b**) $-6.0\cdot10^{-5}$ $s^{-1}$

Fractography also showed the presence of the slip planes escaping to the fracture surface (Fig. 19a), which proves existence of the intragranular plastic deformation and the dislocation mechanism of creep in the temperature range studied. It also can be concluded on the basis of the data obtained that the decisive role in the fracture process is played by a tough (liquid) interlayer, the traces of the deformation of which can be seen in Fig. 19b.

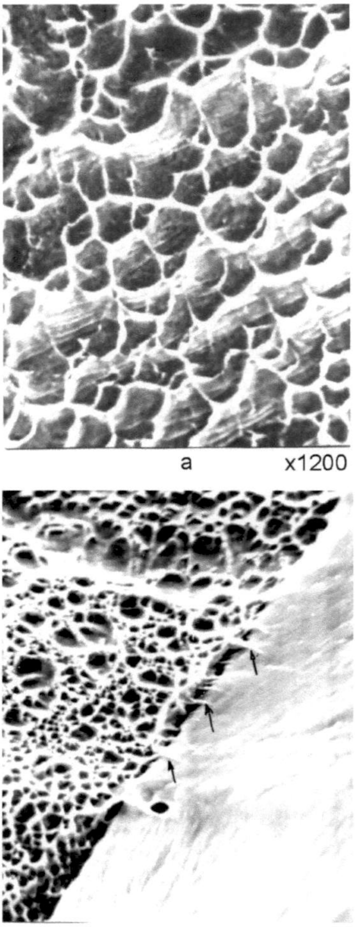


**Fig. 19.** Microplastic deformation in tough component at grain boundaries:
(**a**) – an escape of the slip planes to the fracture surface;
(**b**) – the deformation of a tough (liquid) interlayer.
The traces of deformation are shown arrows

Auger spectroscopy of element composition of the macro brittle fracture surface of specimens deformed at 650°C shows that the surface is rich in sulphur (up to 2.5–3.0 wt.%). Its content stabilises at an ultimate depth (down to 300 nm) from the fracture surface.

Similar data on considerable enrichment of boundaries can be found in study [8]. The authors show that the concentration of sodium at a brittle intergranular fracture of aluminium alloy Al-5%Mg grows by several orders of magnitude (Fig. 20).

This enrichment cannot be realised over a short period of time due to equilibrium segregation and diffusion. In this case the most probable enrichment mechanism is an intensive mass transfer of impurity elements with mobile dislocations [4].

Modelling of thermal-deformation processes in welded joints on stainless steels (Fig. 21a) and heat-resistant nickel superalloys (Fig. 21b) shows that there are real prerequisites for the effect of elastic stresses and localisation of plastic strains on realisation of the processes of redistribution of impurities at the grain boundaries and their embrittlement.

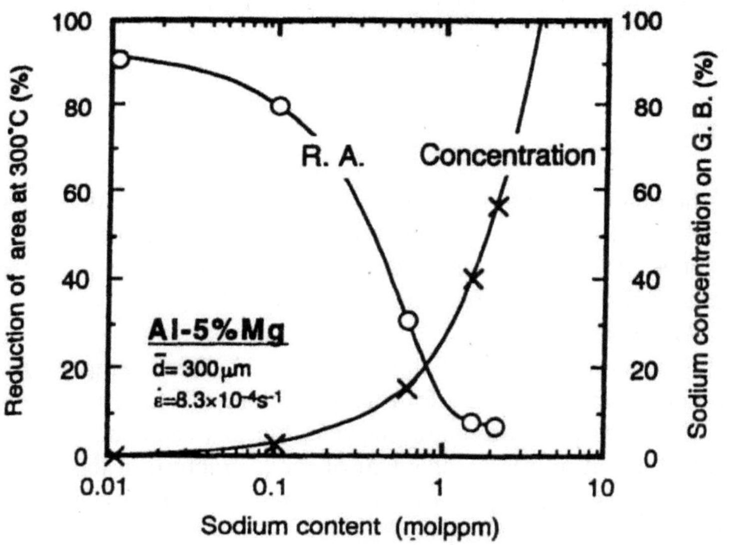


**Fig. 20.** Hot ductility and sodium concentration on grain boundaries of Al-5%Mg alloy as a function of sodium content [8]

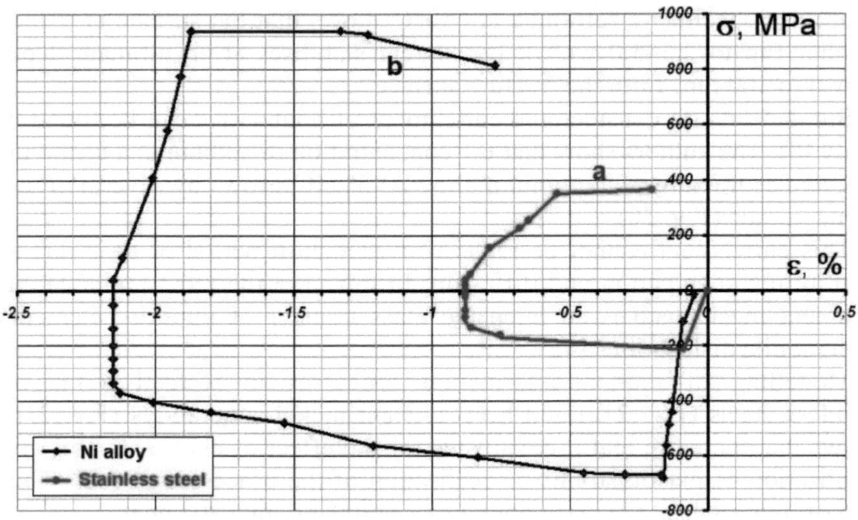


**Fig. 21.** Kinetics of variations in stress-strain state in welding of stainless steel (**a**) and nickel alloy (**b**) at distance of 0.5 mm from fusion line [13]

## Conclusions

1. Analysis of the character and mechanisms of fracture for a welded joint metal with an fcc lattice within a temperature range from ambient to solidification ones proved the presence of characteristic temperature ranges, where cracks were formed, and where ductility dips of the material (weld, HAZ) were fixed.
2. According to the cracking classification suggested, these temperature zones can be characterised by the following fracture mechanisms starting from the $T_L$-$T_S$ temperature range (the first ductility dip zone): The cracks are formed by the mechanism of metal fracture in the intercrystalline interlayers and on the fracture surface. So, these zones are defined as types **D**, **D+F** and **F** [3].
3. The second and third DDC zones, which usually relate to the ductility dip cracking region, are formed in a temperature range of 0.4–0.7 $T_{melt}$, or in a range of $T_{service}$, depending upon the degree of metal purity. One or two mechanisms of intergranular fracture may coexist in this range.

   The first mechanism is characterised by formation of impurity at the grain boundaries. Impurity in the form of metal compounds (S, B, P, O, Na) are formed due to diffusion of impurities to the grain boundaries, which is enhanced by plastic deformation.

The second mechanism is when segregation of elements occurs at the grain boundaries decreasing their cohesive resistance. Fracture in the first zone occurs by the liquation mechanism, and in the second zone by the segregation mechanism.

The third zone shows up at comparatively low temperatures, provided that metal contains chemical elements that have a diffusion mobility in this temperature range (H, He).

Therefore, fractures along the grain boundaries of a liquation character, segregation character, and creep character may coexist at temperature of 0.7 $T_{melt} - T_{service}$.

## References


1. Hemsworth W, Boniszewski T, Eaton NF (1969) Classification and definition of high temperature welding cracks in alloys. Metal Construction and British Welding Journal 1, N25:5–16
2. Ramires AI, Lippold JC (2004) High temperature behavior of Ni-base weld metal, part II – Insight into the mechanism for ductility dip cracking: Materials Science and Engineering, A380, pp 245–258
3. Matsuda F, Nakagawa H, Ogata S, Katayama S (1978) Fractographic Investigation on Solidification Crack in the Varestraint Test of Fully Austenitic Stainless Steel: Studies on Fractography of Welded Zone (III), Transactions of JWRT 7, W1, pp 59–70
4. Yushchenko KA, Savchenko VS, Starushchenko TM (2004) The Role of Segregation of Oxygen in Welding Alloys of the INVAR Type. Hot Cracking Phenomena in Welds, Berlin, pp 59–70
5. Rogberg B (1983) An Investigation on the Hot Ductility of Steels by Performing Tensile Test on “in Situ Solidified” Samples. Scandinavian Journal of Metallurgy 12:51–66
6. Radhakrishnan B, Thompson RG (1993) Kinetics of Grain Growth in the Weld Heat - Affectd Zone of Alloy 718: Metallurgical Trans A 24A, pp 2773–2785
7. Slyško P (1988) Decrease of plasticity in the range of critical recrystallization temperature of fcc metals (in Czech). 2:129–146
8. Horikava K, Kuramoto S, Kauno M (2000) Sources at a trade amount at sodium, and its effect on hot ductility an Al-5 mass%Mg alloy: Light Metals Review, vol 7, pp 18–23
9. Riecke E (1978) Wasserstoff in Eisen und Staul. Archiv für das Eiseuhüttenwesen 11:509–520
10. Ekobori I (1971) Physics and mechanics of fracture and strength of solid bodies (in Russian). J Metallurgiya 1071:264

11. Okrainets PH, Pishchak VK (1978) Stress dependence of creep rate for metals with fcc lattice (in Russian). J Fizika Metallov I Metallovedeniye 46, Issue 3:597–602
12. Sklenička V, Saxl I, Ğadek I (1983) The role of grain boundaries in deformation and fracture at high temperatures (in Czech). Kovove materaly 4, vol 21, pp 364–382
13. Yushchenko KA, Makhnenko VI, Savchenko VS, Velikoivanenko EA, Chervyakov NO (2006) Investigation Of Thermal-Deformation State Of Welded Joints In Stable-AusteniticSteels And Nickel Alloys. IIW Doc.IX-2224-06

# Submerged Arc Welding – A Test for Centreline Cracking

K. Håkansson

Kockums AB, Malmö, Sweden

## Abstract

A newly developed test method for centreline cracking or solidification cracking has been evaluated by adding nickel as cold wire to the weld metal. Welding was performed by using synergic cold wire submerged arc welding with various arc wire diameters to achieve different nickel contents in the weld metal. Welding was done as partial penetration single bevel groove joints in the PB position. This investigation showed that the test method picks up the sensitivity to centreline or solidification cracking. The amount of cracking increased as the nickel content in the weld metal increased.

## Introduction

The present chapter describes briefly the work carried out at Kockums Laboratory, to design a simple test to examine centreline cracking behaviour during submerged arc welding of high strength steels. The welding was done with addition of cold wire containing a high amount of nickel. It discusses in more detail the results of the addition of cold wire containing 98% nickel to the weld. The diameter of the arc wire was changed on purpose to get different amounts of nickel in the weld metal. An increase in the nickel concentration promotes austenitic solidification and therefore should promote solidification cracking, the effect being exaggerated when both the nickel and the sulphur concentrations are large [1]. It is widely accepted that sulphur and phosphorus are harmful impurities with respect to the solidification cracking of steel welds.

An excerpt from "The Fabricators and Erectors Guide to welded Steel Construction" [2] states that centreline cracking is characterized by separation in the centre of a given weld bead. If the weld bead happens to be in the centre of the joint, which is normal for a single-pass weld, centreline cracks will be in the centre of the joint. In the case of multi-pass welds, where several runs per layer may be applied, a centreline crack may not be in the geometric centre of the joint, although it will always be in the centre of the bead.

Centreline cracking is the result of one of the following phenomena:

-segregation induced cracking
-bead shape induced cracking
-surface profile induced cracking

Unfortunately, all three phenomena reveal themselves in the same type of crack, and it is often difficult to identify the cause. Moreover, experience has shown that often two or even all three of the phenomena will interact and contribute to the cracking problem. Understanding the fundamental mechanism of each of these types of centre line cracks will help in determining the corrective solutions [2].

Segregation induced cracking occurs when low melting point constituents such as phosphorous, zinc, copper and sulphur compounds in the admixture separate during the weld solidification process. Low melting point components in the molten metal will be forced to the centre of the joint during solidification, since they are the last to solidify and the weld tends to separate as the solidified metal contracts away from the centre region containing low melting point constituents.

In submerged arc welds, the cracking risk may be assessed by calculating the Units of Crack Susceptibility (UCS) from the weld metal chemical composition (weight %) [3, 4, 7]:

$$UCS = 230C^{*} + 190S + 75P + 45Nb - 12.3Si - 5.4Mn - 1$$

$C^{*}$ = carbon content or 0.08, whichever is higher

Although arbitrary units, a value of $<10$ indicates high cracking resistance whereas $>30$ indicates a low resistance. For fillet welds, runs having a depth to width ratio of about one, UCS values of 20 and above will indicate a risk for cracking [3, 4, 7].

When centreline cracking induced by segregation is experienced, several solutions may be implemented. Since the contaminant usually comes from the base material, the first consideration is to limit the amount of contaminant pick-up from the base material. This may be done by limiting the penetration of the welding process through the use of lower welding currents.

The second type of centreline cracking is known as bead shape induced cracking. This type is associated with deep penetration processes like Submerged Arc Welding (SAW or method 121) and $CO_2$ shielded Flux

Cored Arc Welding (FCAW or method 136). When a weld bead is of a shape where there is more depth than width to the weld cross section, the solidifying grains growing perpendicular to the steel surface intersect in the middle, but do not gain fusion across the joint. To correct for this condition, the individual weld beads must have at least as much width as depth [2]. The total weld configuration, which may have many individual weld beads, can have an overall profile that constitutes more depth than width. If multiple passes are used in this situation, and each bead is wider than it is deep, a crack free weld can be made. [2].

When centreline cracking due to bead shape is experienced, the obvious solution is to change the width to depth relationship. This may involve a change in joint design. Since the depth is a function of penetration, it is advisable to reduce the amount of penetration. This can be accomplished by utilizing lower welding current and larger diameter electrodes. All of these approaches will reduce the current density and limit the amount of penetration.

The third mechanism that generates centreline cracks is surface profile conditions. When concave weld surfaces are created, internal shrinkage stresses will place the weld metal on the surface in tension [2]. Conversely, when convex weld surfaces are created, the internal shrinkage forces will pull the surface into compression. Concave weld surfaces frequently are the result of high arc voltages. A slight decrease in arc voltage will cause the weld bead to return to a slightly convex profile and eliminate the cracking tendency. High travel speeds may also result in this configuration. A reduction in travel speed will increase the amount of fill and return the surface to a convex profile [2].

Hot cracking tests for weldments are given in the standard EN ISO 17641-part 1. The standard divides the tests into two types where part 2 contains a description of Self-Restraint Tests and part 3 contains a description of Externally Loaded Tests [5].

## Experiment

In order to design a simple test for investigating the sensitivity for centreline cracking for submerged arc welding of high strength steels, the following experiments were performed. The joint preparation was a single-bevel groove T-joint, which is regarded as giving the highest shrinkage stresses transverse the weld and showing the highest sensitivity to centreline cracking. The design of the test piece was a 30 mm thick, 300 mm square plate, and a round joint prepared plate with the diameter of 200 mm and the thickness 35 mm was placed in the centre. On the other side of the square plate, a round bar of 25 mm diameter was tack welded in the centre. This bar was used to mount the test piece in a manipulator during welding. Welding was made using a rectifier type constant voltage with 1400A as maximum output and both wires were fed by a synergic cold wire kit (SCW), which feed the cold wire in synergy with the arc wire into the weld pool where it melts. This means that the arc and the cold wire ratio always remain constant after suitable wire diameter is selected. See Fig. 1.

**Fig. 1.** The test set up where the manipulator drives the test piece during the welding of the fillet weld by submerged arc welding

The chemistry of the weld and the deposition rate was controlled and pre-selected [6].

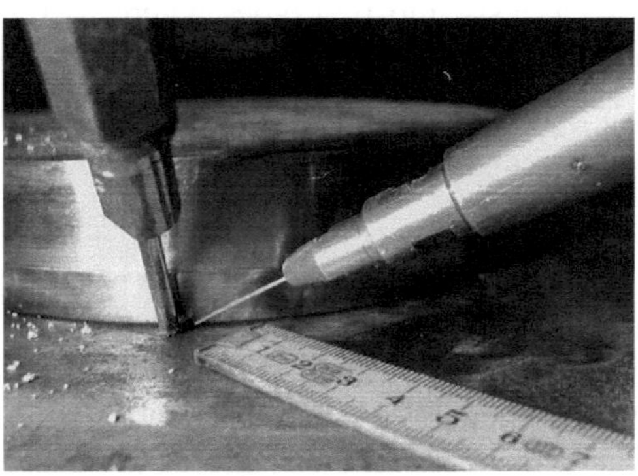

**Fig. 2.** Adjustment of the arc wire and cold wire in the single bevel groove joint. It is essential that the round test piece is aligned with the driving shaft of the manipulator

The arc wire for all welds was a lean alloyed wire with different diameters. This wire gives about 0.5% manganese in the weld.

A basic agglomerated welding flux for welding high strength steels and low temperature steels was used. As cold wire, a wire containing 98% nickel and 2% titanium with a diameter of 0.8 mm was used. Figure 2 shows the alignment of both wires.

The theoretical chemical composition of the weld metal depends on the difference in diameter of the arc wire and the cold wire. See Table 1.

**Table 1.** Area and theoretical composition of nickel of different electrode diameters

| Diameter of wire mm | Area $mm^2$ | Theoretical % Ni in the mixed consumable | Theoretical Ni content in diluted weld metal containing 0.56% base metal and 0.44% of the mixed consumable % |
|---|---|---|---|
| 0.8 Cold wire | 0.503 | – | 0.67 |
| 2.5 Arc wire | 4.909 | 9.2 | 4.64 |
| 3.0Arc wire | 7.069 | 6.5 | 3.51 |
| 4.0 Arc wire | 12.57 | 3.7 | 2.32 |

The weld metal will have a different composition due to dilution from the base material. In submerged arc welding this dilution is assumed to be about 50–60%. The chemical composition of the base material, the arc electrode and the cold wire is given in Table 2.

**Table 2.** Chemical composition base material and wires

| Element | Plate material | Cold wire Ø 0.8 mm | Arc wire |
|---|---|---|---|
| C | 0.112 | 0.031 | 0.080 |
| Si | 0.246 | 0.27 | 0.03 |
| Mn | 0.884 | 0.20 | 0.36 |
| P | 0.009 | 0.000 | 0.021 |
| S | 0.0009 | 0.003 | 0.012 |
| Cr | 0.424 | 0.00 | 0.02 |
| Ni | 1.236 | 96.6 | 0.03 |
| Mo | 0.344 | 0.00 | 0.02 |
| Nb | 0.013 | 0.00 | 0.00 |
| Ti | 0.005 | 2.9 | 0.00 |
| V | 0.007 | 0.01 | 0.02 |
| Cu | 0.198 | 0.01 | 0.08 |
| Fe | Balance | – | Balance |

## Welding Procedures

Welding was performed in flat position without any preheat and in one single pass. The travel speed was set by the rotation performed by the manipulator. The travel speed was aimed at 40–45 cm/min; the welding current was aimed at 350 ampere and the voltage at around 28 V. These figures give a heat input of 1.3–1.5 kJ/mm.

## Results

### Non-Destructive Testing by Liquid Penetrant

Each of the four welds was non-destructive tested by liquid penetrant to reveal any cracks extending to the surface of the weld.

The amount of indication in the form of the length of the indication from the investigation was measured for each of the four tests. The welds containing high amounts of nickel show the largest length of indication for cracks, with the length increasing with an increase in the nickel content of the weld metal. The result from the non-destructive testing is shown in Table 3.

**Table 3.** Result from the non-destructive testing

| Mark | Arc wire diameter mm | Cold wire addition | Total test length mm | Cracked length mm | Percentage cracking as % of the total weld length |
|---|---|---|---|---|---|
| HC1 | 4.0 | Yes | 578 | 10 | 1.7 |
| HC2 | 3.0 | Yes | 578 | 373 | 64.5 |
| HC3 | 3.0 | No | 581 | 0 | 0.0 |
| HC4 | 2.5 | Yes | 580 | 542 | 93.4 |

### Chemical Composition of Weld Metal

Material from the weld metal was removed from each of the welded samples by drilling holes in the weld metal. The chips were used for chemical analysis of the weld metal. The chemical composition was analysed by Inductively Coupled Plasma emission Spectroscopy ICP. The result of the chemical analysis of the weld metal is shown in Table 4.

**Table 4.** Chemical composition of the different weld metals (balance Fe)

| Mark | HC1 | HC2 | HC3 | HC4 |
|---|---|---|---|---|
| Dimension filler wire mm | 4.0 | 3.0 | 3.0 | 2.5 |
| Cold wire NG61 addition | Yes | Yes | No | Yes |
| C | 0.093 | 0.094 | 0.085 | 0.085 |
| Si | 0.21 | 0.21 | 0.19 | 0.22 |
| Mn | 0.70 | 0.75 | 0.68 | 0.65 |
| P | 0.011 | 0.012 | 0.010 | 0.009 |
| S | 0.0075 | 0.0058 | 0.0059 | 0.0068 |
| Cr | 0.22 | 0.21 | 0.23 | 0.22 |
| Ni | 2.54 | 3.41 | 0.70 | 4.77 |
| Mo | 0.18 | 0.17 | 0.20 | 0.18 |
| Ti | 0.02 | 0.02 | 0.01 | 0.05 |
| UCS-value | 16.28 | 15.99 | 14.41 | 14.30 |

UCS value was calculated according to EN 1011-2 appendix E [7], where a value below 10 gives a low risk for cracking and value over 30 represents a high risk.

The nickel content in the weld metal increased as the arc wire diameter was decreased as expected. Figure 3 shows a piece of the surface on the test piece with the highest nickel content. Due to the large penetration into the base material, by the submerged arc welding method, a large part of the base material is mixed together with the consumables to form the weld metal. Calculation shows that the weld metal contains about 50–60% base material.

**Fig. 3.** The test piece with the highest amount of nickel shows a solidification crack almost all around the circumference

## Macro Test of Welds

From each weld two macro test pieces were taken transverse to the weld. A cut divided the test plate into two halves, and the macro tests were removed from each exposed side. The tests were prepared metallographically and etched in Nital. The tests were inspected using a stereo microscope at ten times magnification and the results are shown in Table 5.

**Table 5**. Results of inspection of macro tests transversing the weld

| Mark | Nickel content in weld metal | Remarks | Depth to width ratio |
|---|---|---|---|
| HC3-1 | 0.7% | Acceptable. See Fig. 4a | 0.74 |
| HC3-2 | 0.7% | Acceptable | 0.71 |
| HC1-1 | 2.54% | 1 mm crack in the root. Root gap of 0.4 mm. See Figs. 4b and 5a. | 0.85 |
| HC1-2 | 2.54% | No cracks. Undercut 0.5 mm. Root gap 0.3 mm | 0.70 |
| HC2-1 | 3.41% | Centreline crack 3 mm. Sideline crack 2 mm+1 mm. Root crack 4 mm. Root gap 1 mm. See Fig. 4c. | 1.12 |
| HC2-2 | 3.41% | Centreline crack 1.5 mm. Root crack 3 mm. Root gap 0.5 mm. | 0.88 |
| HC4-1 | 4.77% | Sideline cracks 1 mm + 1 mm + 2 mm. Root crack 1.5 mm. See Fig. 4d. | 0.75 |
| HC4-2 | 4.77% | Centreline crack 1 mm. Sideline cracks 2 mm + 1.5 mm. Root crack 1.5 mm. Root gap 0.2 mm. See Figs. 4e and 5b. | 0.76 |

According to Kazutoshi Ichikawa et al. [1] the probability for solidification cracking is predicted to be dependent on the sulphur and the nickel content in the weld metal, where the probability increases considerably if the nickel content exceeds 2.5% and the sulphur content exceeds 0.010%.

The amount of cracking in the weld increases as the nickel in the weld metal increases. See Fig. 6. All tests with addition of nickel show cracks from the root of the weld. The smaller diameter of wire, which gives the highest amount of nickel, gave deeper penetration. Increased root gap have a tendency to increase the depth of penetration. This deeper penetration increases the depth and decreases the width of the weld, which gives a higher risk for solidification cracking.

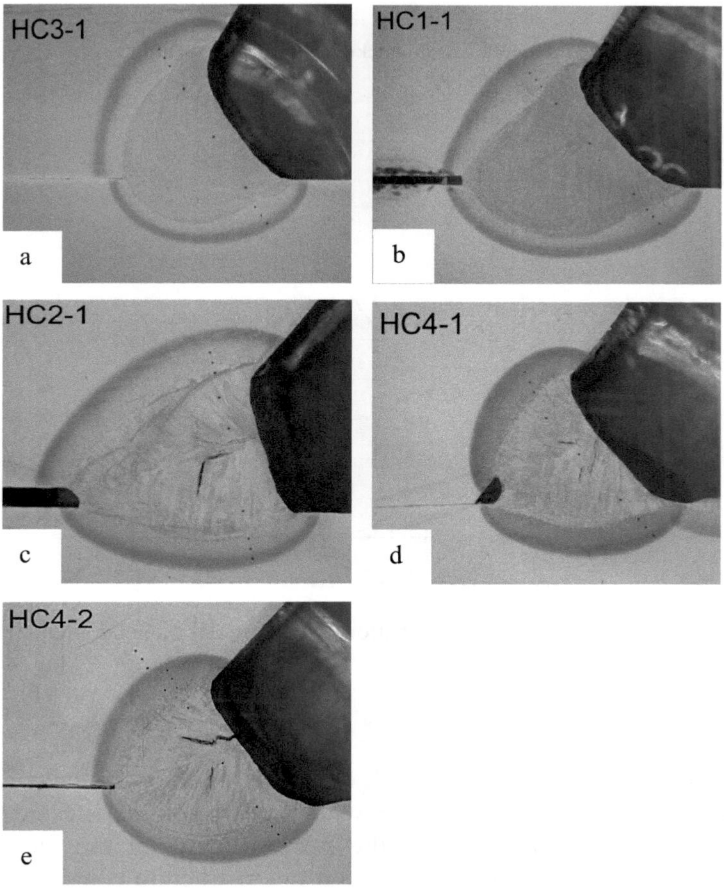


**Fig. 4.** Cross-sections of welds discussed in Table 5 with increasing Ni content: (**a**) 0.7%, (**b**) 2.54%, (**c**) 3.41%, (**d**) 4.77, (**e**) 4.77%

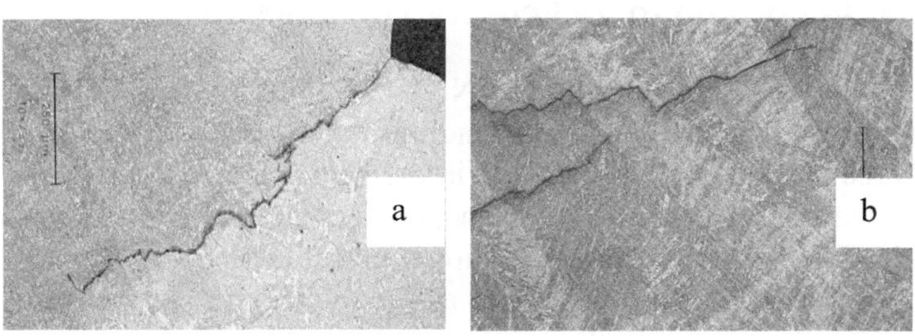


**Fig. 5**. Micrographs of welds (**a**) HC1-1 and (**b**) HC4-2 showing root cracks

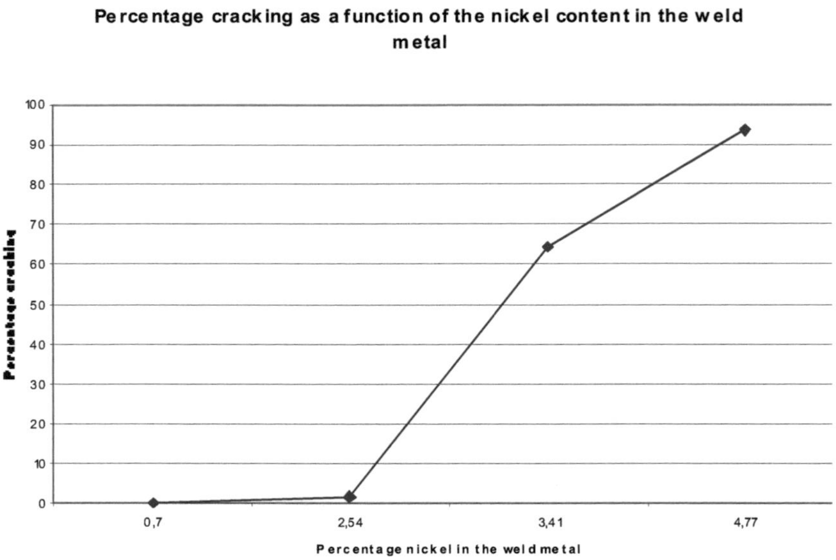


**Fig. 6**. Percentage of cracking shown as a function of weld metal nickel content

## Microstructure

The microstructure of the weld metal with no nickel addition contains a large amount of acicular ferrite together with proeutectoid ferrite side plates. As the nickel content increases the microstructure of the more highly alloyed weld metal shows a change to a microstructure containing more lath martensitic constituents.

## Hardness Test According to Vickers HV5 on Macro Tests

The hardness test was made according to EN 1043-1 with the load of 5 kg, with the exception of that no hardness readings were made in the base material [8]. Three indentations were made in the heat affected zone of the square plate, in the weld metal and in the heat affected zone of the round plate. The result of the hardness survey is shown in Table 6.

The hardness in the weld metal increases as the nickel content increases and this may be due to changes in the microstructure in the weld metal to a more martensitic microstructure. The hardness indicates that the weld metal may have a tensile strength of about 1200 MPa for the two highest nickel levels.

The hardness of the weld metal with no addition of nickel is also rather high, which can be due to the high amount of highly alloyed base material mixed into the weld metal. The fillet weld itself normally experiences a more rapid cooling, which increases the hardness.

**Table 6.** Results from hardness tests on macro tests

| Mark | Nickel content in the weld metal | HV5 in square plate | HV5 in weld metal | HV5 in round plate |
|---|---|---|---|---|
| HC3-1 | 0.7% | 391, 386, 296 | 241, 244, 244 | 396, 391, 376 |
| HC3-2 | 0.7% | 362, 376, 293 | 236, 241, 236 | 376, 376,396 |
| HC1-1 | 2.54% | 371, 381, 353 | 286, 299, 289 | 407, 412, 412 |
| HC1-2 | 2.54% | 401, 386, 283 | 306, 358, 336 | 407, 407, 386 |
| HC2-1 | 3.41% | 396, 371, 289 | 376, 371, 381 | 367, 429, 423 |
| HC2-2 | 3.41% | 391, 391, 277 | 321, 381, 367 | 362, 396, 407 |
| HC4-1 | 4.77% | 396, 396, 299 | 358, 367, 367 | 418, 418, 401 |
| HC4-2 | 4.77% | 391, 386, 286 | 386, 376, 362 | 412, 396, 329 |

## Conclusions

The developed test method uses ordinary welding equipment, such as a manipulator and submerged arc welding equipment, to perform the test.

The design of the test piece in form of a T-butt joint shows that it may be too severe to predict the risk according to the UCS number for solidification cracking in the weld metal.

The amount of centreline cracking in form of solidification cracking increases as the nickel content increases.

The root gap may influence the result by affecting the depth to width ratio.

The microstructure of the weld becomes martensitic for the highest levels of nickel.

The hardness increases as the nickel content increases in the weld metal.

## References

1. Kazutoshi Ichikawa, Bhadeshia HKDH, and MacKay DJC (1996) Modell for solidification cracking in low alloy steel weld metal. Science and Technology of Welding and Joining Vol. 1 no. 1, pp 43–50
2. The James F. Lincoln Arc Welding Foundation. Weld Cracking. An Excerpt from the Fabricators' and Erectors Guide to Welded Steel Construction. pp 1–4

3. TWI World Centre for Material Joining Technology – Defects- solidification cracking
4. Bailey N, Jones SB (1978) The Solidification cracking of ferritic Steel during Submerged Arc Welding. Welding Journal, Aug. pp 217s–231s
5. ISO/FDIS 17641-1 Destructive tests on welds in metallic materials- Hot cracking tests for weldments- Arc welding processes- part 1: General
6. ESAB Brochure "Fluxes and wires for Submerged Arc Welding"
7. EN 1011-2:2001 issue 1 Welding-Recommendations for welding of metallic materials-Part 2: Arc welding of ferritic steels
8. EN 1043-1 (1996) Destructive tests on welds in metallic materials-Part 1: Hardness test on arc welded joints

# Influence of Local Weld Deformation on the Solidification Cracking Susceptibility of a Fully Austenitic Stainless Steel

A. Kromm, Th. Kannengießer

Federal Institute for Materials Research and Testing, Berlin, Germany

## Abstract

For the evaluation of the solidification cracking behaviour of welded structures, the influence of the external boundary conditions needs to be considered, in addition to the metallurgical aspects. Against this background, the solidification crack formation in a fully austenitic stainless steel under variation of external restraint was examined in this study.

For this purpose, a newly developed hot cracking test (CTW test) was used for the first time, which allows application of a defined tensile load transverse to the welding direction during welding. In addition, the strains and strain rates could be determined with the help of a mechanical-electrical measuring device in the near field of the weld pool. These values were examined both under free contraction and varied external load, i.e. under different constant cross head speeds. The Critical strain and strain rate required for the propagation of macroscopic surface cracks were determined. By means of high speed recording the authors succeeded in correlating strain and strain rate with the relative position of the weld pool. In addition, centreline crack initiation and growth were located.

## Introduction

A potential problem when welding austenitic steels is their distinct tendency to hot crack. This is due to the solidification behaviour of these materials. The range between liquidus and solidus temperature, i.e. the mushy zone, is the place of solidification crack formation. Here, liquid metal surrounds the already solidified material. Volume shrinkage during the liquid to solid phase transformation must be compensated for the remaining melt. Otherwise solidification cracks will occur [1, 2].

For the evaluation of the hot cracking resistance of welded components, the influence of the external boundary conditions is to be considered in particular, apart from the metallurgical aspects [3]. According to Prokhorov [1] and Rappaz et al. [2], local strain rates in the vicinity of the weld pool can significantly affect hot crack formation and can thus be regarded as a suitable hot crack criterion. In addition, in the work of several authors [4, 5, 6, 7, 8, 9, 10, 11, 12, 13, 14, 15, 16, 17, 18], it was pointed out that a hot crack formation is usually connected with the occurrence of tensile stresses and strains in hot crack-sensitive areas. In the case of an external restraint, e.g. by the design specific shape of the component itself or by additional external load, these stresses/strains near the weld pool are affected and are directly associated with hot crack formation. Up to now a multiplicity of self restraint or externally loaded tests have been developed in order to evaluate the hot cracking susceptibility of materials. With the help of such tests mostly just material rankings can be provided. Transferability to component welds or quantification of hot cracking susceptibility can be achieved by determining, among other things, the hot crack critical strain and strain rate in the weld pool vicinity [9, 10].

In the present study, the approach of a deformation measurement at pins located near the weld was used [9]. The movements of these pins were transferred to an inductive measuring sensor by means of a mechanical measuring device. The collected data allowed determination of strains and strain rates. Under variation of the external load in the Controlled Tensile Weldability (CTW)-Test, the influence of deformation near the weld pool on the solidification cracking susceptibility was examined. The hot cracking behaviour in the weld pool vicinity was characterized by high speed recording.

## Design Specific Influences on Local Thermo-Mechanics

Solidification cracking in weld metal results from the competition between cracking resistance of the material and the mechanical load. If the load exceeds the cracking resistance, then cracking develops [4]. This means, with regard to the hot cracking behaviour, that there is a close relationship between metallurgical and mechanical influences, which is affected by welding parameters and design boundary conditions. Investigations into this topic have clarified that metallurgical, thermal and mechanical factors can have an influence on the thermo-mechanics in the vicinity of the weld pool. In many investigations the definition of areas is common, in which due to heterogeneous heating and cooling either compressive or tensile stresses are present. Further stresses with opposite algebraic sign develop in other areas, due to the balance of forces [4-8, 12-18].

It is an undisputed fact that in front of the moving weld pool compressive stresses are directly formed. The magnitude of this stress cell is strongly affected by the welding parameters, particularly by the travel speed [4-8]. In reaction to this, tensile stresses in the still cold base material appear. The expansion in front of the weld pool continues directly at its sides. The maximum of the longitudinal extensions is within this range. At high welding speeds, this strain maximum can extend to the region behind the weld pool. This has direct consequences on the solidified material already cooling down there. Normally, contraction due to cooling creates a tensile stress in this region, as with slow welding speeds. However, according to Chihoski [17, 18], the expansion of the surrounding material can thus superimpose the contraction within the cooling range -supported by longitudinal expansions- resulting in a pressure cell. Thereafter, under uninfluenced contraction, a tensile cell develops. The investigations of Feng and Zacharia [4-8] confirm the presence of a pressure cell directly behind the weld pool. Figure 1 illustrates this circumstance. Thus, solidification cracking can be prevented.

The development of compressive and tensile cells around the weld pool crucially depends on the welding parameters, namely on the heat input. From this, it becomes clear that the place of a possible crack initiation can be shifted relative to the weld pool. Whether the mechanisms of crack initiation are the same in different locations, needs to be clarified, particularly since any thermal pre-treatment of the component should have an additional effect [12]. Up to now, the emphasis of most investigations has been on the influence of welding parameters on hot crack formation. The number of investigations into the design specific influences is smaller.

It becomes clear that the influence of external load or geometry itself during the welding procedure acts in a similar way, i.e. with movement of compressive and tensile cells. First, a dependence of the stress/strain distributions of the momentary weld pool position on the specimen is to be emphasized. Consequently, this distribution changes with distance and proximity to boundary regions. Without the supporting effect of these areas, the material behind the weld pool seems clearly higher stressed and more rapidly loaded under tensile stress, which should favour cracking [4, 6].

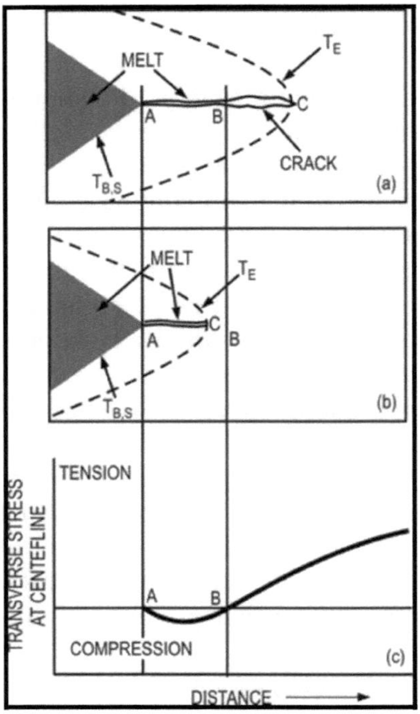


**Fig. 1.** Compressive stresses behind the weld pool [8]

The application of an external, static load has similar effects, as used with the Sigmajig – test [19]. Applied tensile stresses directly affect the level of the locally evolving deformations. These can be increased or decreased [4, 5, 7, 8]. The influence of external restraint was examined recently also by Kannengiesser [20]. Results show that higher restraint intensities can lead to a decrease of the solidification cracking susceptibility.

Hence it follows that potential cracking can either be increased or decreased by the design specific or mechanical boundary conditions. In [17, 18] this statement is also confirmed. It seems that the causes leading directly to cracking (local strains and strain rates, respectively) can alter with the geometry of the material (e.g. round vs. rectangular specimens) [13, 14]. That would mean that crack-critical local values for the same material vary with the geometry (restraint intensity). In the present study a contribution is made to the question of how local strains and strain rates are influenced with respect to external load and how solidification cracking is affected.

# Experimental

## Test Material

The material used in the investigation is the high-alloyed, fully austenitic stainless steel UNS NO. S 34565. The chemical composition is shown in Table 1. The mechanical characteristics determined in the tensile test can be seen from Table 2.

**Table 1.** Chemical composition of S 34565 (wt.%)

| C | Si | Mn | Cr | Ni | Mo | N |
|---|---|---|---|---|---|---|
| 0.03 | 0.111 | 6.639 | 24.1 | 19.04 | 4.972 | 0.551 |

**Table 2.** Mechanical characteristics of S 34565

| Yield Strength, $R_{p0,2}$ | Tensile Strength, $R_m$ | Ultimate Strain, $A_{50}$ | Reduction of Area, Z |
|---|---|---|---|
| 548 MPa | 932 MPa | 50 % | 48 % |

## Hot Cracking Behaviour in the MVT-Test

For the characterisation of the hot cracking susceptibility, preliminary investigations using the MVT-Test [21] were performed. The investigations were accomplished in the Varestraint – mode. The crack evaluation showed predominantly solidification cracks in the weld metal, marginal liquation cracks and no ductility dip cracks. The material was therefore classified by the expression "increasing hot cracking susceptibility".

## Controlled Tensile Weldability Test

The solidification cracking behaviour was examined both under free contraction and under varying external load. For this purpose, a newly developed, externally loaded hot cracking test (CTW-Test) was used for the first time. In this test a defined tensile load can be applied during welding, transverse to the welding direction [11]. The CTW-Test setup is shown in Fig. 2.

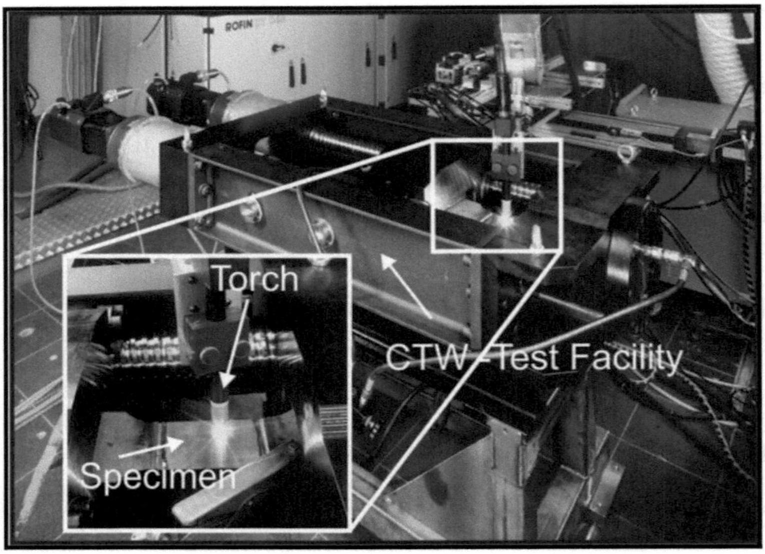


**Fig. 2.** CTW – Test setup

In order to quantify the susceptibility to solidification cracking, a minimum critical cross head speed $v_{a,crit}$ was determined for producing solidification cracks at the specimen surface. This value was determined to be $v_{a,crit} = 4$ mm/min. Based on this value the parameters for the following investigations were chosen and are indicated in Table 3.

**Table 3.** External load parameters

| External Load | Restraint Free | Constant Cross Head Speed [mm/min] | | |
|---|---|---|---|---|
| Parameter Variation | – | 3 | 4 | 5 |
| Solidification Crack | no | no | yes | yes |

The specimen geometry and dimensions are shown in Fig. 3. The welding parameters are summarized in Table 4. In order to avoid additional notch effects and cracking at the weld start, special clamping adapters were used, and the weld start and end were placed 5 mm away from the edge of the sample resulting in a weld length of 45 mm.

**Table 4.** GTA bead-on-plate welding parameters

| Basic parameter | Value |
|---|---|
| Welding Current | 100 A |
| Welding Voltage | 13 V |
| Travel Speed | 5 mm/s |
| Heat Input | 0,26 kJ/mm |

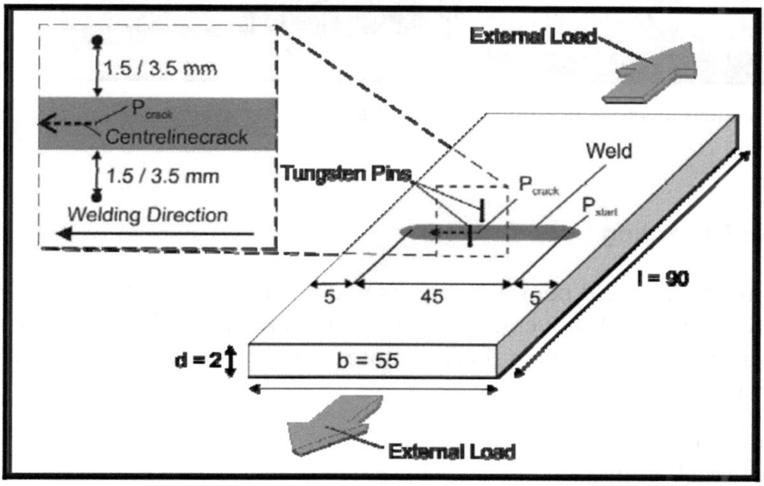


**Fig. 3.** Specimen with weld and centreline crack position

## Strain Monitoring

In order to capture local strains, a mechanical-electrical measuring system was developed. The system is based on direct displacement measurement at defined points on the specimen surface. Via a mechanical leverage system the specimen movements were transferred to an inductive measuring sensor. Its electrical signals were online computed to obtain the respective strains. Figure 4 show the measuring system on a specimen. For the linking of measuring instrument and specimen surface tungsten pins were placed on the surface using induction welding. Reproducible positioning of the pins was accomplished with the help of a positioning device.

Distances from the fusion line of 1.5 mm (near field) and 3.5 mm (far field) were selected. From the measured pin movements over the outside pin distance (see Fig. 3) the transverse strain over the weld seam was determined. The strain rate was calculated as temporal derivative of the strain.

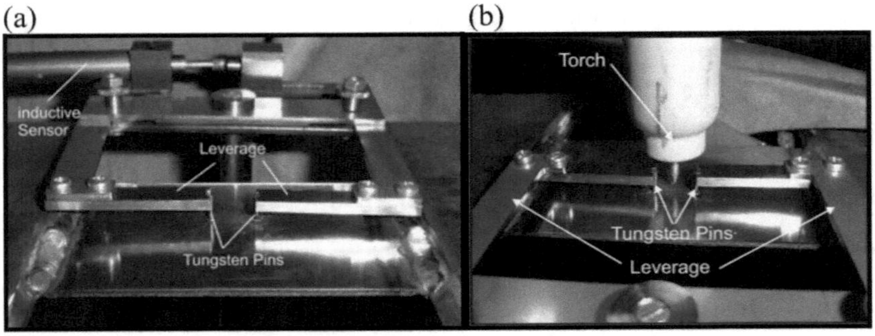


**Fig. 4.** Mechanical leverage positioned on specimen

### High Speed Recording

In order to relate the weld pool dimensions and position to the strain measurements, the weld pool movement was recorded optically. This was realized using a high-speed camera. Furthermore, the records served to determine of the following parameters:

- Time of Solidification
- Time of crack initiation
- Location of crack initiation
- Characteristics of crack initiation

Before the investigations were started, the samples were provided with a mark perpendicular to the welding direction, which was used as point of reference in order to determine the weld pool dimensions. Furthermore, the mark represents the place of the hot crack initiation determined in preliminary tests.

## Results and Discussion

### Weld Pool Dimensions and Crack Initiation

The high speed records helped to determine the dimensions of the weld pool and to observe crack initiation and crack growth. The following Figs. 5–7 illustrate the formation and progress of a solidification crack at a constant cross head speed of $v_{a,crit}$ = 4 mm/min in the CTW-Test. As shown in Fig. 5a, the crack arose at the solidification line. This is at the same point of time as the weld pool begins to narrow.

The hot crack grows perpendicular to the solidification line toward weld pool centre. The growth direction corresponds to the solidification direction. This is represented in Fig. 5b. At the same time, a second hot crack develops on the opposite flank of the travelling weld pool (Fig. 6a). Both cracks grow toward the weld pool centre, until they unify at a certain point in time at the trailing edge of the weld pool, as shown in Fig. 6b.

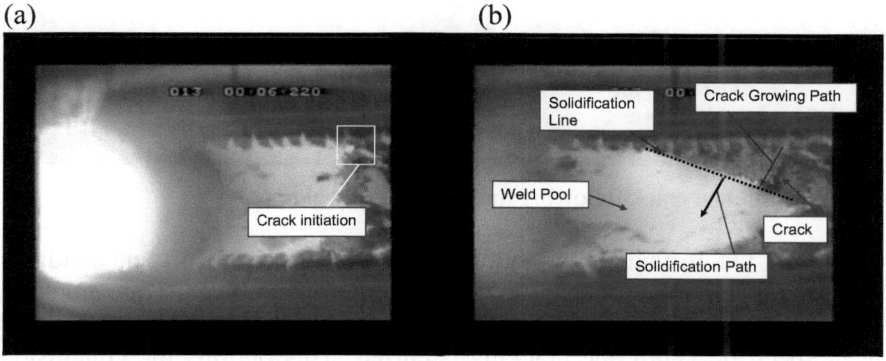


**Fig. 5.** Solidification crack initiation and growth

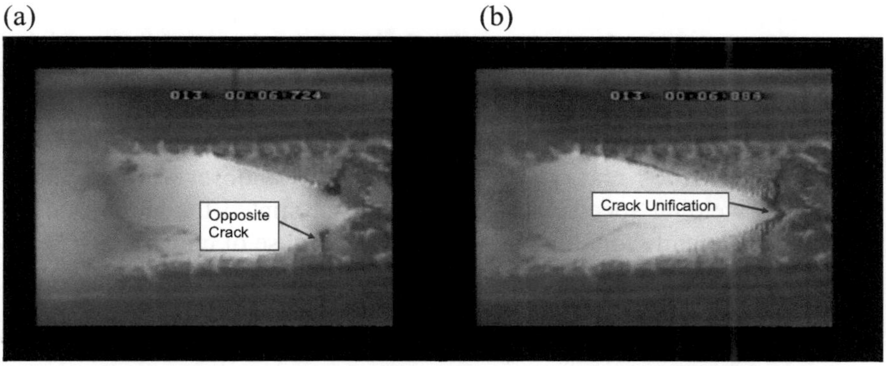


**Fig. 6.** Initiation of opposite crack and crack unification on trailing edge of the weld pool

The described hot cracks obviously unify to become a centreline crack parallel to the welding direction. The centreline crack is always located right behind the weld pool at the trailing edge. Fig. 7 show growth and expansion of the developed centreline crack.

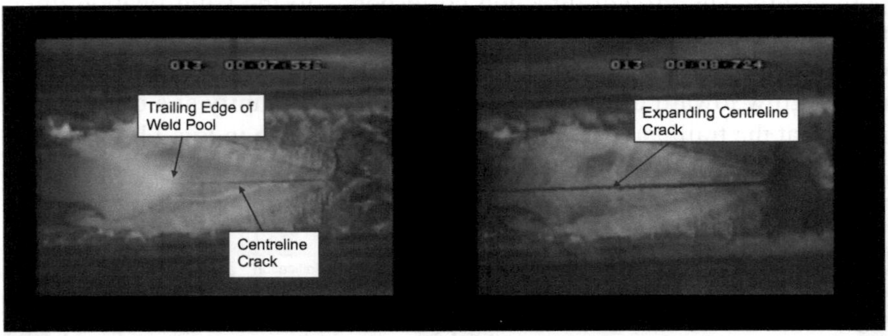


**Fig. 7.** Centreline crack following trailing edge of weld pool

Note that the formation of solidification cracks in the present study is bound to an external load. Without an external load no centreline cracking appeared. Likewise, at an external load below the determined critical cross head speed no centreline cracking occurred. This permits the conclusion that centreline cracking is caused by exceeding a critical external load (strain or strain rate, respectively).

The discussed crack features were observed also by [5] in a similar way. There, transverse cracks arose at the fusion line when welding a nickel base alloy under external preloading. Cracks grew during progressive welding towards the centreline. However, they did not form a centreline crack. By SEM photographs, the observed transverse cracks were identified as solidification cracks. This phenomenon was observed particularly when high welding speeds were applied and prompted the authors to assume strains, mainly longitudinal strains, as the cause of cracking.

## Strain and Strain Rate during Welding

### *Restraint Free*

In the following diagrams, strain and strain rate in the near (Fig. 8) and far field (Fig. 9) of the weld pool are represented. Solidification cracking did not occur with these welding conditions. The weld pool position is highlighted. In the curve of the near field, first a rise in strain rate with approaching weld pool is observed due to the forward-moving heat. The consequence is a slight strain of approx. 0.3%. With the arrival of the weld pool, the strain rate decreases and changes its algebraic sign. The pin movement towards the melt is now accelerated. With narrowing of the weld pool, this movement is slowed and goes finally towards zero during cooling. The cooling-conditioned compression takes a value of approximately –4%.

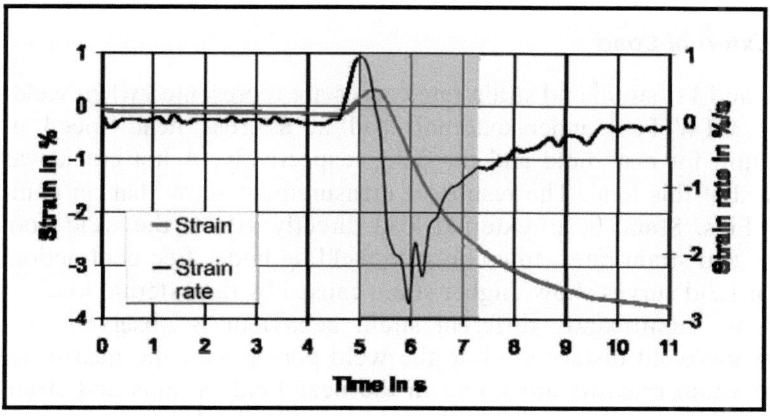


**Fig. 8.** Strain and strain rate in near field of restraint free specimen

A qualitatively similar curve shape is obtained for the far field (Fig. 9). However, the absolute values here are obviously smaller. In [15], similar strain processes were measured and simulated when welding an aluminium alloy.

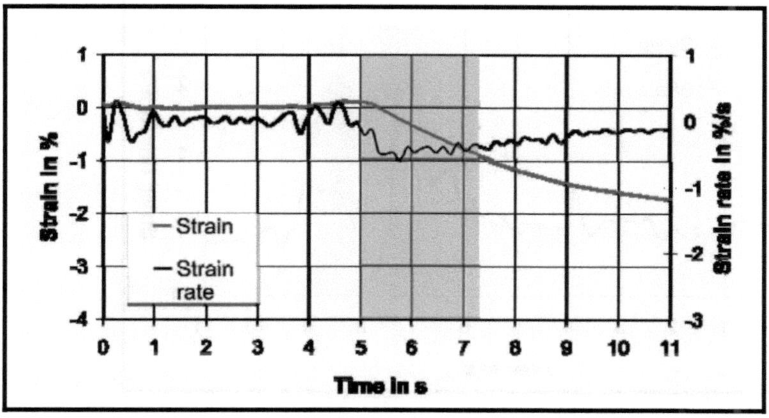


**Fig. 9.** Strain and strain rate in far field of restraint free specimen

It remains to be stated that the material in the near field of the weld pool is exposed to both positive and negative strains. These strains are mainly thermally conditioned, since the material was welded under free contraction. In the near field, the amounts of the arising strains and strain rates are significantly higher than in the far field. The observed strain behaviour agrees in principle to the discussed model conceptions of Zacharia [7, 8] and Chihoski [17, 18].

### *Effect of External Load*

In Figs. 10 and 11, strain and strain rate curves are represented when welding in the CTW-Test under external load at a cross head speed of $v_a = 3$ mm/min for near field and far field, respectively. A hot crack was not observed at this load. The results of measurement show that, in comparison to Figs. 8 and 9, an external load directly affects the weld pool near strains and strain rates. In contrast to welding under free contraction, near and far field curves show higher strain caused by the external load.

In addition, significantly different strain behaviour is observed as a function of the weld distance, while the weld pool passes the measuring point. Two strain maxima are found in the near field. Strains and strain rates are smaller in the far field. It should be emphasized that the already discussed typical strain behaviour in the weld pool vicinity, under free contraction, was registered qualitatively here, although the thermally caused strains were superimposed by an additional external load. Indeed, additional strains acting on the solidifying weld metal are present, but the strain rate does not seem to be high enough to initiate cracking.

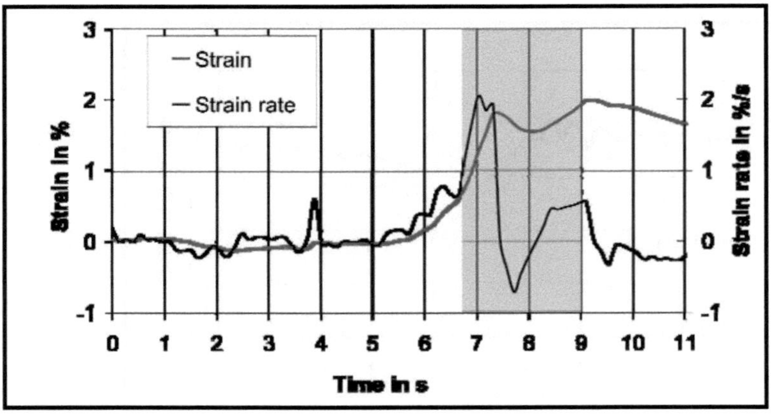


**Fig. 10.** Strain and strain rate in near field applying an external load of $v_a = 3$ mm/min

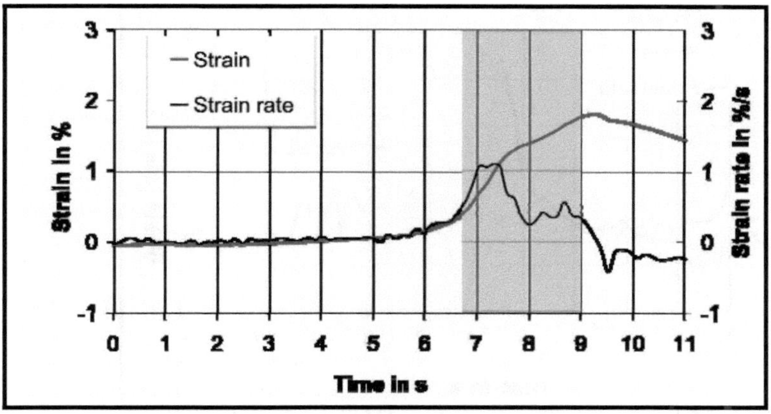


**Fig. 11.** Strain and strain rate in far field applying an external load of $v_a = 3$ mm/min

In Figs. 12 and 13, strain and strain rate are represented when welding in the CTW-Test applying cross head speeds of $v_{a,crit} = 4$ mm/min and $v_a = 5$ mm/min, respectively. These loads triggered solidification crack formation in the form of a centreline crack as mentioned above. Passing of the crack at the measuring point is marked in the Figures.

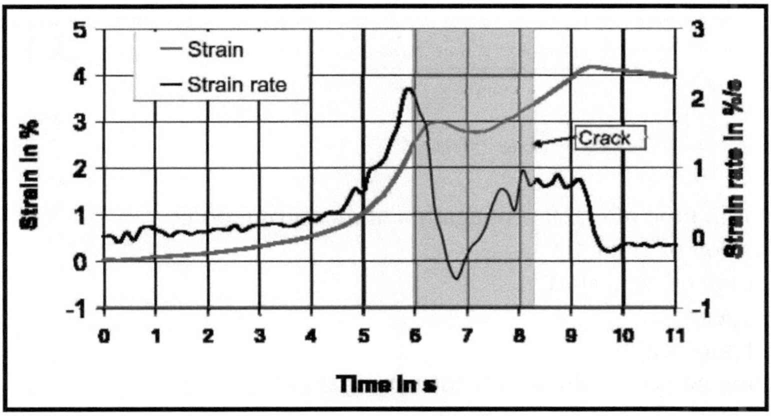


**Fig. 12.** Strain and strain rate in near field applying an external load of $v_{a,crit} = 4$ mm/min (critical cross head speed to induce a centreline crack)

Note that the already mentioned typical strain behaviour in both cases was observed here again. However, the absolute values rise at increased load. The curves of strains and strain rates in the far field show lower values and resemble qualitatively the far field measurements in Fig. 11.

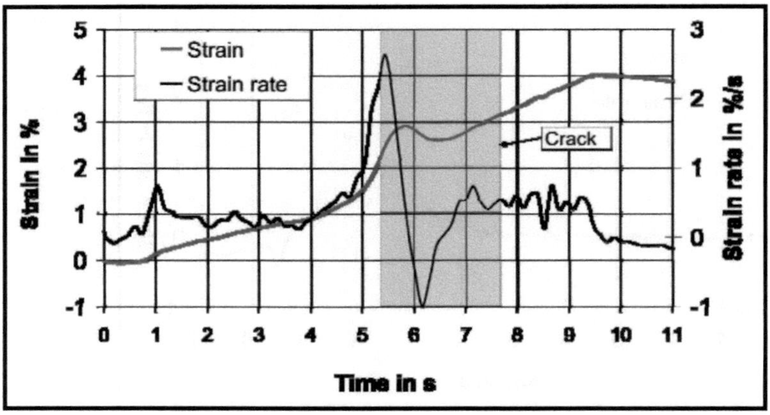


**Fig. 13.** Strain and strain rate in near field applying an external load of $v_a = 5$ mm/min

### *Solidification Crack accompanying Strain and Strain Rate*

Assuming that when a centreline crack follows directly the trailing edge of the weld pool, the time when the centreline crack propagates through the measuring range could be determined with the help of the high speed video records as follows (see Fig. 3).

$$t_{crack} = \frac{P_{crack} - P_{start} + l_{pool}}{v_s} \tag{1}$$

$$l_{pool} = (t_{edge} - t_{arc}) \times v_s \tag{2}$$

$t_{crack}$ point in time of crack propagation at measuring point
$P_{crack}$ location of crack tip
$P_{start}$ location of weld start
$v_s$ travel speed
$l_{pool}$ weld pool length
$t_{edge}$ trailing edge of weld pool at measuring point
$t_{arc}$ arc / electrode at measuring point

Thus, the strains and strain rates available during crack propagation could be determined from the measuring curves. These values are displayed in the diagram represented in Fig. 14. Additionally, strains and strain rates of welds without cracking are also shown in this diagram.

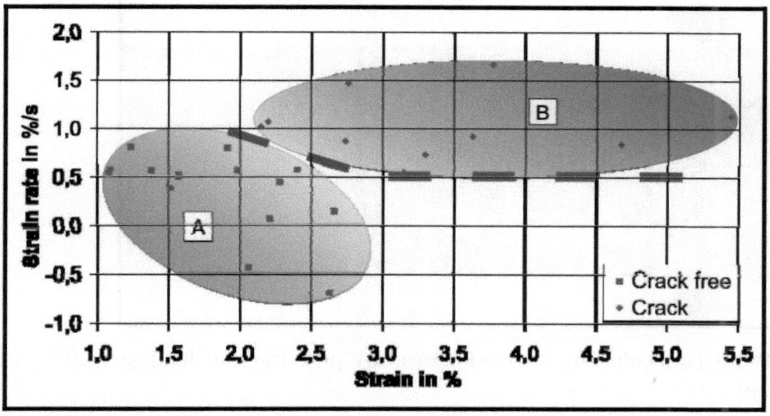


**Fig. 14.** Strain and strain rate for cracked and crack free welds

Although the results are subject to relatively large scatter, it becomes clear that solidification crack propagation requires high critical strain rates. It is obvious from the ranges A and B in Fig. 14 that quantification of the external restraint depending hot crack initiation should be possible based on the local strain rates in the weld pool vicinity.

## Influence of External Load on Local Thermo-Mechanics

Following the established relationship between hot crack formation and tensile stresses present within hot crack-sensitive ranges, as reported in [4-8, 16-18], the effect of an external load on the strain in the direct weld pool vicinity is discussed. First, only thermally conditioned effects are shown schematically on this basis. In Fig. 15, the near field range around the trailing weld pool is represented schematically. Besides heated and cooled areas, arising strains are indicated by arrows.

The temperature gradient in front of the weld pool is extremely high. This entails a sudden material expansion which can run only in transverse or longitudinal direction toward the weld pool edge. In [16-18], such a longitudinal flow of material is mentioned several times. In the transverse direction, the expansion takes place outward, since in front of the weld pool there is also material which must expand. Therefore, a tensile strain is registered in the near field range. In the vicinity of the weld pool the thermal expansion can run also toward the liquid area. The weld pool is able to absorb a certain portion of the expansion [18].

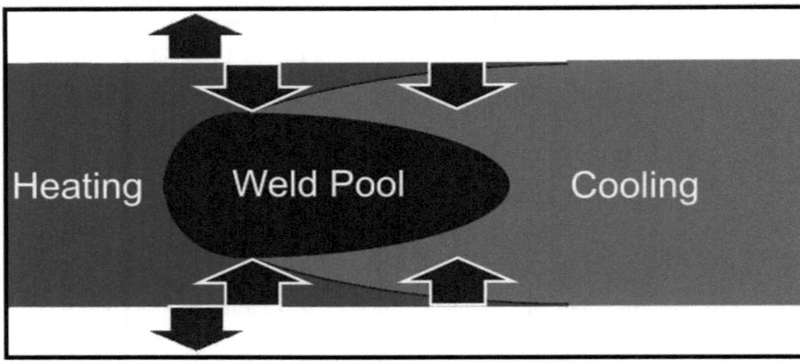


**Fig. 15.** Strains (indicated by arrows) caused by pure thermal load in vicinity of the weld pool

Just behind the arc, when the temperature at the fusion line begins to fall, the contraction by cooling starts in the direct vicinity of the weld pool. The areas outside this region, however, are still heated up, i.e. they still expand. With progressive cooling and contraction of adjacent areas, the heated material here is more and more able to let its expansion proceed inward towards the liquid area. The strain registered changes into a compressive one, although heating (expansion) is still present. Due to progressive cooling, the compressive strain will remain. Figure 15 clarifies this relation. In the literature [17, 18], it is reported that the expansion of areas still heated up can superimpose the contraction of already solidified areas in such a manner that even compressive stresses develop there. According to [4-8], this compressive cell can reach behind the weld pool and may prevent cracking (see Fig. 1).

When an external load is applied, the deformation behaviour (Fig. 16) changes. The described thermo-mechanics are superimposed by external load. In the initial state not yet warmed up material strains, due to the load. With heating of the material these strains become higher, since the rigidity of the material decreases. Additionally, thermal expansion begins. Thermal mechanics and load mechanics interact here in the same direction.

As previously mentioned, it is possible for the heated material to proceed with its expansion towards the weld pool. This thermo-mechanical influence interacts with the external load and at first prevails. With progressive external load; its influence continues to increase until it finally exceeds the thermal mechanics. Now again; a tensile strain is registered. Thermo-mechanics and external load act against each other.

The presence of this tensile cell enables cracking in agreement with the discussed model conceptions (see Fig. 1).

Cracking became possible from the moment when a hot crack-sensitive area (mushy zone) is exposed to a tensile strain and to a corresponding strain rate of a certain level. In this study, this was already possible with narrowing of the weld pool, as shown in the high speed records (Figs. 5 and 6).

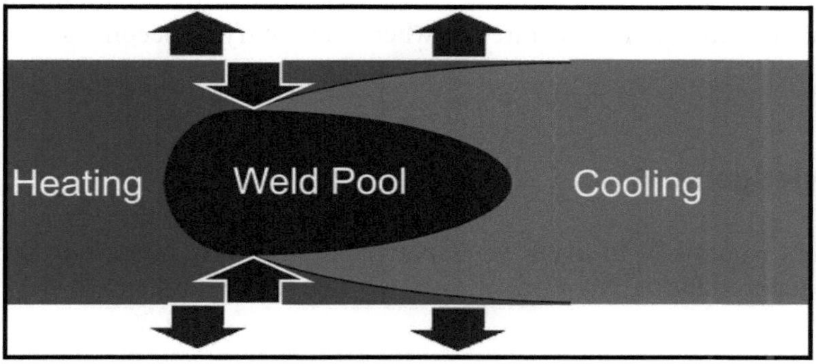


**Fig. 16.** Strains (indicated by arrows) caused by interacting thermal and external load in the vicinity of the weld pool

## Conclusion

1. According to [1, 2], local strain rates in the vicinity of the weld pool can significantly affect hot crack formation, and can be regarded as a suitable hot crack criterion.
2. By means of high speed video technology, the weld pool dimensions as well as the location and point in time of crack initiation and growth could be reconstructed. This permitted a direct correlation with the determined weld pool near strains and strain rates.
3. The observed typical strain behaviour in the weld pool vicinity is in principle in agreement with the model conceptions developed and discussed by Zacharia [7, 8] and Chihoski [17, 18]. An external load causes an additional significant increase in the strains and strain rates. In the near field, the magnitudes of the arising strains and strain rates are clearly higher than in the far field.
4. Contrary to samples under free contraction, additional tensile strains evolved in appropriate areas of the solidifying weld pool. Mechanical and thermal loads superimpose and interact differently in the near and far field.
5. The hot crack formation in the weld pool vicinity is directly affected by the local strain rate, which seems suitable for the quantification of

restraint specific hot crack susceptibility. It could be shown that for hot crack propagation high critical strain rates are necessary.

6. By application of a constant external load, solidification cracks can be initiated with narrowing weld pool on both sides of the solidification line. In the present case, the cracks grow perpendicular to the solidification line towards the weld centreline where they unify to become a centreline crack.

## Acknowledgements

The authors would like to thank A. Hannemann und M. Lammers for the technical support during this work.

## References

1. Prochorow NN, Jakuschin BF, Prochorow N Nikol (1968) Theorie und Verfahren zum Bestimmen der technologischen Festigkeit von Metallen während des Kristallisationsprozesses beim Schweißen (in German). Schweißen und Schneiden 18 1:8–11
2. Rappaz M, Drezet, JM, Gremaud M (1999) A New Hot-Tearing Criterion. Metallurgical and Materials Transactions A vol 30A February:449–455
3. Cross CE (2005) On the Origin of Weld Solidification Cracking. In: Böllinghaus Th, Herold H (eds) Hot Cracking Phenomena in Welds. Springer, Berlin Heidelberg, pp 3–18
4. Feng Z, Zacharia T, David SA (1997) On the Thermomechanical Conditions for Weld Metal Solidification Cracking. In: Cerjak H (ed) Mathematical Modelling of Weld Phenomena 3. Maney Publishing, London, pp 114–148
5. Feng Z, Zacharia T, David SA (1997) Thermal Stress Development in a Nickel Based Superalloy During Weldability Test. Welding Research Supplement November:470–483
6. Feng Z, Tsai CL (1993) Modelling the Thermomechanical Conditions at Weld Pool. In: Zacharia T (ed) International Conference Proceedings on Modelling and Control of Joining Processes, American Welding Society, Florida, U.S.A., pp 525–533
7. Zacharia T (1994) Dynamic Stresses in Weld Metal Hot Cracking. Welding Research Supplement July:164–172
8. Zacharia T, Aramayo GA (1993) Modelling of Thermal Stresses in Welds. In: Zacharia T (ed) International Conference Proceedings on Modelling and Control of Joining Processes, American Welding Society, Miami, Florida, U.S.A., pp 533–540

9. Kannengiesser Th, McInerny T, Florian W, Boellinghaus Th, Cross, CE (2002) The Influence of local weld deformation on hot cracking susceptibility. In: Cerjak H (ed) Mathematical Modelling of Weld Phenomena 6. Maney Publishing, London, pp 803–818
10. Wolf M, Kannengießer Th (2005) Moderne Prüfverfahren zur Bestimmung der Heißrisssicherheit beim Schweißen von hochlegierten Stählen und Nickelbasislegierungen (in German). In: Grellmann W (ed) Tagungsband Werkstoffprüfung. Verlag Stahleisen GmbH, Düsseldorf, pp 419–426
11. Kromm A, Kannengießer Th (2006) Anwendung des neuen Controlled Tensile Weldability (CTW) Tests zur Untersuchung der Heißrissneigung (in German). In: Borsutzki M, Geisler S (eds) Tagungsband Werkstoffprüfung. Verlag Stahleisen GmbH, Düsseldorf, pp 451–456
12. Brooks JA, Dike JJ, Krafcik JS (1993) On Modelling Weld Solidification Cracking. In: Zacharia T (ed) International Conference Proceedings on Modelling and Control of Joining Processes, American Welding Society, Miami, Florida, U.S.A., pp 174–185
13. Dike JJ, Brooks JA, Li M (1998) Comparison of Failure Criteria in Weld Solidification Cracking Simulations. In: Cerjak H (ed) Mathematical Modelling of Weld Phenomena 4. Maney Publishing, London, pp 199–222
14. Brooks JA, Dike JJ (1998) Modelling Weld Solidification Cracking Behavior in Aluminum Alloys-Analysis of Fracture Initiaton. In: Vitek JM, David SA, Johnson JA, Smartt HB, DebRoy T (eds) Trends in Welding Research, Proceedings of the 5th International Conference. Materials Park, Ohio, U.S.A., pp 695–699
15. Dike JJ, Brooks JA, Krafcik JS (1995) Finite Element Modelling and Verification of Thermal-Mechanical Behavior in the Weld Pool Region. In: Smartt HB, Johnson JA, David SA (eds) Trends in Welding Research, Proceedings of the 4th International Conference. ASM International, Materials Park, Ohio, U.S.A., pp 159–164
16. Johnson L (1973) Formation of Plastic Strains During Welding of Aluminum Alloys. Welding Research Supplement July:298–305
17. Chihoski RA (1979) Expansion and Stress Around Aluminum Weld Puddles. Welding Research Supplement September:263–276
18. Chihosky RA (1972) The Character of Stress Fields Around a Weld Arc Moving on Aluminum Sheet. Welding Research Supplement January:9–18
19. Goodwin GM (1987) Development of a New Hot-Cracking Test – The Sigmajig. Welding Research Supplement February:33–38
20. Kannengießer Th, Kromm A (2007) Design-Specific Influences on Local Weld Displacement and Hot Cracking. Proceedings to II. International Conference in Welding and Joining of Materials-ICWJM 2007, Cusco, Peru, 16.-18.4.2007, pp 1–9
21. Wilken K, Kleistner H (1982) Der MVT-Test – ein neues universelles Verfahren zur Prüfung der Heißrissanfälligkeit beim Schweißen (in German). Material und Technik 1:3–10

# Weld Solidification Cracking in Solid-Solution Strengthened Ni-Base Filler Metals

J.C. Lippold[1], J.W. Sowards[1], G.M. Murray[1], B.T. Alexandrov[1], A.J. Ramirez[2]

[1]The Ohio State University, Columbus, Ohio, USA
[2]Brazilian Synchrotron Light Laboratory, Campinas-SP, Brazil

## Abstract

The weld solidification cracking susceptibility of several solid-solution strengthened Ni-base filler metals was evaluated using the transverse Varestraint test. The alloys tested included Inconel 617, Inconel 625, Hastelloy X, Hastelloy W, and Haynes 230W.* Susceptibility was quantified by determining the solidification cracking temperature range (SCTR) which is a direct measurement of the range over which cracking occurs. This temperature range was then compared to the equilibrium solidification temperature range derived from Calphad-based ThermoCalc™ calculations, Scheil-Gulliver solidification simulations, and in-situ measurements using the single sensor differential thermal analysis (SS-DTA) technique.

Good correlation among the simulated and measured solidification temperature ranges, and SCTR values were found for the 617 and 230W filler metals. These two filler metals exhibited the best resistance to weld solidification cracking. Correlation among measured and simulated temperature ranges, and SCTR was poor for Hastelloy alloys X and W. Alloy 625 was found to be the most susceptible to solidification cracking, but this result is in conflict with fabrication experience. This appears to be the result of the inability of the Varestraint test to account for crack "healing" during the final stages of solidification.

---

* Inconel® is a registered trademark of Special Metals Company, a PCC company. Hastelloy® is a registered trademark of Haynes International. In this paper, a shortened version of these alloys will be used that does not include the trademark.

## Introduction

Solid-solution strengthened Ni-base alloys are used in applications where moderate strength and excellent corrosion resistance are required. Nickel-based alloys are widely used in a variety of industries, including pulp and paper, power generation, chemical handling and processing, coal gasification and liquefaction, and turbine engines. In all these industries, the combination of mechanical properties and corrosion resistance of these alloys are required to meet demanding service conditions.

In welded fabrication, one problem that is often encountered with these alloys is weld solidification cracking. Since the Ni-base alloys are austenitic, and solidify as a face centered cubic (FCC) gamma phase, they are inherently susceptible to weld solidification cracking. This form of cracking occurs along solidification grain boundaries in the weld metal due to the presence of liquid films at the boundary at the end of solidification. In this investigation, the susceptibility to weld solidification cracking was quantified using the transverse Varestraint test. The solidification behavior was evaluated using a combination of computational and experimental techniques. A comparison between solidification temperature range and cracking susceptibility measured using the Varestraint test was then conducted.

The Varestraint (VAriable RESTRAINT) test was originally developed by Savage and Lundin in 1965 to measure the susceptibility of an alloy to solidification cracking [1]. The Varestraint test was originally developed to be a direct and relatively simple technique for isolating the metallurgical and compositional factors responsible for cracking from the welding process variables. The test is widely applicable to most materials and the results are quantitative, allowing relative susceptibility to solidification cracking to be easily determined.

The concept of the Varestraint is represented schematically as a beam supported by rollers, as shown in Fig. 1. An autogenous weld is made across the sample as a downward bending motion is initiated by applying force at the loading rollers, which forces the sample to conform to the radius of the die block.

By using die blocks of different radii and by varying the length of the stroke of the hydraulic ram to a pre-calculated distance, the augmented strain in the outer fibers of the test sample can be altered according to the relationship below.

$$\varepsilon = \frac{t}{2R + t} \qquad (1)$$

where: ε = strain
t = sample thickness
R = radius of curvature of die block

By testing over a range of strain, a relationship between strain and cracking susceptibility can be determined. In this investigation, the susceptibility to cracking was determined by measuring the maximum crack distance (MCD) on the surface of the sample tested at a given strain. Duplicate samples are normally run at the same strain under identical testing conditions.

A representative plot of strain versus MCD is shown in Fig. 2. The lowest strain at which solidification cracking is observed is termed the "threshold" strain. The strain level above which the maximum crack distance does not increase is called the "saturated" strain, since at this level sufficient strain has been applied to force the solidification crack to fully propagate within the solidification temperature range.

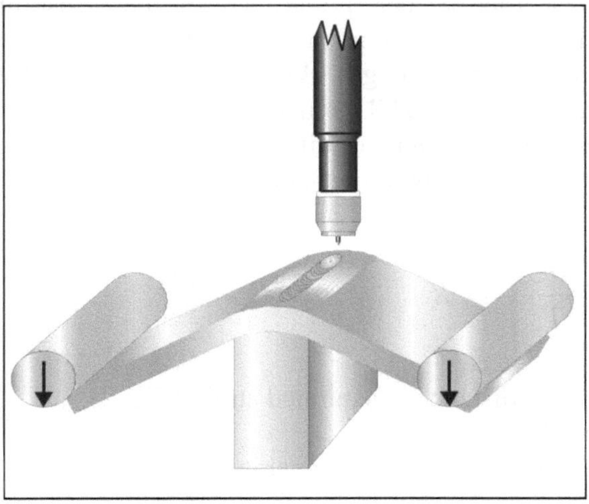

**Fig. 1.** Schematic illustration of the transverse Varestraint test

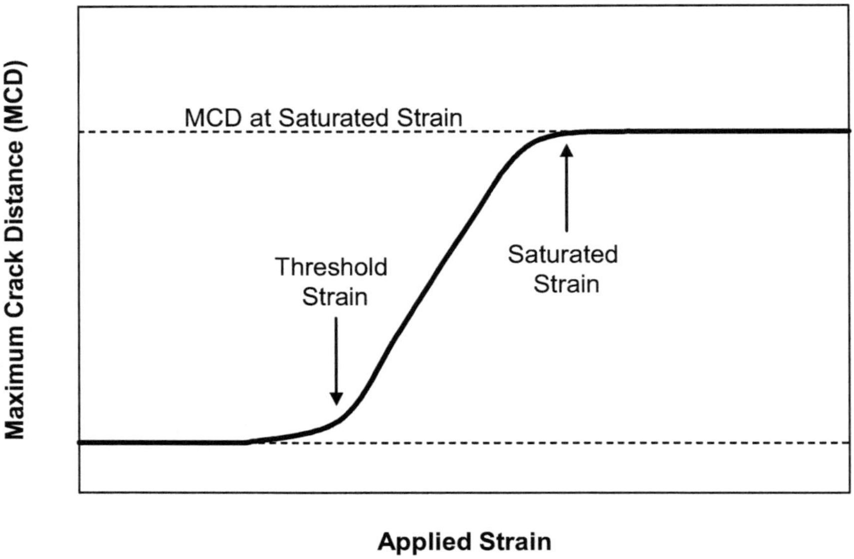


**Fig. 2.** Transverse Varestraint test results plotted as maximum crack distance versus strain

From the MCD versus strain data, the solidification cracking temperature range (SCTR) can then be determined. The SCTR is useful because it quantifies solidification cracking susceptibility in terms of temperature rather than simply a crack length. As a result, it is directly related to the brittle temperature range (BTR) which is commonly used to describe the temperature range over which solidification cracking occurs. The SCTR can be calculated by the following equation:

$$SCTR = \frac{MCD}{V_W} * CR \tag{2}$$

where: MCD = maximum crack distance at saturated strain (Fig. 2),
$V_w$ = weld travel speed,
CR = cooling rate in the solidification temperature range.

By using the SCTR approach both an order ranking and quantitative measure of weld solidification cracking can be obtained. Additional details and SCTR approach using the transverse Varestraint test can be found elsewhere [2].

# Experimental Procedures

## Materials

The Ni-base alloys tested in this investigation were in the form of filler metals. The compositions of the filler metals, as provided by the consumable manufacturers, are shown in Table 1.

**Table 1.** Chemical composition (wt %) of Ni-base filler metals

| Element | 617 | 625 | Hast W | Hast X | Haynes 230W |
|---|---|---|---|---|---|
| Ni | 56.48 | 63.9 | 64.57 | 47.99 | 60.58 |
| Cr | 21.4 | 22.1 | 5.52 | 20.99 | 21.9 |
| Fe | 0.38 | 0.9 | 5.35 | 19.54 | 0.8 |
| Mo | 8.91 | 8.98 | 23.62 | 8.44 | 1.58 |
| Co | 11.03 | 0.04 | 0.6 | 1.5 | 0.5 |
| W | – | – | – | 0.44 | 13.78 |
| Nb | – | 3.21 | – | – | – |
| Ti | 0.3 | 0.24 | – | – | 0.01 |
| Al | 1.2 | 0.2 | – | – | 0.41 |
| Si | 0.03 | 0.06 | 0.32 | 0.16 | 0.42 |
| C | 0.06 | 0.01 | 0.02 | 0.07 | 0.07 |

## Sample Preparation

Samples for Varestraint testing were prepared by depositing the filler metals into grooves machined in Alloy 600 base plates. The base plate was 0.375-inch (~9.5 mm) thick. The groove depth was 0.156-in. (~4 mm) and was 0.75 in. (19 mm) wide at the surface and 0.5 in. (13 mm) wide at the bottom of the groove. Using the gas tungsten arc welding (GTAW) process, a minimum of four layers of weld metal was deposited to minimize dilution between the filler metal and the base plate. In the four plus weld layers, there were approximately 15 weld passes per groove. Dilution of the filler metals by the Alloy 600 base metal in the final layer was determined to be less than 10%.

Upon completion of the groove welds, the weld deposit was machined flush with the base plate. The plates were then sectioned into 3.5 inch wide by 8 inch long samples in preparation for Varestraint testing. A schematic of the sample preparation process is shown in Fig. 3. The compositions of the actual weld metal deposits are provided in Table 2.

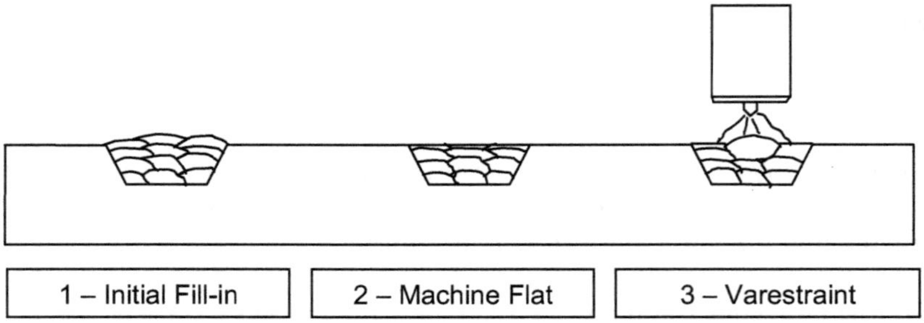


**Fig. 3.** Sample preparation and testing sequence. Grooves in Alloy 600 plate were filled in four layers with approximately 15 total weld passes. The weld reinforcement was then removed and Varestraint testing conducted within the deposited weld metal

**Table 2.** Composition (wt %) of actual weld deposits

| Element | 617 | 625 | Hast W | Hast X | Haynes 230W |
|---|---|---|---|---|---|
| Ni | 60.1 | 65.4 | 63.2 | 52.5 | 63.1 |
| Cr | 19.9 | 20.8 | 8.53 | 20.06 | 20.82 |
| Fe | 2.58 | 1.38 | 4.84 | 17.53 | 1.8 |
| Mo | 7.07 | 8.35 | 16.84 | 7.53 | 1.31 |
| Co | 9.08 | 0.1 | 2.42 | 1.25 | 0.45 |
| W | 0.1 | 0.1 | 0.07 | 0.43 | 11.2 |
| Nb | 0.1 | 3.38 | 0.02 | 0.1 | 0.02 |
| Ti | 0.27 | 0.28 | 0.12 | 0.1 | 0.06 |
| Al | 1.07 | 0.32 | 0.38 | 0.6 | 0.36 |
| Si | 0.1 | 0.1 | 0.3 | 0.1 | 0.33 |
| Mn | 0.14 | 0.1 | 0.41 | 0.39 | 0.47 |
| C | 0.034 | 0.017 | 0.016 | 0.056 | 0.038 |

## Varestraint Testing

The transverse Varestraint test was used to determine the solidification cracking susceptibility of the Ni-base filler metals deposited on Alloy 600. The gas tungsten arc welding (GTAW) parameters were established to develop a weld bead that was approximately 10 mm wide on the surface. The test conditions are listed in Table 3. Prior to testing, the samples were thoroughly cleaned with methanol.

Augmented strains in the range from 0.5 to 7% strain were used to evaluate the cracking susceptibility of the weld deposits. For each sample both the maximum crack distance (MCD) and total number of cracks were determined using a binocular microscope at magnifications up to 50X. Three independent measurements were made for each sample.

The solidification cracking temperature range (SCTR) was determined using the approach described in the Introduction by plunging a Type-C thermocouple into the molten weld metal of a weld made using identical parameters to those listed in Table 3. The cooling rate data collected by this technique was then used to calculate the SCTR.

**Table 3.** Transverse Varestraint Test Conditions

| Variable | Settings |
|---|---|
| Current | 180 amps, DCRP |
| Voltage | ~ 8 V (2.4 mm arc gap) |
| Electrode | W-2$ThO_2$, 2.4 mm diameter |
| Shielding Gas | 100% argon |
| Shielding Gas Flow Rate | 14.2 L/min |
| Gas Preflow Time | 4 s |
| Gas Postflow Time | 10 s |
| Weld Length | 50 mm |
| Bend Position | 44 mm from weld start |
| Bend Orientation | Transverse |
| Bend Rate | 250 mm/s |
| Bend Hold Time | 2 s after arc shut off |
| Weld Bead Width | ~10 mm on surface |

## Characterization

Weld metal samples of 617, Hast X, and Haynes 230W were prepared for metallographic analysis by electrolytic etching with 10% chromic acid at 2.5 V for 30–60 s. The 625 and Hast W samples were etched electrolytically with 10% oxalic acid at 4 V for 30 s. Scanning electron microscopy (SEM) and EDS analysis was performed on a Sirion SEM/FEG and Phillips XL-30 ESEM/FEG at 15 kV.

## Calphad Calculations

Equilibrium phase diagrams and Scheil-Gulliver solidification simulations were calculated using Calphad-based software Thermocalc® and Ni-base alloys database developed by Thermotec®. The measured chemical compositions of the actual weld deposits (Table 2) were used as input for the calculations. The phases used for the calculations were liquid, δ, γ, γ′, Laves, η, μ, σ, R-phase, MC, $M_6C$, $M_7C_3$, and $M_{23}C_6$.

## Results and Discussion

### Calculated Phase Diagrams

Calphad-based calculated equilibrium phase diagrams for Ni-base alloys based on Inconel 625 and Hastelloy X systems as a function of Mo content are shown in Fig. 4. The vertical lines in these diagrams represent the specific Mo content of Inconel 625 and Hastelloy X. As can be seen from these diagrams, Ni-base systems are very complex and contain multiple phases.

Since cooling from the solidification temperature in arc welds is relatively rapid, these alloys do not reach equilibrium conditions. Therefore, the phase diagrams and phase molar fraction evolution presented as examples in Figs. 4 and 5 are just a first approximation to the actual phase transformation behavior under non-equilibrium cooling conditions. Despite this limitation, these diagrams are extremely important when evaluating the material behavior under slower cooling conditions, postweld heat treatment, and material service at elevated temperature.

According to the equilibrium calculations, all the filler metals in this study undergo fully austenitic solidification with the exception of Hastelloy X, which forms $M_6C$ at the end of the solidification. Table 4 summarizes the calculated *solidus* ($T_S$) and *liquidus* ($T_L$) temperatures, equilibrium solidification temperature ranges and sequences for the weld deposit compositions listed in Table 2.

During solidification there is a strong segregation that results in the liquid enrichment in some elements and impoverishment of others. As a result of this segregation, there are significant changes in the solidification temperatures and path relative to the equilibrium diagrams. There is a simple model that approximates the solidification path under non-equilibrium conditions, the so-called Scheil-Gulliver solidification simulation. This model is based on local equilibrium at the solid/liquid interface, complete mixing of solute within the liquid, and no solid diffusion following solidification (once formed the solid composition does not change). Scheil-Gulliver solidification simulations were performed using Thermocalc™ software and a summary of the results is presented in Table 4.

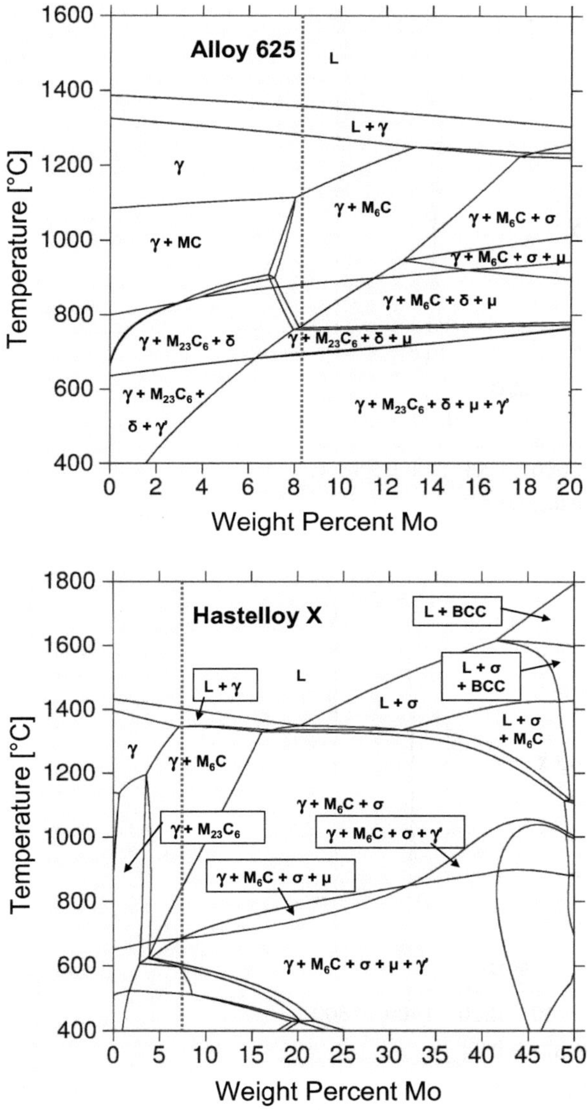


**Fig. 4.** Calculated phase diagrams for Alloy 625 (*top*) and Hastelloy X (*bottom*) systems as a function of Mo content. The vertical line represents the nominal Mo content of each alloy

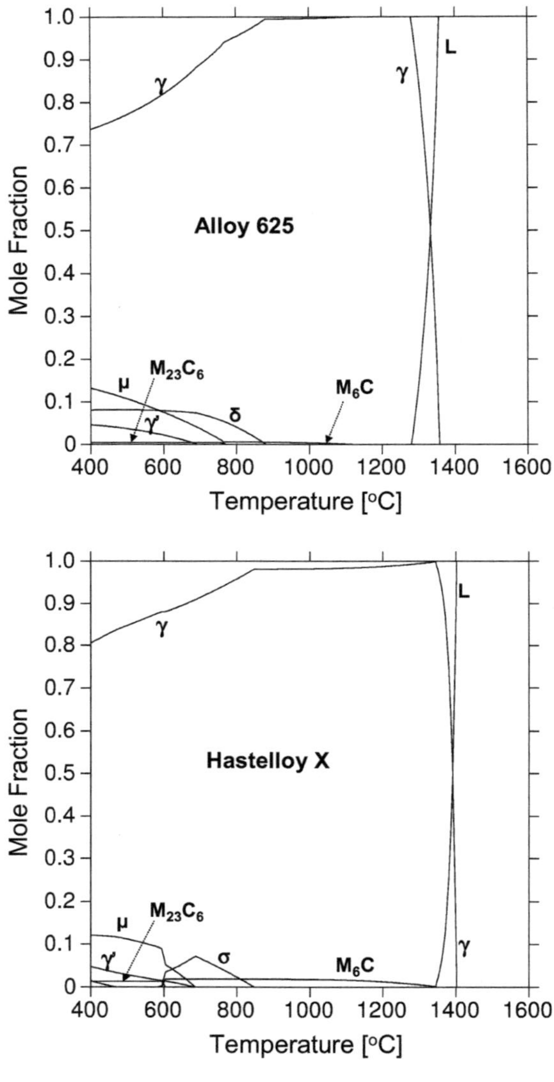


**Fig. 5.** Calculated equilibrium evolution of molar fraction of phases of Alloy 625 and Hastelloy X at the Mo concentrations indicated in Fig. 4

The comparison of the equilibrium and non-equilibrium solidification behavior in Tables 4 and 5, respectively, shows the effect of solidification segregation on the both the solidification temperature range and solidification path of the studied alloys. There are a number of eutectic reactions that occur during the solidification of these alloys, causing a significant reduction in the *solidus* temperature relative to equilibrium solidification. This expansion of the solidification temperature range has an important impact on solidification cracking susceptibility.

**Table 4.** Equilibrium solidification temperature ranges and solidification sequence determined using Calphad-based ThermoCalc™ calculations

| Alloy | Equilibrium Solidification | | | |
|---|---|---|---|---|
| | $T_S$ (°C) | $T_L$ (°C) | $\Delta T$ (°C) | Predicted Solidification Sequence |
| Alloy 617 | 1339 | 1386 | 47 | $L \rightarrow L+\gamma \rightarrow \gamma$ |
| Alloy 625 | 1275 | 1355 | 80 | $L \rightarrow L+\gamma \rightarrow \gamma$ |
| Hastelloy W | 1344 | 1385 | 41 | $L \rightarrow L+\gamma \rightarrow \gamma$ |
| Hastelloy X | 1333 | 1396 | 64 | $L \rightarrow L+\gamma \rightarrow L+\gamma+M_6C \rightarrow \gamma+M_6C$ |
| Haynes 230W | 1365 | 1402 | 37 | $L \rightarrow L+\gamma \rightarrow \gamma$ |

**Table 5.** Solidification Temperature Ranges and Solidification Sequence determined using ThermoCalc™ Scheil-Gulliver Prediction

| Alloy | Scheil-Gulliver Simulation | | | |
|---|---|---|---|---|
| | $T_S$ (°C) | $T_L$ (°C) | $\Delta T$ (°C) | Predicted Solidification Sequence |
| Inconel 617 | 1226 | 1386 | 160 | $L \rightarrow L+\gamma \rightarrow L+\gamma+M_6C \rightarrow L+\gamma+M_6C+\delta \rightarrow \gamma+M_6C+\delta$ |
| Inconel 625 | 1112 | 1355 | 243 | $L \rightarrow L+\gamma \rightarrow L+\gamma+MC \rightarrow L+\gamma+MC+\delta \rightarrow L+\gamma+MC+\delta+\sigma \rightarrow \gamma+MC+\delta+\sigma$ |
| Hastelloy W | 1060 | 1385 | 325 | $L \rightarrow L+\gamma \rightarrow L+\gamma+M_6C \rightarrow L+\gamma+M_6C+\mu \rightarrow \gamma+M_6C+\mu$ |
| Hastelloy X | 1236 | 1396 | 160 | $L \rightarrow L+\gamma \rightarrow L+\gamma+M_6C \rightarrow L+\gamma+MC+M_6C+\sigma \rightarrow \gamma+MC+M_6C+\sigma$ |
| Haynes 230W | 1277 | 1402 | 125 | $L \rightarrow L+\gamma \rightarrow L+\gamma+M_6C \rightarrow L+\gamma+M_6C+M_{23}C_6 \rightarrow L+\gamma+M_{23}C_6 \rightarrow L+\gamma+M_6C+M_{23}C_6 \rightarrow +\gamma+M_6C+M_{23}C_6$ |

## Weld Metal Characterization

The alloys examined in this study solidify as austenite (face centered cubic crystal structure). Austenite solidification promotes strong segregation resulting in local variations in composition in the weld metal. Due to this segregation, many of these weld metals form a terminal eutectic constituent (or constituents), as predicted by the Scheil-Gulliver simulations. Figs. 6–8 show the solidification structure of the alloys examined in this study. Included with the micrographs are composition spectra (measured with SEM/XEDS) showing the composition of both the austenite matrix and interdendritic eutectic constituent.

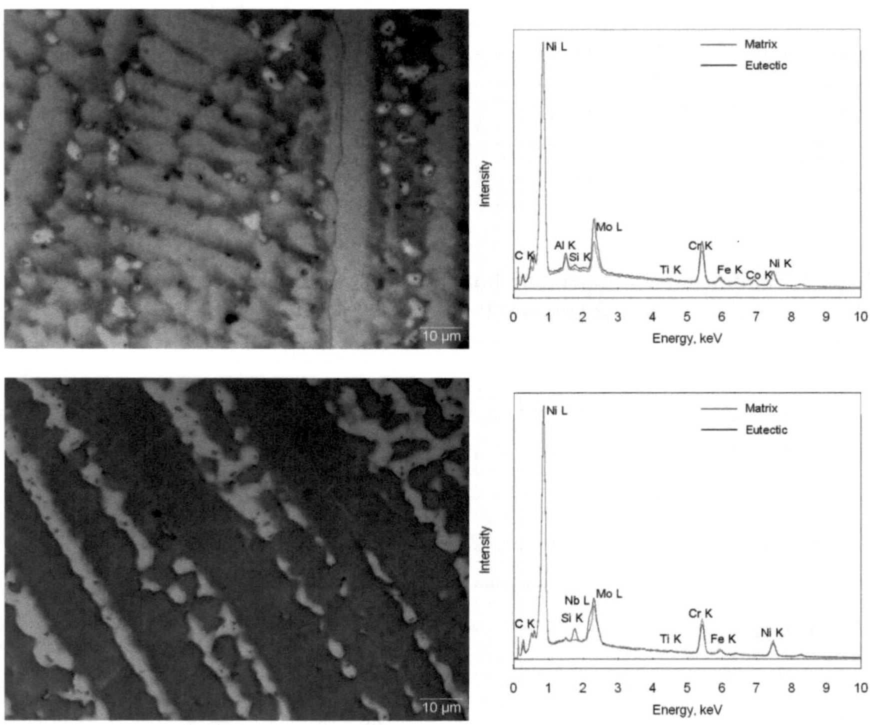


**Fig. 6.** Micrographs showing Alloy 617 (*top*) and Alloy 625 (*bottom*) solidification structure and XEDS spectra of matrix and eutectic compositions

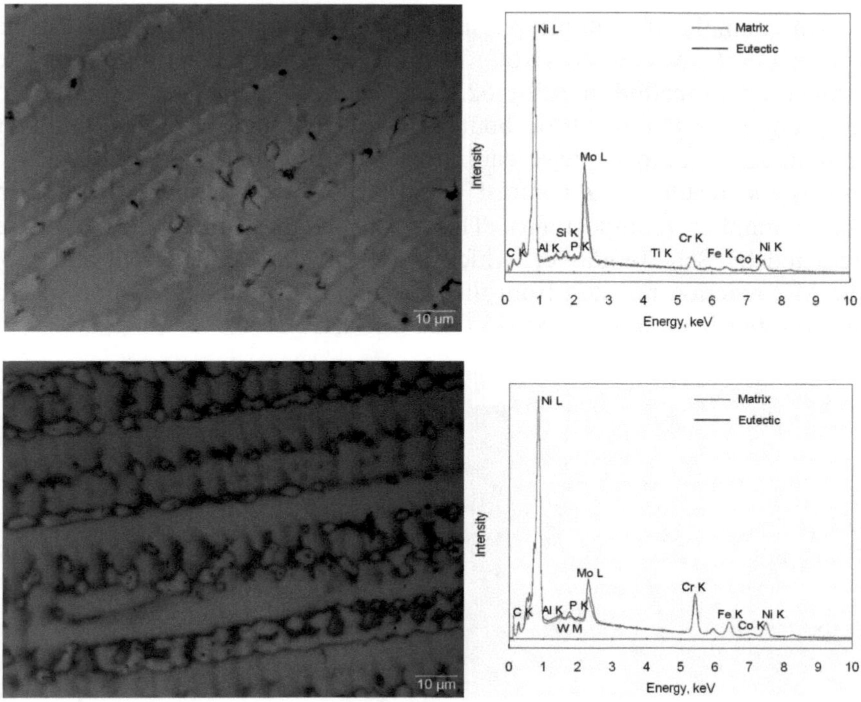


**Fig. 7.** Optical micrograph showing Hastelloy W (*top*) and Hastelloy X (*bottom*) solidification structure and XEDS spectra of the matrix and interdendritic regions

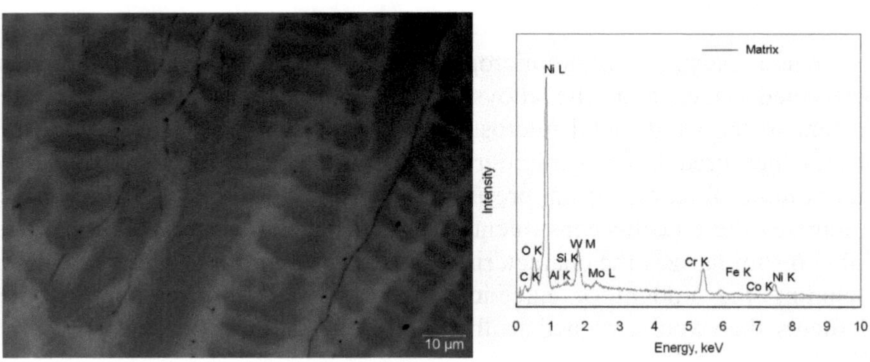


**Fig. 8.** Optical micrograph showing Haynes 230W solidification structure and XEDS spectra of matrix composition

An example of multiple phases forming during solidification can be seen in Fig. 9, where MC (NbC) forms by a eutectic reaction during the terminal solidification of Alloy 625. The MC eutectic phase is the first to form along the interdendritic boundaries as predicted by Scheil-Gulliver simulations. A second phase can be seen surrounding the MC particles. This is the result of "Z-contrast" in the SEM due to the difference in atomic number (composition). This phase is most likely the δ phase predicted by Scheil-Gulliver, which is rich in the elements Nb and Mo. The MC reaction resulted from the presence of C and segregation of Nb during solidification of Alloy 625 (Nb = 3.4 wt%).

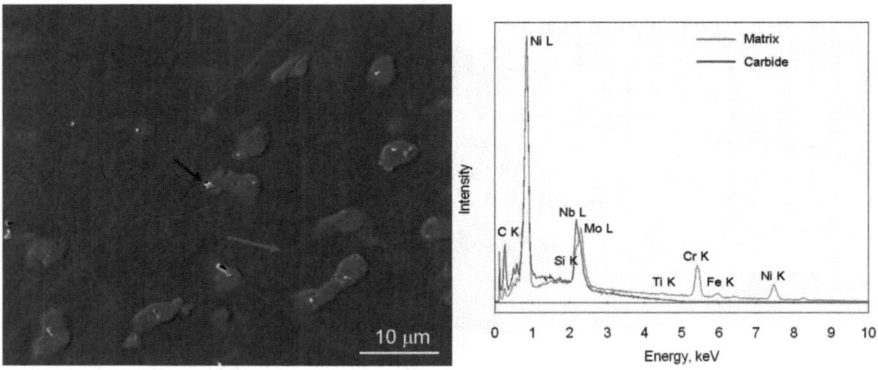


**Fig. 9.** Back-scattered electron SEM image (1000X) and XEDS spectra of NbC and matrix composition of Alloy 625

Image analysis of photomicrographs taken at 400X magnification was performed on each of the alloys to rank the amount of second phase present in the weld metal microstructure. The area fraction of secondary phases measured in each micrograph was assumed to be proportional to the volume fraction. It is presumed that most of this second phase comprises the eutectic constituents that form at the end of solidification. Table 6 summarizes the characteristics of eutectic formation for each alloy including the amount of eutectic formed during solidification and the elements that were enriched in the eutectic constituents as measured by XEDS.

Weld metal produced during Varestraint testing of all five filler metals solidified with a cellular dendritic solidification structure. Primary dendrite arm spacing (DAS) and secondary dendrite arm spacing (SDAS) were measured at locations near the weld centerline of samples tested at 3% strain. Solidification rate along the weld centerline can be approximated by the travel speed. The measured values of DAS and SDAS (shown in Table 6) were comparable for each of the alloys indicating similar solidification behavior.

## Transverse Varestraint Results

Both the maximum crack distance (MCD) and total number of cracks (TNC) were determined from the tranverse Varestraint test samples for the five filler metals. The results from this analysis are shown in Figs. 10–12. Except for Hastelloy W, cracking was observed at even the lowest applied strain (0.5%). Based on these results, the 617 and Haynes 230W filler metals exhibited the lowest cracking susceptibility. The MCD results for all five filler metals are compiled in Fig. 13.

**Table 6.** Summary of weld metal microstructure characteristics

| Alloy | Area Fraction of 2nd Phase | Elements Enriched in Eutectic | DAS (μm) | SDAS (μm) |
|---|---|---|---|---|
| 617 | 0.045 | Al, Si, Mo, Ti, Cr | 17.9 | 7.4 |
| 625 | 0.13 | C, Si, Nb, Mo | 16.8 | 6.4 |
| Hast X | 0.082 | Al, W, P, Mo, Cr | 17.6 | 6.3 |
| Hast W | 0.011 | C, Si, P, Mo | 17.8 | 6.7 |
| 230W | 0.002 | Not evaluated | 17.6 | 5.9 |

Using the technique described in the Introduction, the solidification cracking temperature range (SCTR) was determined based on weld metal cooling rate data collected by plunging thermocouples into the molten weld metal. Based on this data, the SCTR values for the five filler metals are plotted in Fig. 14. The SCTR data from an autogenous GTA weld in Type 304L stainless steel with Ferrite Number (FN) 6 is included for comparison. This weld metal is known to be extremely resistant to weld solidification cracking. In general, SCTR values below 50°C are consistent with weld metals that are very resistant to weld solidification cracking.

Based on SCTR values below 100°C, Alloy 617 and Haynes 230W would be the most resistant to weld solidification cracking. The SCTR results for Alloy 625 and Hastelloy X would predict that these filler metals have the highest susceptibility. The results for Alloy 625 are not consistent with actual fabrication experience with this filler metal, since it is generally considered to have good resistance to cracking. This inconsistency will be discussed later.

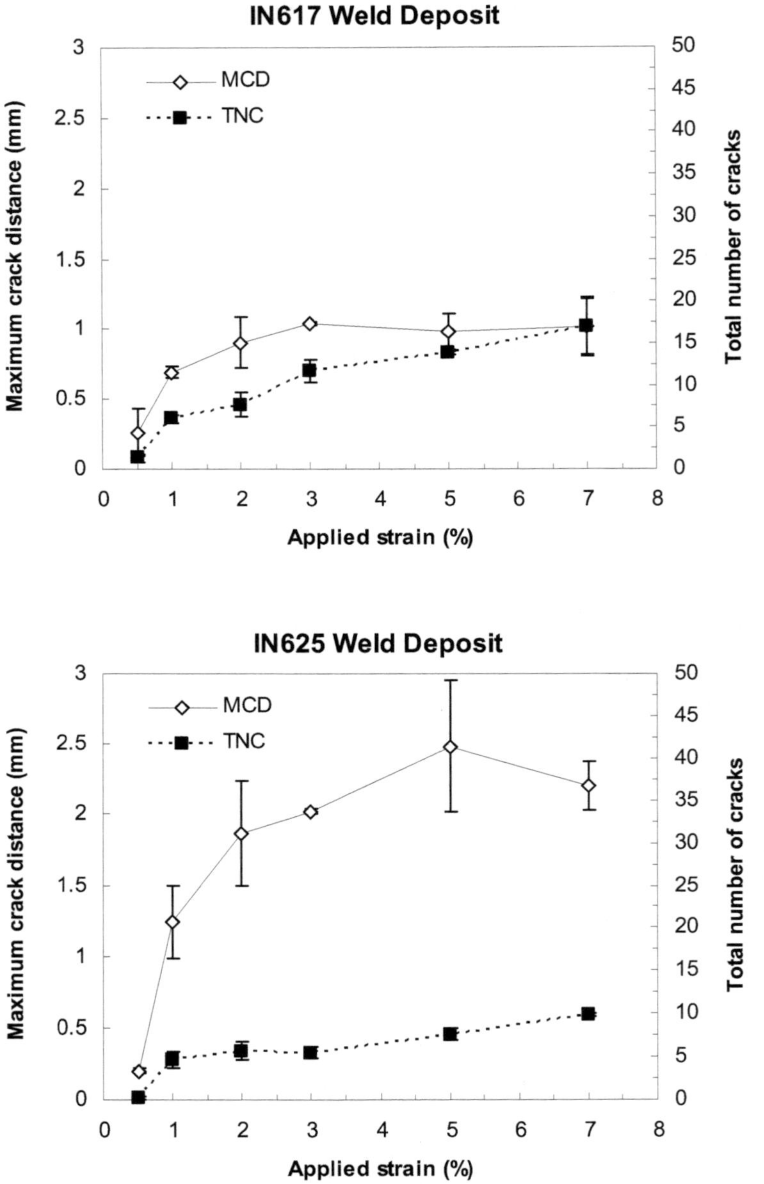


**Fig. 10.** Transverse Varestraint results for Alloys 617 and 625 filler metals

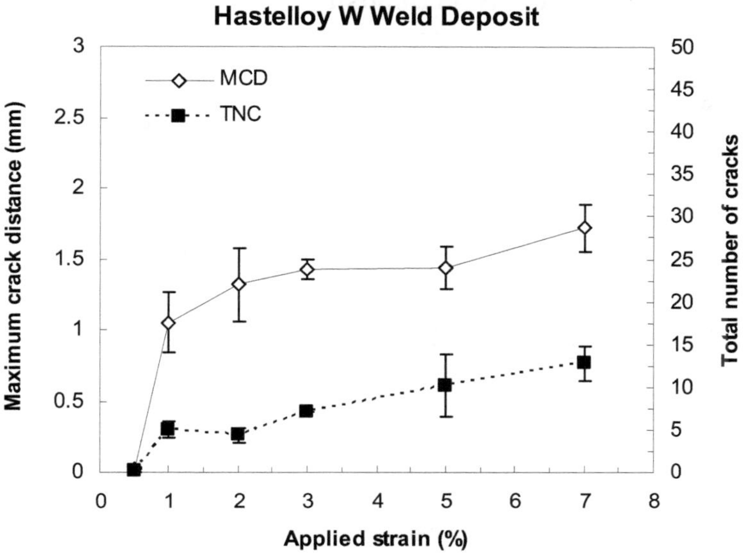


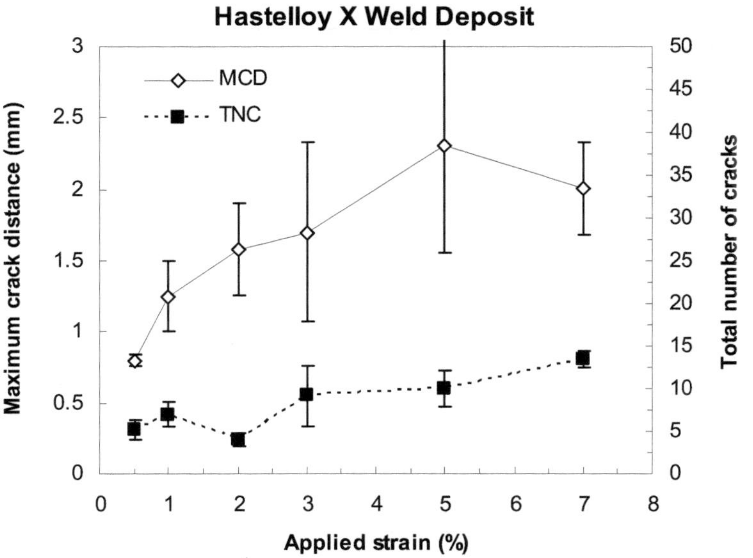


**Fig. 11.** Transverse Varestraint results for Hastelloy W and Hastelloy X filler metals

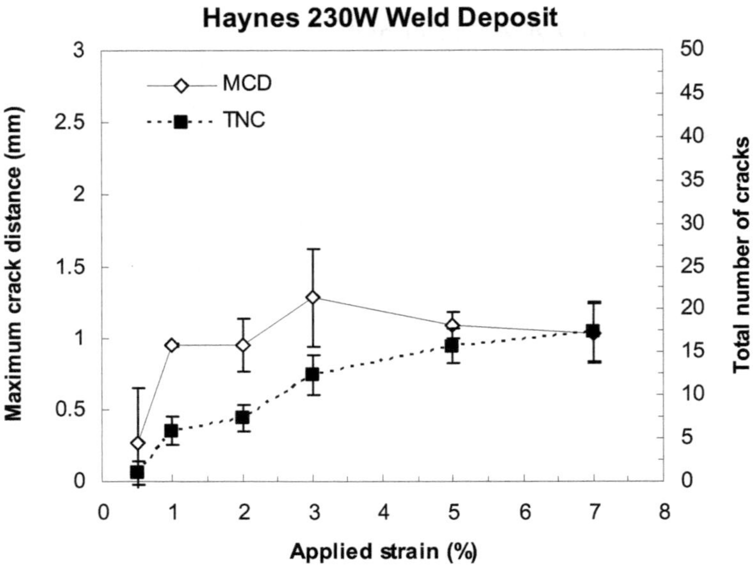


**Fig. 12.** Transverse Varestraint results for Haynes 230W filler metal

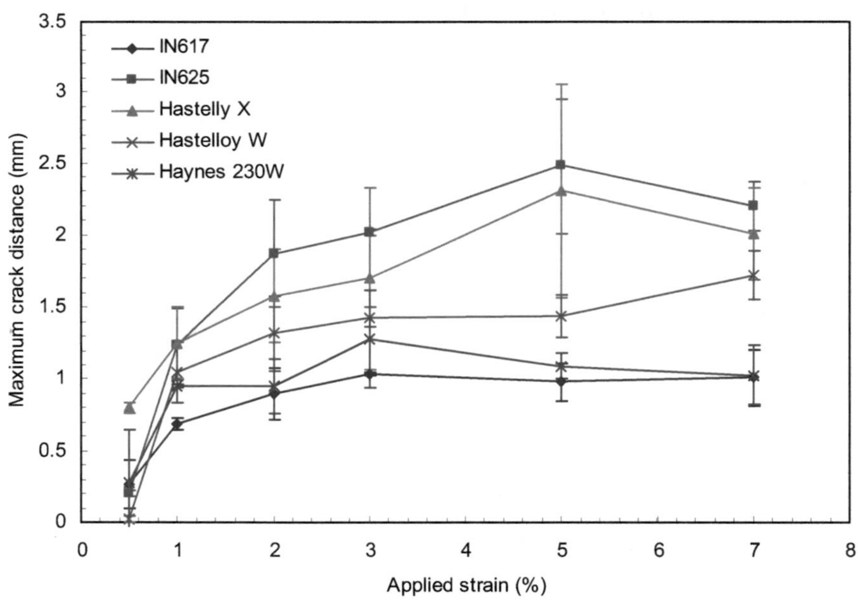


**Fig. 13.** Summary of maximum crack distance (MCD) versus strain for all the filler metals tested

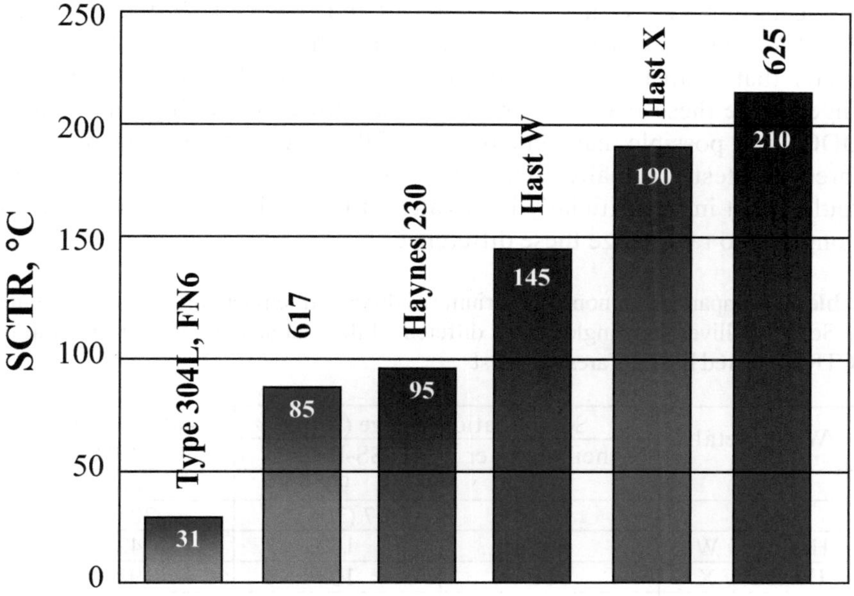


**Fig. 14.** Solidification cracking temperature range for the filler metals tested. Type 304L with Ferrite Number (FN) 6 is listed for comparison

## Comparison of Non-equilibrium Solidification and SCTR

Actual weld solidification temperature ranges were measured using the SS-DTA technique [3, 4, 5]. This technique uses thermocouples plunged directly into the weld pool to measure the cooling rate from above the liquidus to room temperature. The thermal history is then analyzed to determine the liquidus and solidus temperatures, and the solidification temperature range is calculated.

The comparison of the Scheil-Gulliver simulations, SS-DTA measurements, and Varestraint SCTR are shown in Table 7. The Scheil-Gulliver and SS-DTA values are in reasonably good agreement for Alloys 617, 625, and Haynes 230W. For Hastelloy alloys X and W, the simulation overestimated the measured solidification temperature range.

Based on solidification cracking theory, the SCTR should be a fraction of the solidification temperature range, since cracking will occur only when substantial solid has formed. Thus, the SCTR values for Alloy 625, and Haynes 230W seem most appropriate when compared to both SS-DTA and Scheil-Gulliver values. For Alloy 617 and Hastelloy W, the SCTR and SS-DTA values are nearly equivalent, while the SCTR for Hastelloy X is greater than both the simulated and measured solidification temperature range. Since these materials are also susceptible to ductility dip cracking (DDC), it is possible that some portion of the crack length measured in the Varestraint test is actually DDC extension off the solidification crack. This would result in an artificially high value of the SCTR. Additional analysis is ongoing to rationalize these differences.

**Table 7.** Comparison of non-equilibrium solidification temperature range determined by Scheil-Gulliver and single sensor differential thermal analysis (SS-DTA) with the SCTR measured by the Varestraint test

| Weld Metal | Solidification Range (°C) | | Varestraint SCTR (°C) |
|---|---|---|---|
| | Scheil-Gulliver | SS-DTA | |
| 617 | 160 | 93 | 88 |
| 625 | 243 | 97 (306)* | 205 |
| Hastelloy W | 325 | 162 | 142 |
| Hastelloy X | 160 | 108 | 191 |
| Haynes 230W | 125 | 139 | 98 |

* Value in parentheses includes end of eutectic solidification

## Effect of Backfilling on Cracking Susceptibility

The SCTR determined using the Varestraint test is based on the MCD at saturated strain. In the case of the weld metals tested here, the saturated strain at which SCTR was determined was 5%. At such high applied strains, the effect of crack backfilling by excess liquid at the end of solidification can be reduced or eliminated. Under normal welding conditions, where actual strains in the solidifying weld metal may be much lower, backfilling may in fact be effective in reducing cracking susceptibility.

Photomicrographs taken near the crack tips of Varestraint samples from Alloy 625 and Haynes 230W are shown in Fig. 15. Note that in Alloy 625 considerable residual liquid (arrows in Fig. 15A) is observed near the crack tips. This represents liquid that was along the solidification grain boundaries at the end of solidification and is drawn into the tips of the cracks by capillary attraction during testing at high strain levels. In contrast, virtually no residual liquid is observed near the crack tips of the Haynes 230W weld metal (Fig. 15B).

This residual liquid effect is also observed when examining the fracture surface of the solidification cracks in these weld metals, as shown in Fig. 16. In both cases, the characteristic dendritic nature of solidification cracking is evident. However, in the case of Haynes 230W, the dendritic fracture morphology is much more distinct. This occurs since there is little residual liquid at the end of solidification and the solidifying dendrites are covered by only a thin layer of liquid. In contrast, the large amount of terminal liquid during solidification of Alloy 625 acts to coat the dendritic structure and make it less distinct.

A)

B)

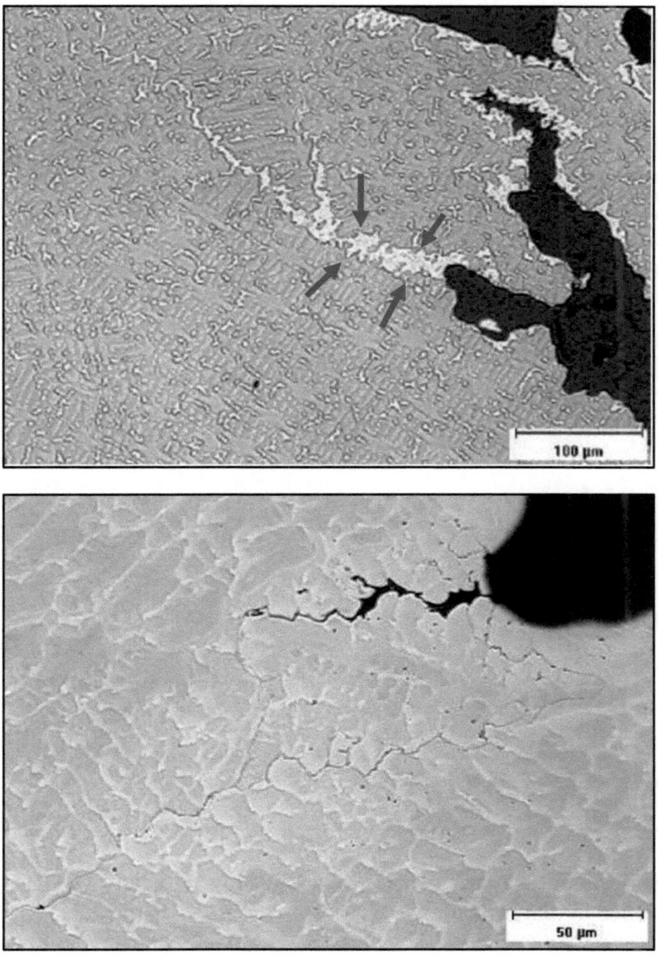


**Fig. 15.** Microstructure near the crack tip of a solidification crack in (**A**) Alloy 625, and (**B**) Haynes 230W. Note difference in magnification

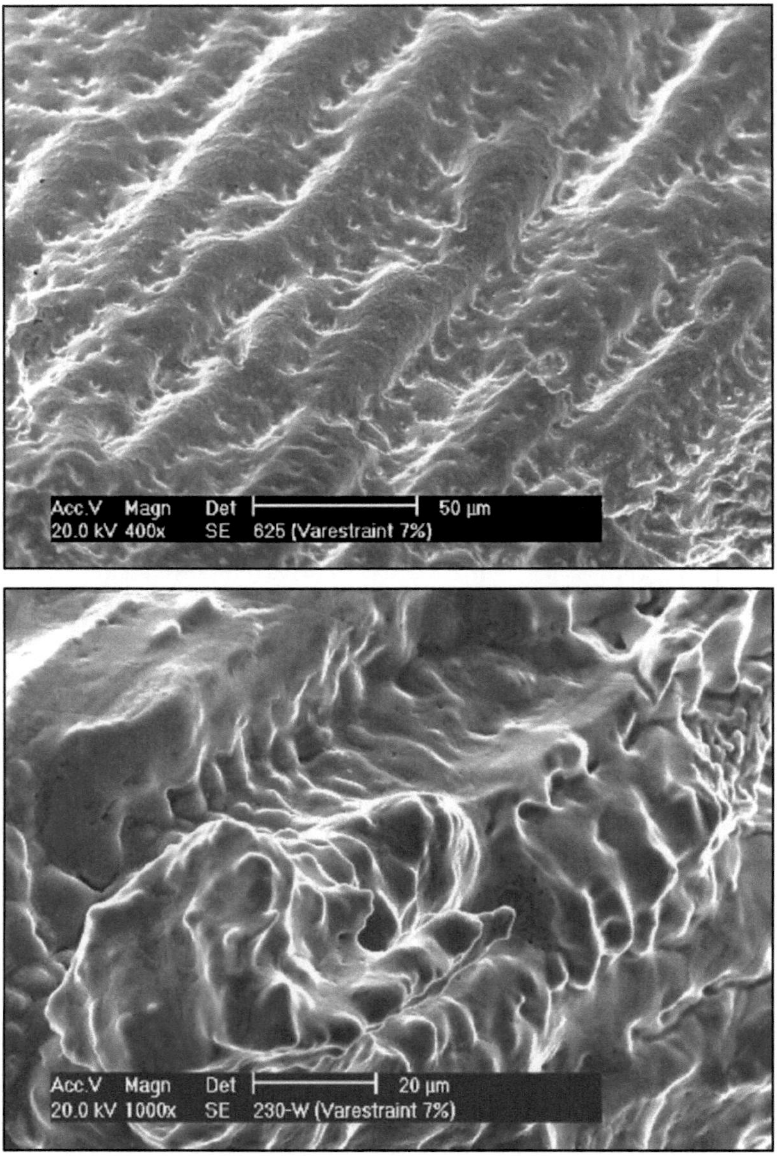


(A)

(B)

**Fig. 16.** Solidification crack surface of (**A**) Alloy 625, and (**B**) Haynes 230W. Note difference in magnification

Under lower applied strain conditions, the terminal liquid in the Alloy 625 weld metal would be available to heal any solidification cracks that would form, improving the weld solidification cracking resistance relative to what the SCTR ranking predicts. This may explain why Alloy 625 is generally considered to have good resistance to weld solidification cracking in actual practice, particularly at low to moderate restraint levels where backfilling can be effective in healing cracks that form..

For Alloy 625, differences between the solidification cracking susceptibility determined using the Varestraint test and actual fabrication experience has been reported by other investigators [6, 7]. This suggests that for weld metals that achieve resistance to solidification cracking through a crack backfilling mechanism, the Varestraint test may not accurately reflect their susceptibility to cracking. The high applied strains used during this test force cracking to occur even in cases where considerable terminal liquid is present. The use of a threshold strain for cracking rather than MCD at saturated strain may be a more appropriate measure of susceptibility for these weld metals.

## Conclusions

1. Solidification cracking susceptibility of the five Ni-base filler metals as determined by the solidification cracking temperature range (SCTR) is as follows (lowest to highest): Alloy 617, Haynes 230W, Hastelloy W, Hastelloy X, and Alloy 625.
2. Good correlation was found among the SCTR, simulated Scheil-Gulliver solidification range, and SS-DTA values for Alloy 617 and Haynes 230W. These filler metals exhibited the highest susceptibility to solidification cracking.
3. Correlation between the simulated solidification range and the range measured by SS-DTA was poor for both Hastelloy alloys W and X. In both cases the simulation overestimated the actual range.
4. Alloy 625 weld metal was determined to be the most susceptible to solidification cracking, even though fabrication practice has shown that this filler metal is quite resistant to cracking. It appears that the crack "healing" mechanism may be responsible for this contradiction.
5. Additional research is needed to relate calculated and measured solidification temperature ranges, and to effectively use these to predict solidification cracking susceptibility.

## Acknowledgements

The authors wish to thank Siemens-Westinghouse for providing the filler metals used in this investigation. The Varestraint testing and characterization was carried out as part of a senior project at OSU. We extend our thanks to Jeff Bernath and Matthew Slate (along with Jeff Sowards) who participated in that project.

## References

1. Savage, W.F. and C.D. Lundin, The Varestraint Test, *Welding Journal,* October 1965, pp. 433s–442s.
2. J.C. Lippold, 2005. Recent Developments in Weldability Testing, 1st International Workshop, *Hot Cracking Phenomena in Welds,* Berlin, March 2004, publ. by Springer-Verlag, pp. 271–290.
3. Alexandrov B.T. and J.C. Lippold, Single Sensor Differential Thermal Analysis of Phase Transformations and Structural Changes during Welding and Postweld Heat Treatment, IIW Doc. IX-2199-06, 59th Annual Assembly of IIW, Quebec City, Canada, August 2006, accepted for publishing in Welding in the World.
4. Alexandrov B.T. and J.C. Lippold, Relationship Between the Solidification Temperature Range and Weld Solidification Cracking Susceptibility of Stainless Steels and Ni-base Alloys, IIW Doc. IX-2163-05, 58th Annual Assembly of IIW, Prague, July 2005, www.iiw-iss.org.
5. Alexandrov B.T. and J.C. Lippold, In-Situ Weld Metal Continuous Cooling Transformation Diagrams, Welding in the World, Vol. 50, No. 9/10, 2006, pp. 65–74 (Doc. IIW 1744-06, ex-doc. IX-2162-05).
6. Lingenfelter, A.C., Varestraint Testing of Nickel Alloys, *Welding Journal*, September 1972, pp. 430s–436s.
7. Karlsson, L., E.L. Berquist, S. Rigdal, and N. Thalberg, Evaluating Hot Cracking Susceptibility of Ni-base SAW Consumables for Welding of 9% Ni Steel. Proceeding of 2nd International Workshop on Hot Cracking, Berlin, March 2007.

# Hot Cracking Susceptibility of Ni-Base Alloy Dissimilar Metal Welds

H. Hänninen, A. Brederholm, T. Saukkonen

Helsinki University of Technology, Espoo, Finland

## Abstract

Hot cracking susceptibility of Ni-base alloy dissimilar metal welds in nuclear power plant (NPP) applications was studied by preparing mock-ups of various relevant dissimilar metal welds of the NPPs. The motivation for the study was the need for repair welding of Alloy 182 welds of safe ends in BWR plants and possible welding problems with new Alloy 152/52 weld metals for new reactors. Weldability of the Ni-base alloys (Alloys 182/82 and Alloys 152/52) was evaluated based on their composition and weld metal microstructures. Susceptibility to hot cracking was examined by Varestraint tests. The studied dissimilar metal welds showed clear segregation of Nb, Si, P and Mn to the last liquid to solidify at dendrite boundaries. Hot cracking occurred along the dendrite boundaries and the susceptibility to hot cracking in the weld mock-up samples was observed to follow the order: Alloy 152 > Alloy 52 > Alloy 182 > Alloy 82, while in pure (undiluted) weld metal hot cracking tests the susceptibility followed the order: Alloy 182 ≥ Alloy 152 > Alloy 52 ≥ Alloy 82. The differences are thought to be related to dilution effects in mock-up welds because Fe, Si, and C enhance the eutectic phases and expand the solidification temperature range.

## Introduction

The Ni-base alloy dissimilar metal welds are typically made using Alloy 182 and Alloy 82. Recently, Alloy 52 has been used both in new constructions as well as in repair welding. The trend towards alloys with higher contents of chromium is driven by the observed environment-assisted cracking (EAC) in Alloy 182, and recently also in Alloy 82. One driving force towards the more EAC resistant alloys is also the challenges and costs related to non-destructive examination of dissimilar metal welds in nuclear power plant applications.

Several studies have been made to investigate hot cracking susceptibility of filler metal Alloys 82, 182, 52 and 152 and base material Alloy 690 [1, 2, 3]. Hood and Lin [1] studied hot cracking susceptibility of Ni-base alloy filler metals (Alloys 82, 182, 52 and 152) using the Varestraint and spot-Varestraint tests. The results showed that the hot cracking resistance of the four filler metals deposited on Alloy 690 and carbon steel A285 is similar or better than that for 1¼Cr-½Mo steel and Alloy 690-AISI 316L base metal combinations. Alloy 52 exhibited the best resistance to both weld solidification cracking and weld metal liquation cracking, followed by Alloys 82, 152 and 182. The AISI 316LN steel/Alloy 52 combination exhibited better resistance than Alloy 690/Alloy 52. Wu and Tsai [2] investigated the hot cracking susceptibility of Alloys 82 and 52 with Alloy 690 by Varestraint tests. Alloy 82 showed greater total crack length (TCL) than Alloy 52. Liquation cracks and ductility dip cracks were found in the HAZ, but an accurate distinction of the crack type was difficult. These results indicate that hot cracking susceptibility of Alloy 82 is greater than that of Alloy 52 in Varestraint testing.

Nishimoto et al. [4] investigated the microcracking behaviour in TIG-welded multi-pass Alloy 690 weld metal by spot- and trans-Varestraint tests using three different filler metals of Alloy 52 by varying the impurity contents of P and S. The results of the study showed that the microcracking susceptibility decreased with decreasing impurity element content. The microcracking susceptibilities of ductility-dip, liquation and solidification cracking were much improved by adding 0.01–0.02 mass% La to the weld metal. However, excessive La addition to the weld metal led to liquation and solidification cracking. The ductility-dip cracking susceptibility was improved by the amelioration of hot ductility of reheated weld metal. The excessive La addition resulted in the formation of Ni-La intermetallic compound, and therefore, liquation and solidification cracks can occur in the weld metal attributed to the enlargement of brittle temperature range (BTR) and/or the occurrence of local liquation during the reheating. It was concluded that the microcracking susceptibility in multi-pass welding of Inconel 690 was improved drastically by La addition to the filler metal [4].

Recently a Gleeble-based test technique, termed strain-to-fracture (STF), has been developed by Nissley et al. [5], and employed for evaluation strain-to-fracture of ductility-dip cracking (STF DDC) susceptibility curves for filler metals of Alloys 82 and 52 [3, 5]. The filler metal Alloy

52 was more susceptible to DDC than Alloy 82, which also exhibited heat-to-heat variation in susceptibility. Thus, because of varying observations in various tests there is a major need to rank the Ni-base alloy weld metals for their hot cracking susceptibility and to understand the mechanisms of solidification cracking properly.

## Materials and Experimental Methods

In this study the hot cracking tendency of six different dissimilar metal weld joints was tested by Varestraint testing [6]. The Varestraint equipment, Fig. 1, was built and the obtained hot cracking and weld microstructures were studied by optical and scanning electron microscopy (FEG-SEM/EDS). The studied materials and welds are typical for nuclear power plant and oil and gas industry applications. In Table 1 the chemical compositions of the base materials and filler metals used in the manufacturing of the studied dissimilar metal welds are presented.

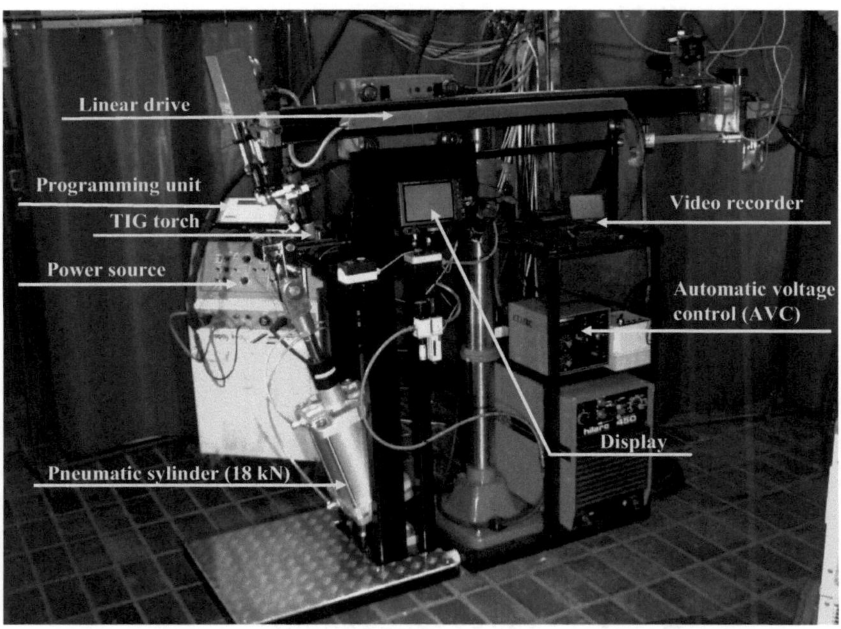


**Fig. 1.** General view of Varestraint testing equipment built for hot cracking testing

**Table 1.** Chemical compositions of the base materials and filler metals used in the dissimilar metal welds (wt.-%)

| | C | Si | Mn | P | S | Cr | Mo | Ni | Nb | Ti | Fe | Al | Cu |
|---|---|---|---|---|---|---|---|---|---|---|---|---|---|
| **Base materials** | | | | | | | | | | | | | |
| 42CrMo4 | 0,42 | | 0,7 | | 0,4 | 1,05 | 0,225 | | | | | | |
| AISI 321 | < 0,08 | | | | | 17-19 | | 9-12 | | 5*C-0,7 | | | |
| AISI 316 NG | < 0,03 | | | | | 16,5-18,5 | 2,5-3 | 11-14 | | | | | |
| SA508 (Grade 2) | 0,24 | 0,21 | 1,38 | | 0,009 | 0,16 | 0,54 | 0,58 | | | | 0,015 | 0,12 |
| Inconel 600 | 0,07 | 0,19 | 0,21 | 0,007 | 0,001 | 16,3 | | 72,86 | Nb+Ta 0,10 | 0,28 | 9,44 | 0,229 | 0,13 |
| Inconel 690 | 0,02 | 0,04 | 0,16 | | 0,001 | 29,46 | | 59,82 | | 0,33 | 9,96 | 0,2 | <0,01 |
| **Filler metals** | | | | | | | | | | | | | |
| E 347 (OK 61.81) | 0,07 | 0,7 | 1,7 | 0,021 | 0,006 | 19,8 | 0,1 | 10 | 0,61 | | | | 0,14 |
| AISI 347 (OK Tigrod 16.11) | 0,04 | 0,85 | 1,4 | 0,02 | 0,01 | 19 | 0,02 | 9,5 | 0,55 | | | | 0,03 |
| E 347 (OK Band 11.62) | < 0,025 | 0,4 | 1,8 | | | 20 | | 10 | 0,7 | | | | |
| AISI 308L (OK Band 11.61) | < 0,025 | 0,4 | 1,8 | | | 20 | | 10 | | | | | |
| AISI 309L (OK Band 11.65) | < 0,025 | 0,4 | 1,8 | | | 24 | | 13 | | | | | |
| Inconel 182 (OK 92.26) | 0,03 | 0,8 | 6,5 | 0,01 | 0,003 | 15,7 | | 68 | 1,8 | 0,1 | 6,7 | | < 0,01 |
| Inconel 82 (OK Tigrod 19.85) | 0,039 | 0,03 | 2,98 | 0,001 | 0,004 | 19,94 | | 72,6 | Nb+Ta 2,47 | 0,34 | 1 | | 0,01 |
| Inconel 52 | 0,03 | 0,13 | 0,24 | <0,001 | <0,001 | 29,2 | 0,03 | 59,28 | Nb+Ta <0,02 | 0,51 | 9,8 | 0,72 | 0,04 |
| Inconel 52M | 0,02 | 0,09 | 0,8 | 0,003 | 0,001 | 30,06 | 0,01 | 59,54 | Nb+Ta 0,83 | 0,224 | 8,22 | 0,11 | 0,02 |
| Inconel 52MS | 0,014 | 0,12 | 0,68 | 0,004 | 0,0007 | 29,53 | 0,02 | 60,14 | 0,78 | 0,19 | 8,33 | 0,13 | 0,03 |
| Inconel 152 | 0,048 | 0,41 | 3,48 | 0,003 | 0,003 | 28,74 | 0,01 | 55,2 | Nb+Ta 1,54 | 0,09 | 10,39 | 0,06 | <0,01 |
| Inconel 152M | 0,0293 | 0,3276 | 3,247 | <0,001 | 0,0042 | 28,98 | <0,005 | 57 | Nb+Ta 1,533 | 0,039 | 8,75 | <0,005 | <0,005 |

Figure 2 presents the groove geometry, materials and cross-section of Neste Oil mock-up dissimilar metal weld. Inconel 182 buttering side of the test specimen was heat treated at 700°C for 7 h. Figures 3–6 present the groove geometries, materials and cross-sections of mock-up TV, 1, 2 and 3 dissimilar metal welds for nuclear applications. The clad pressure vessel steel SA 508 side of specimen TV (with Alloy 182 buttering) was heat treated at 610°C for 6 h and mock- up 3 (with Inconel 52 and Inconel 152 buttering) was heat treated at 610°C for 4 h before welding. Whole welded test specimens of mock-ups 1 and 2 were heat treated at 610°C for 16 h after welding.

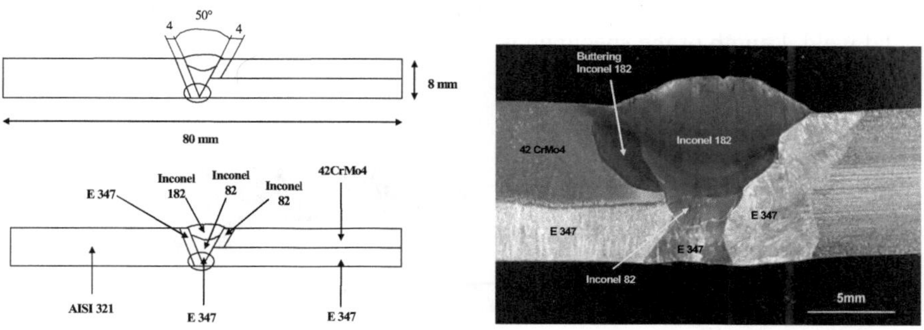


**Fig. 2.** Groove geometry, materials and cross-section of Neste Oil mock-up dissimilar metal weld. Length of the specimen is 300 mm

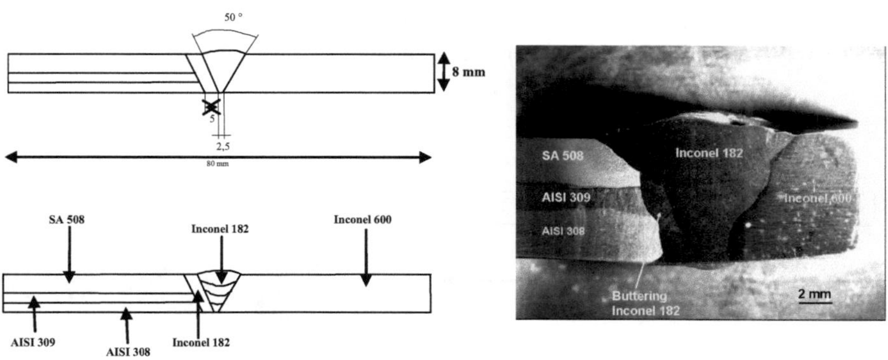


**Fig. 3.** Groove geometry, materials and cross-section of mock-up TV dissimilar metal weld. Length of the specimen is 300 mm

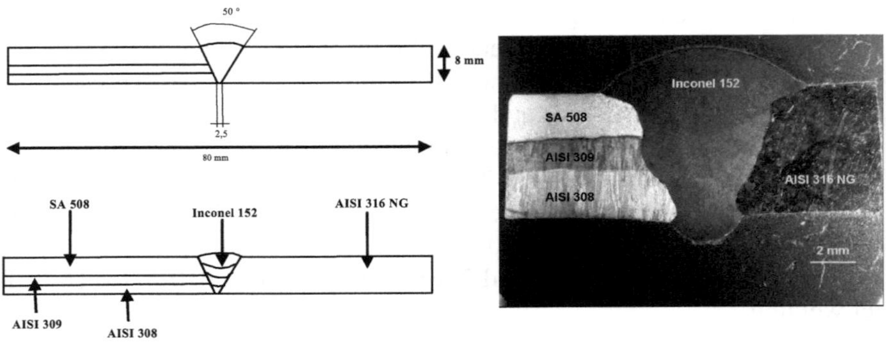


**Fig. 4.** Groove geometry, materials and cross-section of mock-up 1 dissimilar metal weld. Length of the specimen is 300 mm

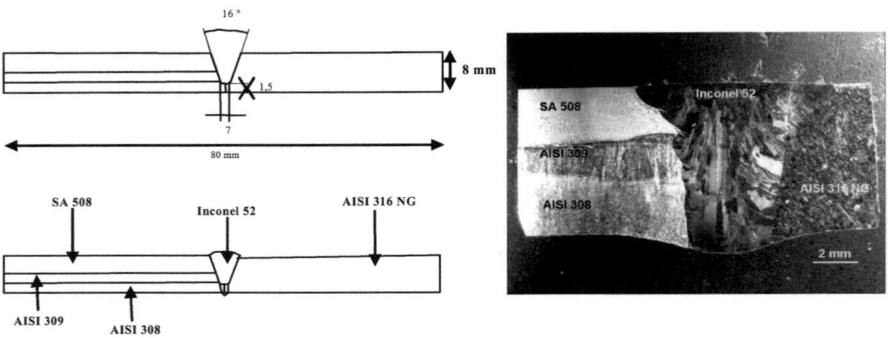


**Fig. 5.** Groove geometry, materials and cross-section of mock-up 2 dissimilar metal weld. Length of the specimen is 300 mm

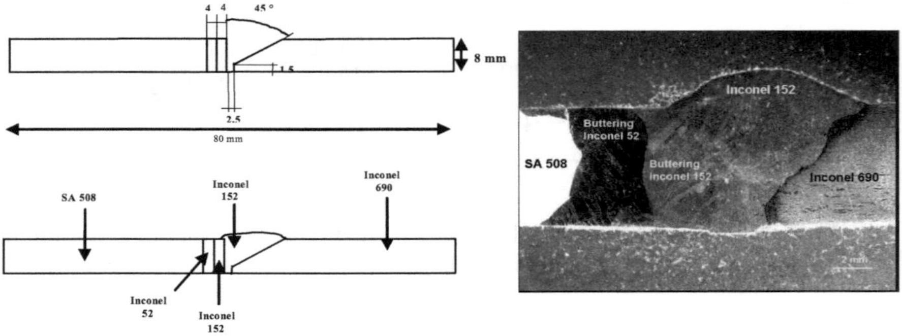


**Fig. 6.** Groove geometry, materials and cross-section of mock-up 3 dissimilar metal weld. Length of the specimen is 300 mm

Figure 7 presents the test piece for Varestraint testing of pure weld metals and weld pass sequence of the test specimen welded with Alloy 82. Base material was AISI 304 plate into which a 20 mm wide, 60 mm long and 4 mm deep groove was machined and then welded with different filler metals. The surface of the weld was ground to the level of the base metal so that the surface quality was equal in all Varestraint tests. Filler metals (supplier in brackets) used in the pure weld metal tests were: Alloy 182 (Esab OK 92.26), Alloy 82 (Esab Tigrod 19.85), and Alloys 152, 152M, 52, 52M, and 52MS (Special Metals). Table 2 presents the welding parameters used in the Varestraint testing.

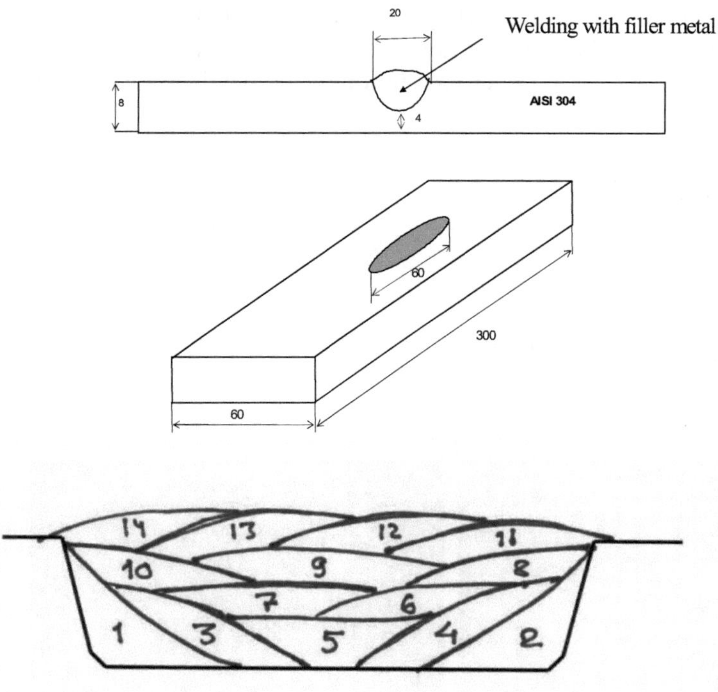


**Fig. 7.** Test piece for Varestraint testing of pure weld metals and weld pass sequence of the test specimen welded with filler metal Alloy 82

**Table 2.** Summary of the welding parameters used in the Varestraint tests

| | Current [A] | Voltage [V] | Welding speed [cm/min] | Heat input [kJ/cm] |
|---|---|---|---|---|
| Neste Oil mock-up (Inconel 182) | 250 | 12,5 | 12 | 9,4 |
| Mock-up TV (Inconel 182) | 150 | 10,5 | 12 | 4,7 |
| Mock-up 1 (Inconel 152) | 150 | 10,5 | 12 | 4,7 |
| Mock-up 2 (Inconel 52) | 150 | 10,5 | 12 | 4,7 |
| Mock-up 3 (Inconel 52/152) | 150 | 10,5 | 12 | 4,7 |

## Experimental Results

Figure 8 (a) presents the surface (Alloy 182 weld, augmented strain 4%) of Neste Oil mock-up with hot cracks after Varestraint test. Optical microscopy of a hot crack (dark areas) in the cross-section of the weld (b) and scanning

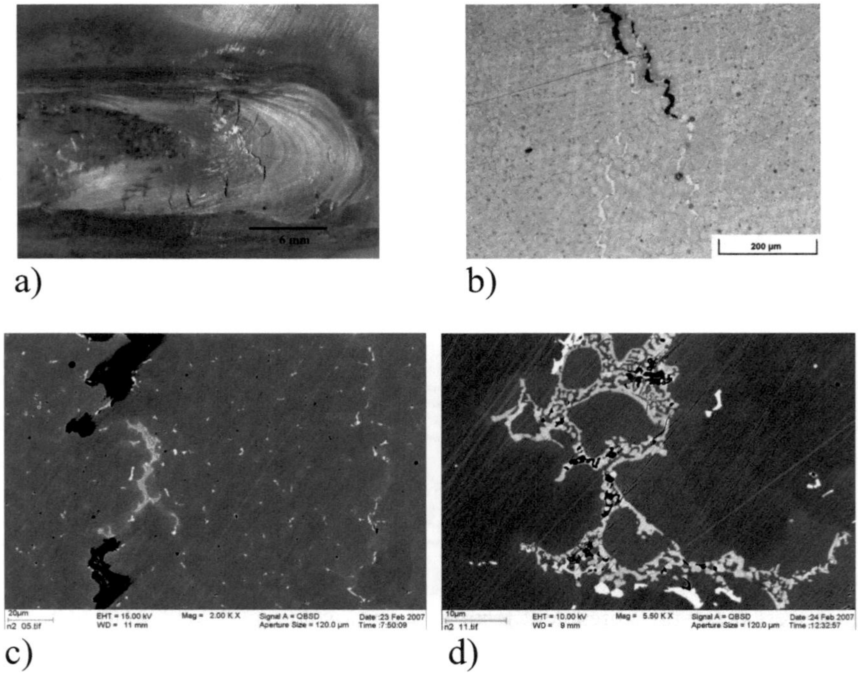


**Fig. 8.** (**a**) Surface (Alloy 182 weld, augmented strain 4%) of Neste Oil mock-up with hot cracks after Varestraint test. (**b**) Optical micrograph of the hot cracks (dark areas) in a cross-section of the weld (note the white phase ahead of the hot crack tip). (**c**), (**d**) Scanning electron micrographs (SEM) of weld hot crack tips showing details of the irregular shaped white phase

electron microscopy (FEG-SEM) of weld hot crack tips (c) and (d) show details of the irregular shaped white phase connected to hot cracks of this alloy. The composition of the white phase areas was examined by EDS analysis. Figure 9 presents X-ray element maps of the white phase area in a dendrite boundary region and Fig. 10 shows a SEM image and EDS analyses of the white phase showing marked Si, Nb, Mn and P enrichment, which is compensated by depletion of Cr, Fe and Ni. Based on the high magnification BSE-images and the EDS analyses the observed white phase consists of two phases, white and grey, with varying Nb, Si, Mn and Ti contents. Inside and along the dendrite/grain boundaries white Nb-rich particles, Nb(C, N), are present.

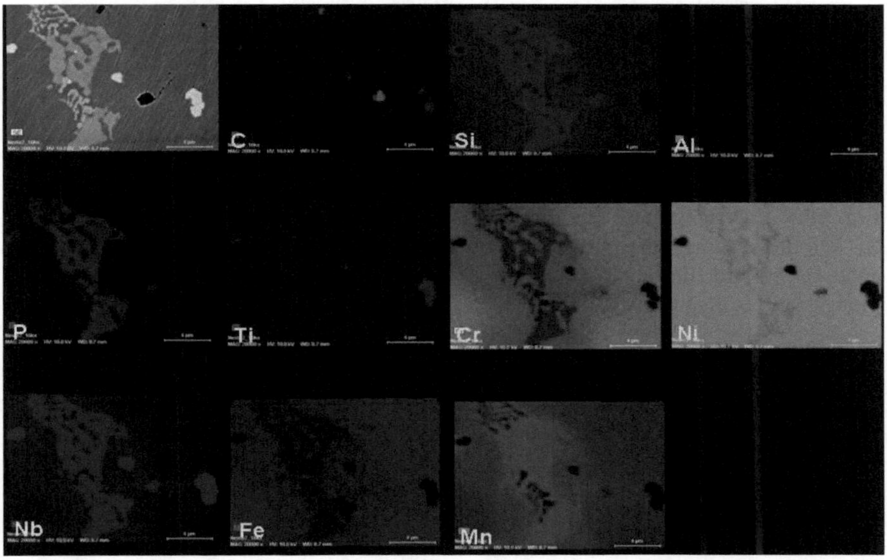


**Fig. 9.** X-ray element maps of the white phase area in a dendrite boundary region. Note marked Nb enrichment in the white phase and less enrichment in a wider area along the dendrite boundary. Marked Si, Mn and P enrichment compensated by depletion of Cr, Fe and Ni can also be seen

| Element | Filler metal wt.% | Grey area wt.% | Enrichment vs. filer metal | Light white area wt.% | Enrichment vs. filer metal | White area wt.% | Enrichment vs. filer metal |
|---|---|---|---|---|---|---|---|
| Si | 0.8 | 3.4 | 4.3 | 6.5 | 8.1 | 0 | 0 |
| P | 0.01 | 1.3 | 130 | 1.3 | 130 | 0.2 | 20 |
| Ti | 0.1 | 0.04 | 0.4 | 0.7 | 7 | 2.8 | 28 |
| Cr | 15.7 | 5.4 | 0.3 | 4.9 | 0.3 | 1.4 | 0.1 |
| Mn | 6.5 | 17.7 | 2.7 | 11.8 | 1.8 | 1.1 | 0.2 |
| Fe | 6.7 | 4.7 | 0.7 | 4.9 | 0.7 | 1.5 | 0.2 |
| Ni | 68 | 51.4 | 0.8 | 49.4 | 0.7 | 0 | 0 |
| Nb | 1.8 | 15.2 | 8.4 | 20.6 | 11.4 | 89.6 | 49.8 |

**Fig. 10.** SEM image and EDS analyses of the white and grey phase showing Nb, Si, Mn and P enrichment and Nb rich particles, Nb(C, N). The spot size in EDS analysis has an interaction diameter about 1 µm, and therefore the surrounding matrix affects also the result

The hot crack fracture surfaces were opened for fractography and phase analysis with an FEG-SEM/EDS system. Figure 11 shows an opened fracture surface of Alloy 182. With a low acceleration voltage of 7 kV the information comes from close to the surface. The wavy morphology of the fracture

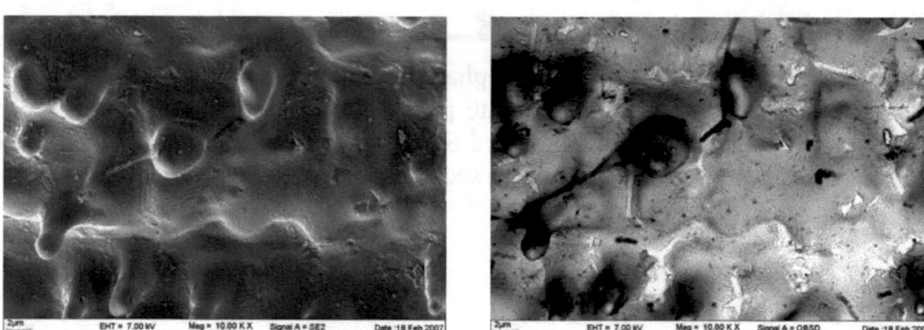

a) b)

**Fig. 11.** Fracture surface of Alloy 182 hot crack showing the white phase in the valleys between dendrites

surface can be clearly seen, which is an evidence of the molten phase along the dendrite boundaries in which the separation has occurred.

Study of the fracture surfaces of the Neste Oil mock-up weld (Alloy 182) hot cracks revealed marked segregation of the same elements on fracture surfaces as seen in the white phase in the cross-sections of the hot cracks. On the fracture surface Nb-rich particles, Nb(C, N), and the lamellar white phase are present intermixed in the final microstructure. The white phase exhibits similar enrichment of Nb, Si, Mn, and P as observed in the cross-sections of the weld metal.

Pure Alloy 82 weld metal hot cracking took place only at higher heat input (Q = 9,4 kJ/cm, augmented strain 4%). The formed hot cracks were markedly oxidized, probably because of the high heat input and also because of element segregation to the final melt. Figure 12 presents a thick oxide layer on the fracture surface of Alloy 82 weld metal hot crack. Table 3 presents EDS analyses of three areas of the oxide layer on the hot crack surface presented in Fig. 12. The oxide layer is enriched in Mn, Nb, Fe, and Ti.

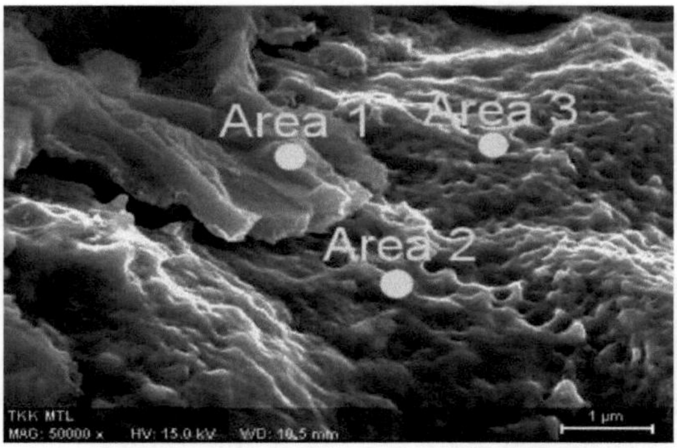


**Fig. 12.** Oxide layer on the hot crack fracture surface of Alloy 82 weld metal

**Table 3.** EDS analyses of three areas of the oxide layer on the hot crack surface presented in Fig. 12 (wt.-%)

| Element | O | Al | Si | Ti | Cr | Mn | Fe | Ni | Nb |
|---|---|---|---|---|---|---|---|---|---|
| Area 1 | 23,4 | 0,02 | 0,18 | 1,2 | 8,6 | 13,4 | 11,7 | 30,1 | 11,6 |
| Area 2 | 1,4 | 0,03 | 0,17 | 0,3 | 8,9 | 2,5 | 5,9 | 77,2 | 3,6 |
| Area 3 | 1 | 0,04 | 0,15 | 0,3 | 10 | 2,2 | 5,2 | 79,2 | 1,9 |

Figure 13 (a) presents SEM image and (b) X-ray element maps of a cross-section of hot cracks exhibiting a strong Nb segregation and less marked Si, P and Ti segregation to the dendrite boundary at the hot crack tip. Figure 14 shows SEM image and X-ray element map of fracture surface of Alloy 82 exhibiting distribution of Nb and Ti.

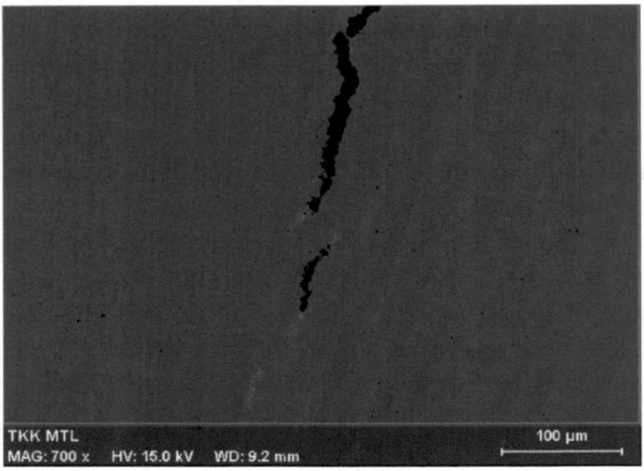


a)

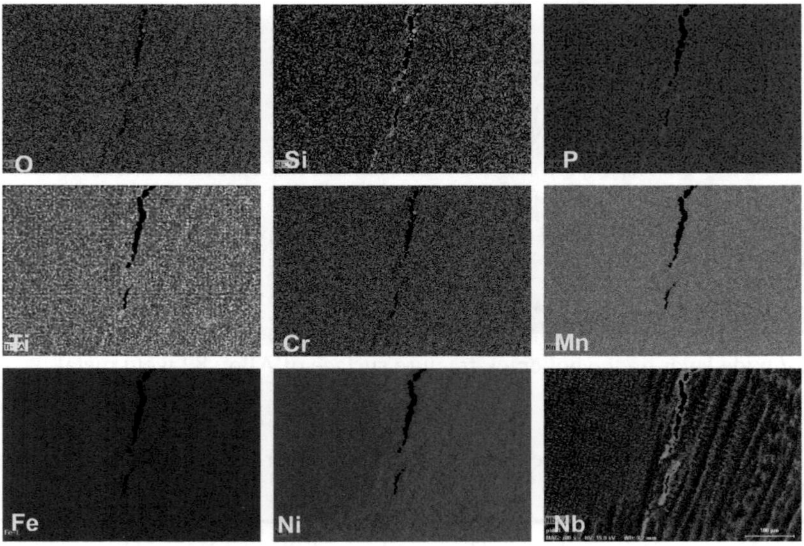


b)

**Fig. 13.** (**a**) SEM image and (**b**) X-ray element maps of a cross-section of hot crack exhibiting strong Nb segregation and additional segregation of Si, P and Ti to the dendrite boundary along and ahead of the hot crack

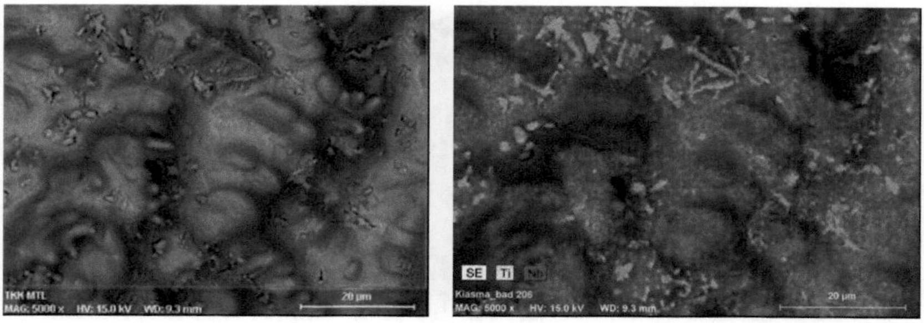


a) b)

**Fig. 14.** (**a**) SEM image and (**b**) X-ray element map of fracture surface of Alloy 82 showing distribution of Nb and Ti

The chemical composition of the oxide layer on the hot crack fracture surface of Alloy 82 shows marked increase of Nb, Mn and Fe contents, which are the elements of a high tendency for oxidation and are available on the fracture surface of a hot crack. Under the oxide layer the white phase rich in Nb, Mn, Si and P as well as the Nb- and Ti-rich (Ti content of Alloy 82 0,34 wt.-%) particles, Nb, Ti(C, N), are also clearly visible. In the pure weld metal tests it was also observed that cracking took place under the weld bead made in the Varestraint test by a liquation cracking mechanism following the boundaries of lower weld passes (see Fig. 7).

Figure 15 presents SEM images of a cross-section of hot cracks in mock-up 1 weld metal (Alloy 152 weld, augmented strain 4%) after Varestraint test and Fig. 16 shows SEM images of fracture surface, where the presence of the white phase in the valleys between the dendrites is clearly visible. SEM image and X-ray element map of the hot crack fracture surface of Alloy 152 exhibits high enrichment of Nb to the white phase (see Fig. 17).

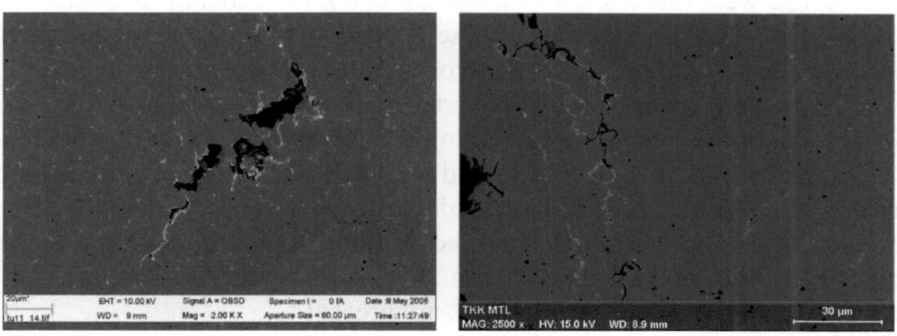


a) b)

**Fig. 15.** SEM images of a cross-section of hot cracks in Alloy 152 weld metal exhibit the connection of the white phase to hot cracks

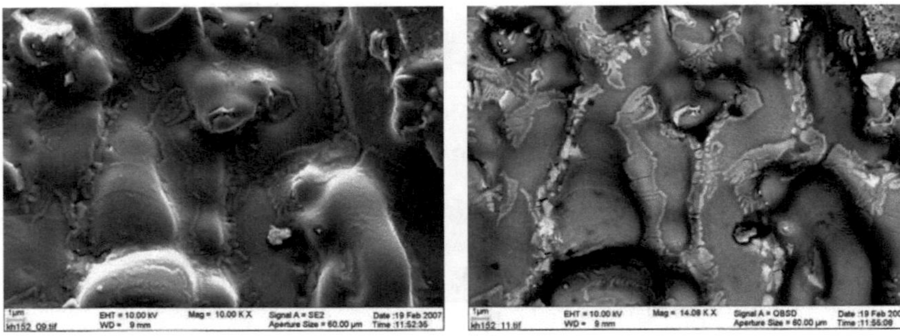

a) b)

**Fig. 16.** SEM images of hot crack fracture surface of Alloy 152 weld metal. Note the brittle nature of the lamellar white phase

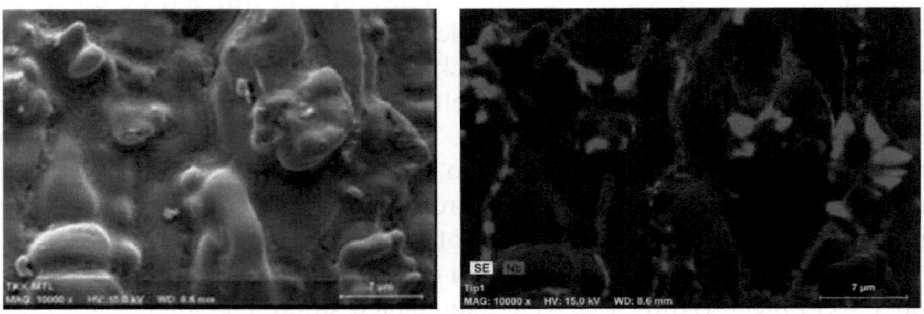

a) b)

**Fig. 17.** (**a**) SEM image and (**b**) X-ray element map of the fracture surface of Alloy 152 exhibit high enrichment of Nb to the white phase

Alloy 152 weld metal shows similar hot cracking tendency as compared to Alloy 182. The Nb-rich white phase is connected to the hot cracks and the phase can be easily found in the cross-sections of the hot cracks as well as on the fracture surfaces (local coverage can be as high as 30%, see Fig. 16). The dendrite boundaries containing the white phase show in addition to Nb a marked increase of Si, Mn and P contents both in the white phase, but also in the wider zone along the dendrite boundaries. Simultaneously depletion of Cr, Ni and Fe is observed in these zones. In addition to the white phase both in the cross-section as well as on the fracture surfaces plenty of small particles, Nb, Ti(C, N), are present.

Filler metal Alloy 52 contains Ti (0,51 wt.-%), but does not contain Nb as Alloys 182, 82 and 152 and, thus, hot cracks form along the dendrite boundaries without the presence of the Nb-rich white phase. Figure 18 presents SEM images of mock-up 2 weld metal (Alloy 52, augmented strain 4%) hot cracks after Varestraint test. A lot of small particles (TiN) are present along the dendrite boundaries. Figure 19 presents a SEM image and X-ray element maps of the hot crack fracture surface.

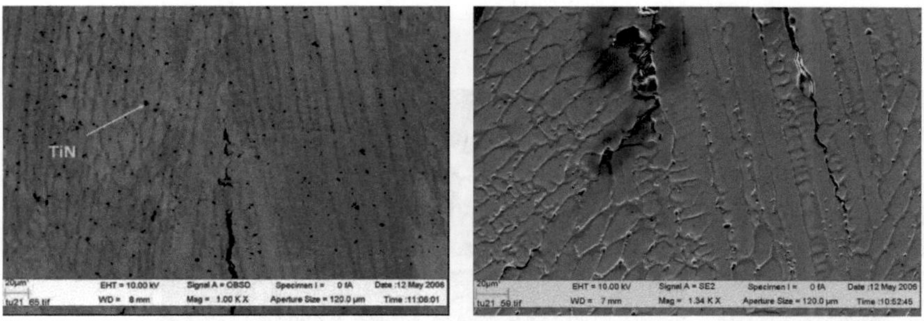


**Fig. 18.** SEM images of mock-up 2 (Alloy 52, augmented strain 4%) weld hot crack cross-sections after Varestraint test

Hot cracking of Alloy 52 does not show any indications of the Nb-rich white phase observed in the earlier cases with Alloys 182, 82 and 152, since Alloy 52 does not contain Nb. Alloy 52 contains Ti and therefore a lot of precipitation of TiN(C) was observed both in the cross-sections along the dendrite boundaries as well as on the fracture surfaces of the hot cracks, where the coverage of the hot crack fracture surface by TiN(C) can be quite extensive. The dendritic large TiN(C) phase particles are presumed to have formed in the melt prior to final solidification of the weld. On the fracture surface also MnS phase particles are observed as separate phases. Thus, in Alloy 52 the only marked phase on the hot crack surfaces is TiN(C).

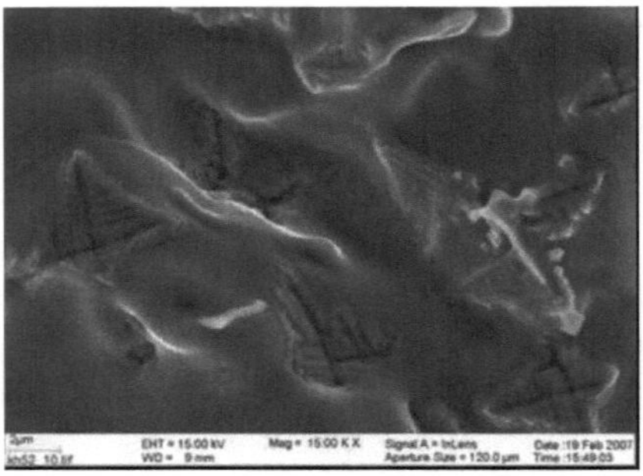

a)

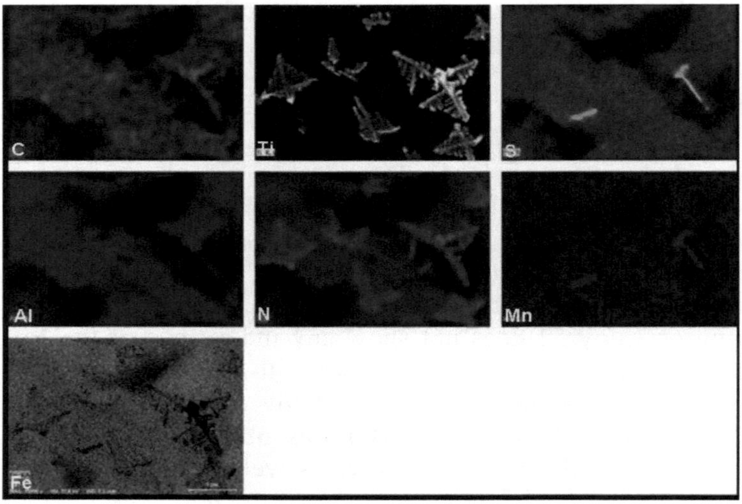


b)

**Fig. 19.** (**a**) SEM image and (**b**) X-ray element maps of fracture surface of Alloy 52 revealing the dendritic phase of TiN(C), and separate MnS particles are also present

## Results of Varestraint Tests

Total crack lengths (TCL) of Neste Oil mock-up, mock-up TV and mock-ups 1, 2 and 3 are shown in Fig. 20 (a) and maximum crack lengths (MCL) in Fig. 20 (b), respectively, with augmented strains of 0%, 0.5%, 1%, 2%, 3% and 4%. In the legend box markings are as follows: case, tension side and filler metal. Table 4 presents summary of Varestraint tests (augmented strain 4%) made for pure Ni-base alloy weld metals (see, Table 1 and Fig. 7).

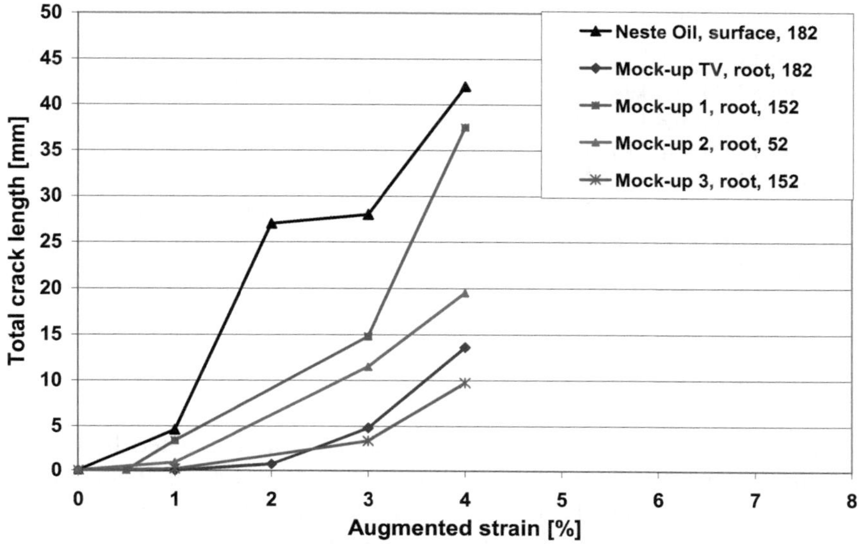


a)

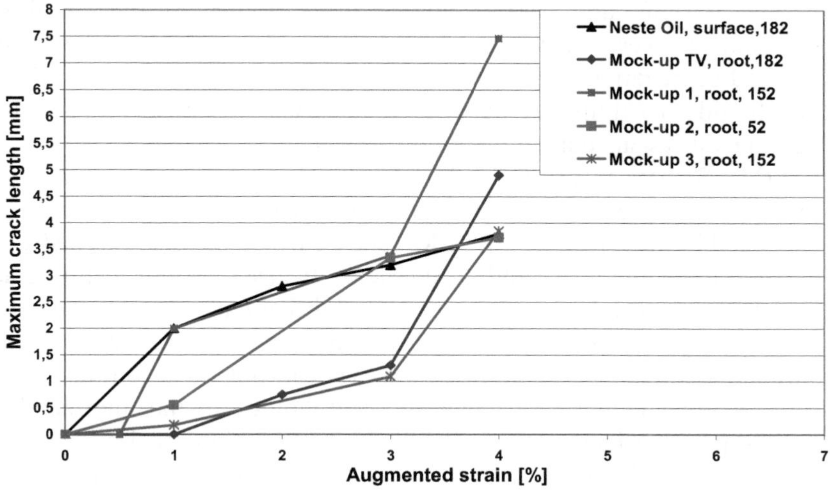


b)

**Fig. 20.** Total solidification crack lengths (TCL) (**a**) and maximum crack lengths (MCL) (**b**) as a function of augmented strain of Neste Oil mock-up, mock-up TV and mock-up 1, 2 and 3 samples in Varestraint testing

**Table 4.** Summary of Varestraint tests made for pure Ni-base alloy weld metals (augmented strain 4%, heat input 4,7 kJ/cm, except for Alloy 82)

| | |
|---|---|
| Inconel 182 (Q=4,7 kJ/cm) | TCL = 19,3 mm / 2nd test -> cracks |
| Inconel 82 (Q=4,7 kJ/cm) | no cracks / 2nd test -> no cracks |
| Inconel 82 (Q=9,4 kJ/cm) | TCL = 32,9 mm |
| Inconel 152 (Q=4,7 kJ/cm) | TCL = 19,7 mm / 2nd test -> cracks |
| Inconel 152M (Q=4,7 kJ/cm) | TCL = 8,1 mm |
| Inconel 52 (Q=4,7 kJ/cm) | no cracks / 2nd test -> no cracks |
| Inconel 52M (Q=4,7 kJ/cm) | TCL = 3,1 mm / 2nd test -> no cracks |
| Inconel 52MS (Q=4,7 kJ/cm) | no cracks / 2nd test -> no cracks |

## Discussion

Weldability of the studied Ni-base materials was evaluated based on the results obtained with various mock-up welds of real components or anticipated weld designs of future plants and weld metals of different chromium contents. The susceptibility to hot cracking was examined based on the Varestraint test results by ranking the materials. The mechanistic understanding of the hot cracking susceptibility of Ni-base weld metals is still largely missing and careful metallurgical studies of the solidification mechanisms were carried out by a modern FEG-SEM/EDS instrument. Marked segregation of alloying elements such as Nb, Si, P and Mn to the last liquid to solidify at dendrite boundaries was observed and in Nb-bearing alloys an eutectic Nb-rich Laves phase (white phase), $(Ni, Cr, Fe)_2(Nb, Si, Mn, P, Ti)$, formed in the interdendritic regions containing Nb as much as 15–30 wt.-%. The formation of the Laves phase is probably enhanced by the presence of Si, Mn and P, which are also markedly segregated to the white phase and/or interdendritic areas. Solidification cracking occurs during the terminal stages of solidification, when a liquid film is distributed along the interdendritic regions as a continuous film and the shrinkage strains across these boundaries can not be accommodated. The solidification temperature range and morphology of the interfacial liquid are the primary factors controlling the solidification cracking susceptibility. Solute redistribution affects the solidification temperature range and amount of terminal liquid (small amount of terminal liquid is most detrimental).

Minor variations in Nb, Si, and C content in Ni-base alloys have a strong influence on the solidification temperature range, type and amount of secondary phases, which form during terminal stages of solidification and affect the solidification cracking susceptibility [10]. These alloy additions decrease the liquidus and solidus temperatures and increase the solidification temperature range, Nb being most effective in this. Of the two types of the eutectic-type constituents, γ/NbC (or TiN(C)) and γ/Laves, known to form as the Ni-base alloy welds solidify, the γ/Laves constituent is more deleterious in terms of hot cracking as it forms at a lower temperature and extends thus the solidification temperature range. In dissimilar metal welds the studied alloys are used to join low alloy steels to stainless steels and due to dilution the weld metal can become significantly enriched in Fe, Si and C, which may significantly change the solidification behavior and associated cracking tendency. It is known from Fe-base alloys and based on Fe additions to Nb-bearing Ni-base alloys that Fe promotes the formation of γ/Laves constituent. Silicon addition has similar effect. Niobium addition promotes higher amounts of total eutectic and both eutectic-type structures (Laves and NbC) are highly enriched in Nb. [10]

Solidification temperature ranges were measured by differential thermal analysis (DTA) by Wu and Tsai [2] for Alloy 82 (44°C) and Alloy 52 (13°C). However, in the studied Ni-base alloys the γ/Laves constituent forms in very small amounts, which probably precludes the measurement of the L => (γ + Laves) reaction temperature by DTA. For Alloys 718 and 625 weld overlay a wide solidification temperature range (around 160°C and 170°C, respectively) and presence of interdendritic Laves phase is typical. The γ within the γ/Laves mixture in Alloy 625 weld overlay has composition about 9,3 wt.-% Nb for the eutectic composition of 18,9 wt.-% Nb and Laves phase composition of 22.1 wt.-% Nb and respectively for Alloy 718 9,3, 19,1 and 22,4 wt.-% Nb were found [7, 8, 9, 10]. In this study the composition of γ close to the white (Laves) phase in the interdendritic region was 5,5 wt.-% Nb and in the Laves phase even 30,8 wt.-% Nb content (Alloy 152) was measured.

Hot cracking occurred along the dendrite boundaries and the susceptibility to hot cracking in the weld mock-up samples was observed to follow the order based on TCL: Alloy 152 > Alloy 52 > Alloy 182 > Alloy 82. In pure weld metal hot cracking tests the susceptibility was observed to follow the order: Alloy 182 ≥ Alloy 152 > Alloy 52 ≥ Alloy 82. The differences between these two test types are thought to be related to dilution effects because Fe, Si, and C enhance the eutectic phases and expand the solidification temperature range.

In order to avoid solidification cracking in Nb-bearing Ni-base alloys the solidification temperature range has to decrease and the terminal secondary constituents have to form in small quantities. Carbon addition may improve the resistance to hot cracking by decreasing the amount of Laves phase (promotes formation of γ/NbC), but especially Nb and Si seem to be detrimental. Niobium segregation is governed by the prevailing solidification conditions: the low weld metal cooling rate and high heat input increase

Nb segregation and the amount of Laves phase [11]. Thus, for controlling the Laves phase in the Nb-bearing Ni-base weld metals lowest possible heat input with high cooling rates have to be used. Additionally, control of dilution by Fe, Si, and C is very important, since these elements lower the liquidus and solidus temperatures, increase the solidification temperature range and favour eutectic-type solidifications.

## Conclusions

The following conclusions can be drawn based on the results of this study:

- Dissimilar metal weld joints of Ni-base weld metals show in their microstructure marked segregation of Nb, Si, P and Mn to the last liquid to solidify wetting the dendrite boundaries. In the Nb-bearing alloys eutectic Laves and Nb(C, N) phases form along the dendrite boundaries and in Ti-alloyed Alloy 52 TiN(C) phase forms, respectively.
- Hot cracking occurred along the dendrite boundaries and the susceptibility to hot cracking in the weld mock-up samples was observed to follow the order: Alloy 152 > Alloy 52 > Alloy 182 > Alloy 82.
- In pure weld metal hot cracking tests the hot cracking susceptibility was observed to follow the order: Alloy 182 ≥ Alloy 152 > Alloy 52 ≥ Alloy 82.
- The differences are probably related to the dilution effects because Fe, Si, and C enhance the eutectic phases and expand the solidification temperature range.

## References


1. Hood B, Lin W (1995) Weldability Testing of Inconel Filler Materials. In: Seventh International Symposium on Environmental Degradation of Materials in Nuclear Power Systems – Water Reactors. Breckenridge, Colorado, USA, 7–10 August 1995. pp. 69–79.
2. Wu W, Tsai C (1999) Hot Cracking Susceptibility of Fillers 52 and 82 in Alloy 690. Met. Mat. Trans., 30A, pp. 417–426.
3. Collins M, Lippold J (2003) An Investigation of Ductility Dip Cracking in Nickel-Based Filler Materials – Parts I-III. Welding Research, pp. 288–295.
4. Nishimoto K, Saida K, Okauchi, H (2004) Microcracking Susceptibility in Multi-Pass Weld Metal of Inconel 690 Alloy. IIW Doc. No. IX-2097-04. 15 p.

5. Nissley N, Collins M, Guaytima G, Lippold J (2002) Development of the Strain to Fracture Test for Evaluating Ductility Dip Cracking in Austenitic Stainless Steels and Ni-base Alloys. Welding in the World, 46, 7/8, pp. 32–40.
6. CEN ISO/TR 17641-3 (2005) Destructive Tests on Welds in Metallic Materials. Hot Cracking Tests for Weldments. Arc Welding Processes. Part 3: Externally Loaded Test. 17 p.
7. Knorovsky GA, Cieslak MJ, Headley TJ, Romig Jr AD, Hammetter WF (1989) Inconel 718: A Solidification Diagram. Met. Trans., 20A, 10, pp. 2149–2158.
8. Cieslak MJ, Headley TJ, Kollie T, Romig Jr AD (1988) A Melting and Solidification Study of Alloy 625. Met. Trans., 19A, 9, pp. 2319–2331.
9. Dupont JN (1996) Solidification of an Alloy 625 Weld Overlay. Met. Trans., 27A, 11, pp. 3612–3620.
10. Dupont JN, Robino CV, Marder AR (1998) Solidification and Weldability of Nb-Bearing Superalloys. Welding J., 77, pp. 417s–431s.
11. Radhakrishna CH, Prasad Rao K (1997) The Formation and Control of Laves Phase in Superalloy 718 Welds, J. Mater. Sci., 32, pp. 1977–1984.

# Evaluation of Weld Solidification Cracking in Ni-Base Superalloys Using the Cast Pin Tear Test

B.T. Alexandrov[1], J.C. Lippold[1], N.E. Nissley[2]

[1]The Ohio State University, Columbus, Ohio, USA
[2]Exxon Mobil Upstream Research, Houston, Texas, USA

## Abstract

A second-generation cast pin tear test (CPTT) that is capable of ranking the weldability of Ni-base superalloys has been developed at the Ohio State University. The CPTT utilizes an optimized testing procedure and apparatus design that provide controllable and repeatable testing conditions, and yield reproducible and reliable test results.

The CPTT provided a weldability ranking of four highly alloyed stainless steels that is in very good correlation to the results of a round robin study on six externally restrained weld hot-cracking tests. The solidification microstructure of the test samples (cast pins) closely simulates the microstructure of low to medium heat input welds, such as those made by the GTAW process. The CPTT fracture surfaces exhibit dendritic "eggcrate" fracture morphology, which is a characteristic of solidification cracking. This fracture morphology is similar to the high and medium temperature regime of solidification cracks generated using the Varestraint test. The nature of crack nucleation and propagation, and the crack healing phenomenon in CPTT closely resemble weld solidification cracking.

The CPTT technique was applied for evaluating the solidification cracking susceptibility of a series of high performance turbine engine alloys. These alloys were ranked in order of decreasing susceptibility to solidification cracking as follows: René alloys 142, 125, and 77, alloy 718, René 80, Waspaloy, and alloy 600. These rankings are in good correlation to field experience with the solidification cracking susceptibility of the tested alloys.

The second generation CPTT proved to be capable of ranking the solidification cracking susceptibility of both "difficult-to-weld" alloys and "standard" alloys, and to differentiate their solidification cracking behavior. It also provides an efficient and inexpensive tool for weldability testing in the process of alloy and consumable development.

## Introduction

The cast pin tear test was originally developed by Hull [1] and used in the 1970's as a method for evaluating the susceptibility of alloys to hot cracking during welding and casting. It consisted of levitation melting and casting of small charges of material in copper molds to produce conical cast pins with varying geometry. The charge mass in this test was constant (19 g). The test sensitivity was controlled by the pin length and geometry, and by the mold diameter. The pins were examined for circumferential cracks and the total crack length was plotted versus the mold number. Based on the percent of circumferential cracking and mold number, the CPTT generated arbitrary ranking of the susceptibility to hot cracking. Compared to the other hot-cracking tests, it required only several hundred grams of material to develop a cracking susceptibility curve and was suitable for alloy development purposes.

The CPTT of Hull was modified by the Welding and Joining Metallurgy Group at The Ohio State University [2]. The costly and complex levitation melting technique was replaced by an arc melting technique. A standard gas tungsten arc welding (GTAW) torch and power supply are used to melt a small charge of material on a water-cooled copper hearth under an inert (argon) atmosphere. The hearth was equipped with a simple delivery system that transferred the molten charge into a copper mold positioned below the hearth. Although a successful system was initially developed using this approach, problems with complete mold filling and reproducibility of the cracking results were encountered.

Because of the problems with mold filling, a second-generation cast pin tear test that is more robust and capable of ranking the weldability of Ni-base superalloys has been recently developed at the Ohio State University [3]. The reproducibility and reliability of test results was improved by further development of the CPTT apparatus and by optimizing the melting and mold filling procedures. The test sensitivity was improved by introducing smaller increments in cast pin length. The capacity of testing various alloys was expanded by using mold materials with different thermal conductivity and by extending the range of mold lengths.

This chapter presents the development and verification of the second-generation CPTT, and its application for ranking the susceptibility to solidification cracking of a series on Ni-base superalloys for turbine engine application.

## Second-Generation Cast Pin Tear Test

The general design of the second-generation CPTT apparatus is shown on Fig. 1. A charge of the tested alloy is melted in a gas-shielded chamber over water cooled copper hearth by a GTAW torch. The hearth has a central opening that connects the gas-shielded chamber to a copper mold. The molten charge is transferred through the hearth opening into the mold where it solidifies as a cast pin. The parameters of the CPTT procedure are provided in Table 1.

**Fig. 1.** Cast Pin Tear Test apparatus

The CPTT apparatus is equipped with vacuum and overpressure systems that are controlled by electromagnetic valves and pressure and vacuum gauges. These are used to both provide an inert atmosphere in the melting chamber and mold before testing, and facilitate the transfer of molten charge to the mold.

**Table 1.** Cast Pin Tear Test procedure

| Parameter | Value | |
|---|---|---|
| Mold Material | Cu; Cu-Be | |
| Shielding Gas | Ar | |
| Gas Flow Rate, ml/s | 70–80 | (9–10 cf $h^{-1}$) |
| Arc Current, A | 250 | |
| Arc Length, mm | 15 | (0.6 in) |
| Arc Time, s | 5–7 | |
| Over-Pressure, MPa | 0.007–0.014 | (1–2 psi) |
| Cast Pin Diameter, mm | 9.525 | (0.375 in) |
| Cast Pin Length, mm | 12.7–50.8 | (0.5–2 in) |
| Length Increments, mm | 3.175 | (0.125 in.) |

The hearth design is presented in Fig. 2. The hearth opening limits the heat extraction from the charge during its melting, thus allowing for sufficient superheating of the latter and facilitating its quick transfer into the mold. The mold cavity is designed to produce a cylindrical cast pin with restraining head and foot, Fig. 3. The CPTT test utilizes molds and cast pins with constant diameters and varying lengths. The charge mass corresponds to the cast pin length, Table 2.

**Fig. 2.** Open hearth

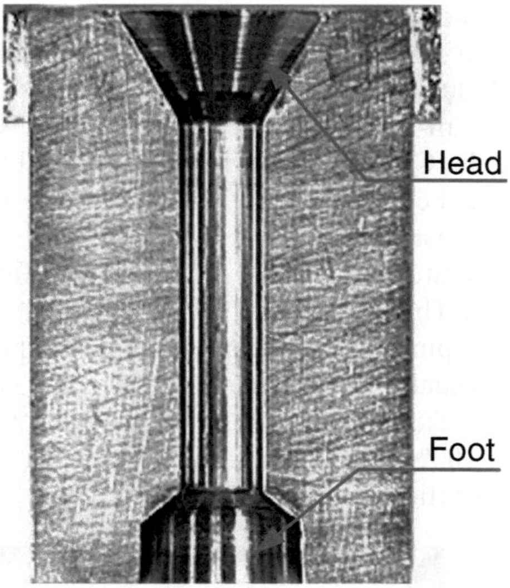


**Fig. 3.** Cast pin mold

**Table 2.** Range of mold (cast pin) lengths and correspondent pin/charge volume and mass (for pure Ni samples)

| Mold Length, mm | (in.) | Volume, $mm^3$ | Mass, g |
|---|---|---|---|
| 12.700 | (0.500) | 1122.6 | 10.0 |
| 15.875 | (0.625) | 1178.7 | 10.5 |
| 19.050 | (0.750) | 1234.8 | 11.0 |
| 22.225 | (0.875) | 1291.0 | 11.5 |
| 25.400 | (1.000) | 1347.1 | 12.0 |
| 28.575 | (1.125) | 1403.2 | 12.5 |
| 31.750 | (1.250) | 1459.4 | 13.0 |
| 34.925 | (1.375) | 1515.5 | 13.5 |
| 38.100 | (1.500) | 1571.6 | 14.0 |
| 44.450 | (1.750) | 1683.9 | 15.0 |
| 50.800 | (2.000) | 2806.5 | 16.0 |

The rate of heat extraction from the molten metal is controlled by the thermal conductivity of mold material. In this study, two mold materials were used: high purity copper (C10100) and a Cu-Be-Co alloy (C17000) with thermal conductivities of 391 W/m-K and 118 W/m-K respectively. The lower rate of heat extraction of the Cu-Be-Co molds made it possible to test some Ni-base alloys that had been previously problematic with respect to filling of copper C10100 molds.

Longitudinal tensile stress is generated in the cast pin as it solidifies and cools down to room temperature. The stress level increases with the pin length. At the critical stress level (pin length) for particular alloy, hot cracking occurs at the pin surface usually just below the pin head, as indicated on Fig. 4. With further increase in the cast pin length, the total surface crack length increases to 100% circumferential cracking and eventually results in complete pin separation, Fig. 5.

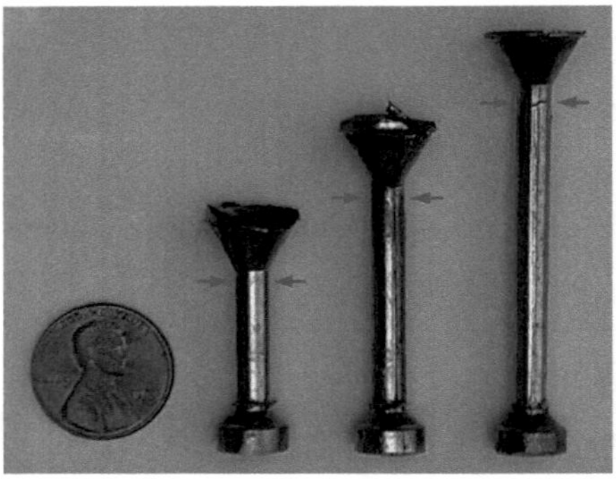

**Fig. 4.** Cast pin samples of alloy 625. Arrows show the typical location of solidification cracks

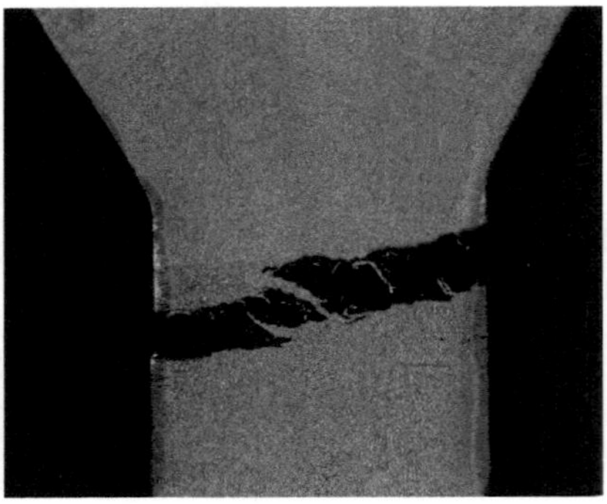

**Fig. 5.** Complete separation in 1.25 in. long pin of alloy 800H

The cast pins are examined for cracks using a binocular microscope at magnifications from 10x to 70x. During this examination the cast pins are rotated around their longitudinal axes by a specialized device. The projected crack length in a plane that is perpendicular to the pin axis is measured in degrees. The percentage of cracking is calculated as follows:

$$\%\,Cracking = \frac{L_T}{360} \times 100\,,\ \%,$$

where $L_T$ is the total length in degrees of all cracks measured on the pin surface.

The fracture surface of each completely cracked sample is examined to ensure that no casting defects contributed to cracking. The partly cracked cast pins are longitudinally sectioned and also checked for internal casting defects that might have affected the test results. Based on the valid test results, a response curve of maximum circumferential cracking is plotted as a function of pin length, Fig. 6. The maximum pin length of no cracking and the minimum pin length of 100% circumferential cracking are used as quantitative criteria for ranking the susceptibility to solidification cracking.

The sensitivity of the CPTT depends on the increments in cast pin length. Pin lengths between 12.7 mm and 25.4 mm with increments of 3.175 mm proved to be sufficient to differentiate and rank the susceptibility to solidification cracking in highly susceptible Ni-base alloys. The reproducibility of the CPTT results and the reliability in determination of the maximum pin length of no cracking and the minimum pin length of 100% cracking have been demonstrated by extensive testing of alloy René 77 (Fig. 6).

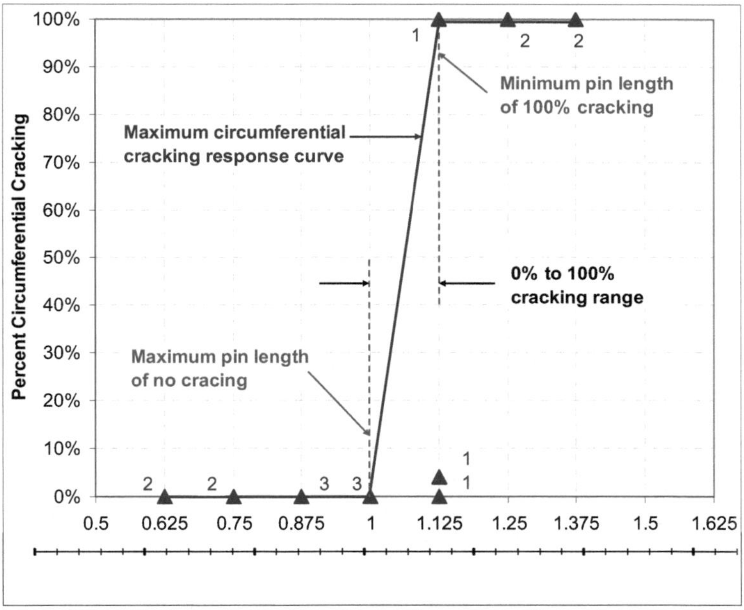


**Fig. 6.** Quantitative criteria for evaluation of susceptibility to solidification cracking by the Cast Pin Tear Test (CPTT). The circumferential cracking response curve for René 77 is shown here. The numbers on the plot indicate the number of samples tested for each mold length

## Comparison of CPTT to Other Hot Cracking Tests

In order to compare the CPTT to other "hot" cracking test techniques, a series of high-alloy austenitic stainless steels were tested by the CPTT, including alloys 800H, 926, AC66, and 825. The chemical composition of these alloys is given in Table 3.

**Table 3.** Composition of the IIW round robin study alloys, wt.%

| Alloy | 800H | 926 | AC66 | 825 |
|---|---|---|---|---|
| Ni | 30.55 | 24.80 | 31.45 | 39.15 |
| Cr | 20.40 | 20.85 | 27.35 | 22.25 |
| Fe | Bal | Bal | Bal | 31.30 |
| C | 0.068 | 0.010 | 0.072 | 0.006 |
| Si | 0.38 | 0.32 | 0.21 | 0.32 |
| Mn | 0.70 | 0.82 | 0.50 | 0.69 |
| P | 0.011 | 0.017 | – | 0.015 |
| S | 0.002 | 0.003 | – | 0.003 |
| Mo | – | 6.38 | – | 3.16 |
| Ti | 0.33 | – | – | 0.77 |
| Al | 0.280 | – | 0.014 | 0.090 |
| Other | – | 0.91 Cu<br>0.196 N | 0.83 Nb<br>0.085 Ce | 1.90 Cu |

These alloys had previously been the subject of an IIW round robin study of externally restrained weld hot cracking tests [4, 5]. The study was conducted by ten participating organizations and included the Varestraint, Trans-Varestraint, PVR, MVT, MIS-1, and LTP-1-6 tests [4]. Each participating laboratory applied its own testing procedures.

That study produced a total of twelve separate rankings of the hot cracking susceptibility of the tested alloys. Most of the rankings, generated by the different hot cracking tests and by different laboratories using same tests, were conflicting. This poor correlation was related to the different aspects of hot cracking evaluated by the different tests, and to the different testing procedures used with the same hot cracking tests [4, 5]. A summary ranking of the tested alloys that is based on the results from that study is presented in Table 4.

The CPTT results for these alloys, in terms of maximum pin length for no cracking and response curve for circumferential cracking, are summarized in Fig. 7. The CPTT ranked alloy 800H as the most susceptible, followed by alloys 926, AC66 and 825. These results are in fairly good correlation with the general results of the IIW round robin test. A very good correlation was found between the CPTT and PVR test rankings. This is not surprising since both the CPTT and the PVR test restrain the test samples in the direction of solidification.

**Table 4.** Hot cracking susceptibility ranking of the IIW round robin study alloys

| Test Type and Laboratory | | 800H | 926 | AC66 | 825 |
|---|---|---|---|---|---|
| CPTT | | 1* | 2 | 3 | 4 |
| PVR | Laboratory 1 | 1 | 2 | 4 | 3 |
| | Laboratory 2 | 1 | 2 | 3 | 4 |
| | Laboratory 3 | 1 | 2 | 4 | 3 |
| | Laboratory 4 | 1 | 3 | 2 | 4 |
| IIW ranking (out of 12 labs) | | 1 (8) | 2 or 3 (7) | 2 or 3 (9) | 3 or 4 (7) |

* 1 = highest susceptibility

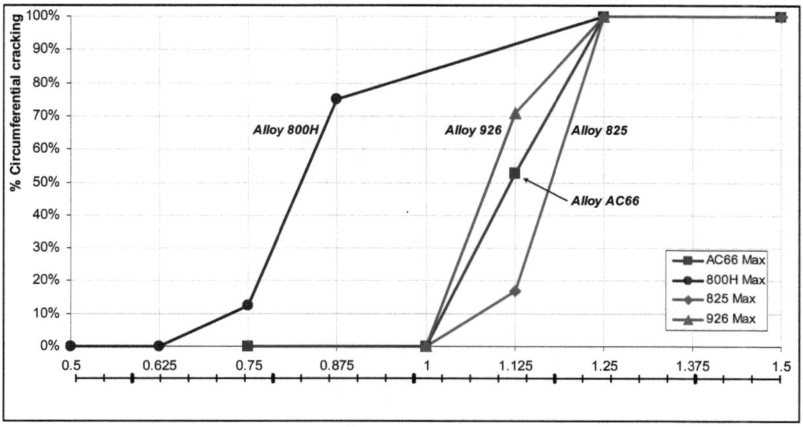


**Fig. 7.** Susceptibility to solidification cracking of the IIW round robin study alloys evaluated by CPTT

The fracture morphology of the solidification cracks produced by the CPTT and the transverse Varestraint test was examined using scanning electron microscopy (SEM), as shown in Fig. 8. The fracture surface in transverse Varestraint samples has a dendritic appearance that varies from flat at the low temperature region of the crack to the dendritic "eggcrate" morphology at medium and high temperatures (Fig. 8a). The cracks generated by CPTT have a dendritic fracture morphology that is similar to the high and medium temperature region in the Varestraint cracks.

The cracking in the CPTT samples initiates at the pin surface and propagates along the solidification grain boundaries and solidification subgrain boundaries, as shown in Fig. 9.

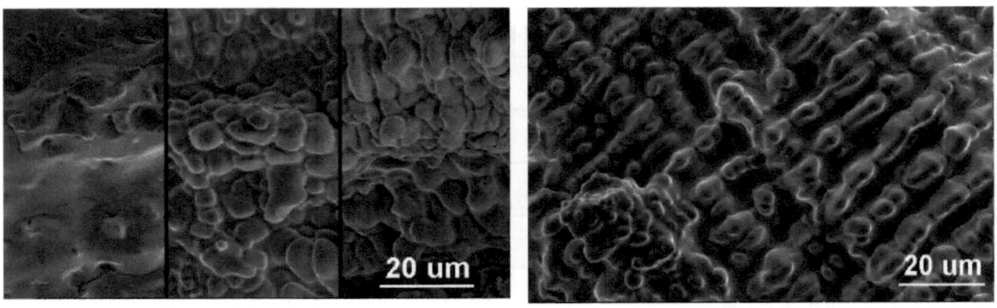


a) b)

**Fig. 8.** Alloy 926 fracture surface tested with Varestraint (**a**) and CPTT (**b**)

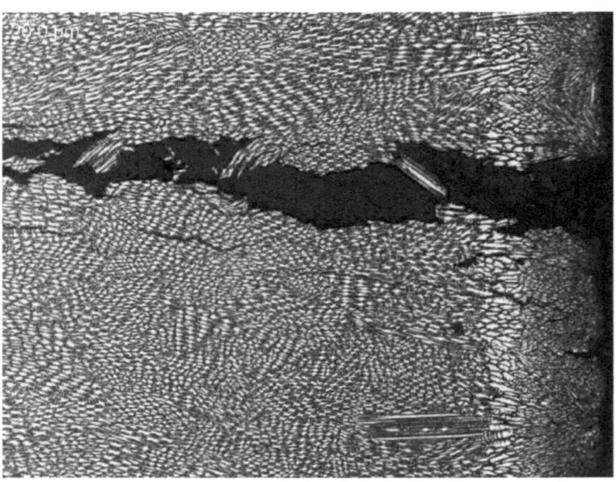

**Fig. 9.** Solidification cracking at the cast pin surface of alloy A66

## Testing of High Performance Ni-base Superalloys

A solidification cracking susceptibility of seven high performance turbine engine alloys was also determined by the CPTT. These alloys included René alloys 77, 80, 125, and 142, and alloys 600, 718, and Waspaloy. The chemical compositions for these materials are provided in Table 5.

**Table 5.** Nominal chemical composition of the tested Ni-base superalloys in wt.%

| Alloy | Inconel 600 | Inconel 718 | Wasp-aloy | René 77 | René 80 | René 125 | René 142 |
|---|---|---|---|---|---|---|---|
| Ni | 76 | 53 | 58 | 57 | 60 | 59 | 57 |
| Co | – | – | 13.5 | 15.0 | 9.5 | 10.0 | 12.0 |
| Fe | 8.0 | 18.5 | – | 0.5 | – | – | – |
| Cr | 15.5 | 19.0 | 19.5 | 14.6 | 14.0 | 8.9 | 6.8 |
| Al | – | 0.50 | 1.40 | 4.30 | 3.00 | 4.80 | 6.15 |
| Ti | – | 0.90 | 3.00 | 3.35 | 5.00 | 2.50 | – |
| Ta | – | – | – | – | – | 3.80 | 6.35 |
| Nb | – | 5.10 | – | – | – | 0.10 | – |
| W | – | – | – | – | 4.00 | 7.00 | 4.90 |
| Mo | – | 3.00 | 4.30 | 4.20 | 4.00 | 2.00 | 1.50 |
| C | 0.08 | 0.08 | 0.07 | 0.07 | 0.17 | 0.11 | 0.12 |
| Hf | – | – | – | – | – | 1.55 | 1.50 |
| B | – | – | 0.006 | 0.015 | 0.015 | 0.015 | 0.015 |
| Zr | – | – | 0.09 | 0.04 | 0.03 | 0.05 | 0.02 |
| Cu | 0.15 | 0.15 | – | – | – | – | – |
| Other | – | – | – | 0.20 Si | – | – | 2.80 Re |

## Solidification Cracking Susceptibility

The solidification cracking susceptibility of the tested alloys was evaluated by the CPTT procedure given in Table 1 using Cu-Be-Co molds. At least three samples were tested at each pin length in the transition range from no cracking to 100% cracking. The results of this testing is summarized in Table 6 and plotted as function of percent circumferential cracking versus pin length in Fig. 10.

**Table 6.** Solidification cracking parameters in Ni-base superalloys tested by CPTT

| Alloy | Max. pin length for no cracking, mm | (in) | Min. pin length 100% cracking, mm | (in) | Zero to 100% cracking range, mm | (in) |
|---|---|---|---|---|---|---|
| Inconel 600 | 38.100 | 1.500 | 50.800 | 2.000 | 12.700 | 0.500 |
| Waspaloy | 38.100 | 1.500 | 50.800 | 2.000 | 12.700 | 0.500 |
| René 80 | 34.925 | 1.375 | 38.100 | 1.500 | 3.175 | 0.125 |
| Inconel 718 | 25.400 | 1.000 | 38.100 | 1.500 | 12.700 | 0.500 |
| René 77 | 25.400 | 1.000 | 28.575 | 1.125 | 3.175 | 0.125 |
| René 125 | 15.875 | 0.625 | 19.050 | 0.750 | 3.175 | 0.125 |
| René 142 | 12.700 | 0.500 | 15.875 | 0.625 | 3.175 | 0.125 |

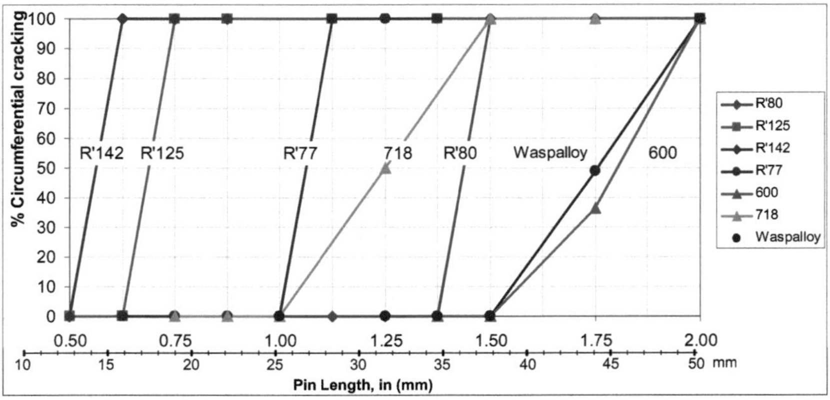


**Fig. 10.** Relative susceptibility to solidification cracking of high performance Ni-base superalloys tested by the CPTT

The alloys were ranked in order of decreasing susceptibility to solidification cracking as follows: R'142, R'125, R'77, alloy 718, R'80, Waspaloy, and alloy 600. Alloys 718 and René 77 had equal maximum pin lengths of no cracking, but the minimal pin length of 100% in alloy 718 was three increments (0.375 in.) higher, thus showing lower susceptibility. Alloys 600 and Waspaloy had coinciding cracking ranges with somewhat lower extent of cracking in the former at the 1.750 in. pin length (Fig. 10). The zero to 100% cracking ranges of all the René alloys were one increment of pin length (0.125 in.) wide. The cracking ranges of alloys 718, Waspaloy and 600 were four increments wide (0.5 in.). These rankings are in generally good correlation to the field experience with the solidification cracking in the tested alloys.

## Solidification Microstructure

Longitudinal sections of tested cast pins were mounted, polished, and etched electrolytically in a chromic acid solution at 2.5 V for 30 s. The cast microstructure was investigated by optical microscope at magnifications up to 400X. Although some variations were observed among the tested alloys, the general solidification macrostructure of the cast pins consisted of a thin layer of equiaxed grains at the mold surface, followed by an area of extensive growth of columnar and cellular dendrites (Fig. 11a), and a central area of equiaxed dendritic growth (Fig. 11b). Alloys R'142, R'125 and R'80 exhibited all of these areas, with comparatively finer equiaxed dendrites in alloy R'142, Fig. 11c. In alloys R'77 and 718 narrow equiaxed areas with large equiaxed dendrites were present along the cast pin axis (Fig. 11d) mixed with areas of center contact of columnar dendrites, Fig. 11e. The macrostructure of alloy 600 and Waspaloy consisted of columnar dendrites with small isolated areas of equiaxed dendrites in the cast pin central areas, Fig. 11f.

A series of measurements were conducted to evaluate the primary dendritic arm spacing (DAS) of the investigated alloys under the solidification conditions of the CPTT. The measurements were performed on grains of columnar dendrites that were located close to the fracture surface and whose primary arm axes were normal to the surface of measurement. The results have shown that the DAS of the investigated alloys under the specific conditions of CPTT is in the range from 13 to 30 μm, as summarized in Table 7. This range of DAS corresponds to what is observed during solidification of low to medium heat input welds using the gas-tungsten arc welding (GTAW) process.

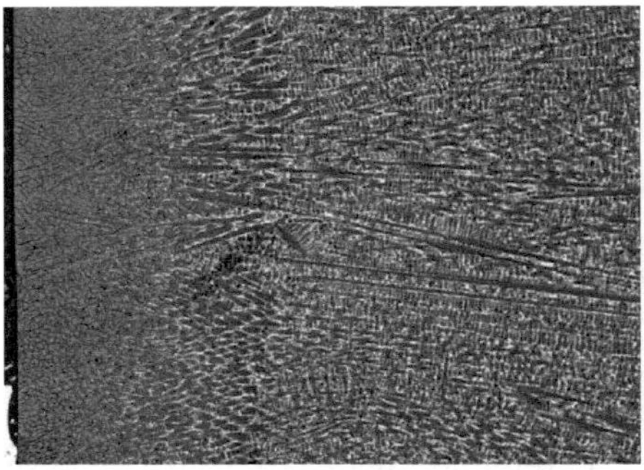

**a)** Crack initiation area. Fine grained surface region followed by dendritic growth in alloy R'80

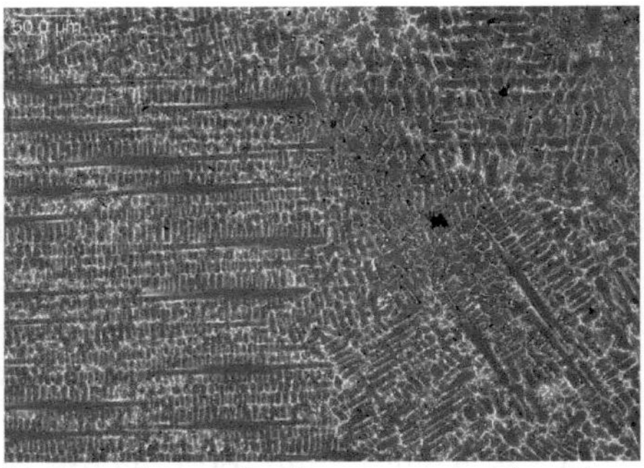

**b)** Transition of columnar to equiaxed dendritic growth in cast pin of alloy R'125

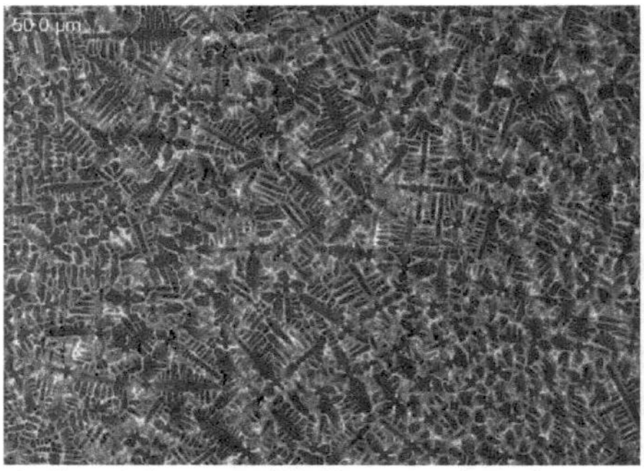


**c)** Central area with finer equiaxed dendrites in cast pin of alloy R'142

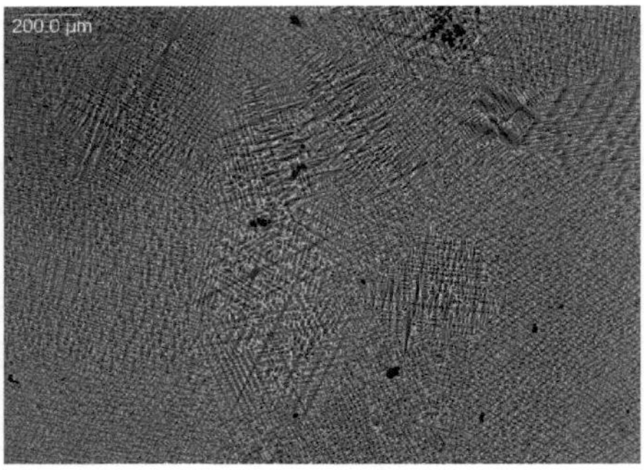


**d)** Isolated central area with large equiaxed dendrites in cast pin of alloy R'77

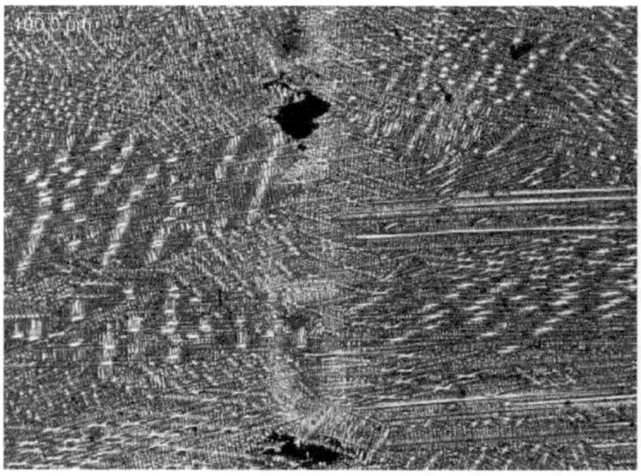

**e)** Central contact of opposite growing dendrites at cast pin center line, alloy 718

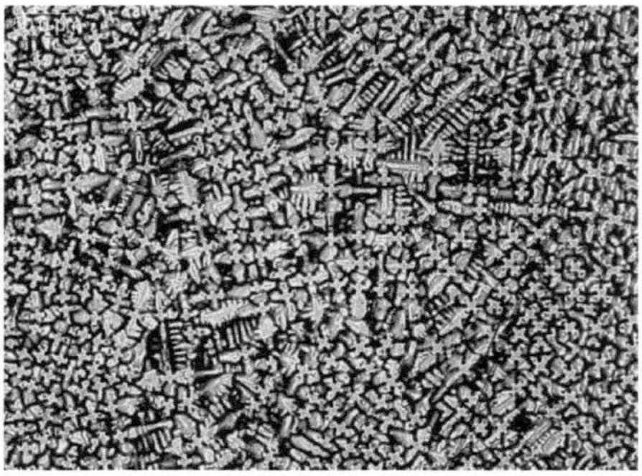

**f)** Isolated equiaxed dendrites in the central of area cast pin of Waspaloy

**Fig. 11.** Solidification structure in cast pins of the tested Ni-base superalloys

**Table 7.** Primary dendritic arm spacing in the tested Ni-base alloys in the conditions of CPTT (average of 30 measurements)

| Alloy | Average DAS, [μm] | Standard Deviation |
|---|---|---|
| René 142 | 18.2 | 3.1 |
| René 125 | 26.0 | 5.0 |
| René 80 | 16.0 | 2.6 |
| René 77 | 16.9 | 2.2 |
| Inconel 718 | 13.3 | 2.0 |
| Inconel 600 | 30.1 | 5.6 |
| Waspalloy | 14.7 | 2.2 |

## Shrinkage Porosity

Some shrinkage porosity was found along the cast pin center line in the alloys forming large equiaxed dendrites in the last region of the pin to solidify. Alloys René 125 and René 77 (Fig. 11d) had more significant center-line shrinkage porosity, followed by alloy 718 (Fig. 11e) and René 80 with some moderate porosity. The shrinkage porosity was typically found in the longer pins and decreased to an insignificant level with decreasing the pin length. In general, the shrinkage prosity was not related to solidification cracking but instead increased the percent of invalid samples at the higher mold lengths.

Alloys 600, Waspaloy, and René 142 exhibited very little dispersed shrinkage porosity that was not related to solidification cracking. The dispersed shrinkage porosity typically forms in the cast pin central area, it is evenly distributed, and the size of the separate pores is smaller then the dendrite arm spacing. Therefore, this dispersed porosity is not expected to affect the general stress distribution in the cast pin, and the processes of crack nucleation and propagation, which occur in the pin surface area. This kind of porosity was considered acceptable and not affecting the test results.

From the total of 27 sectioned cast pin samples, only three had large shrinkage pores that caused cracking. The test results from these samples were not included in the results. Most of the tested samples had either insignificant or no shrinkage porosity and, in general, porosity did not affect the test results.

## Fracture Morphology

Optical microscopy of longitudinal cast pin sections revealed that solidification cracking occurred along solidification grain boundaries (SGBs) and solidification subgrain boundaries (SSGBs), as shown in Fig. 12. Figures 12a,b represent a clear example of the process of crack nucleation and propagation under the conditions of CPTT. The solidify-cation cracks appear to form initially in the fine grained surface area that solidifies at the surface of the pin and then propagate along the SGBs (Fig. 12b). Propagation along the SSGB's is also possible during the final stage of fracture (Fig. 12c,e,f). Cracking often occurred in multiple planes within the samples (Figs. 12c,e,f) and sometimes joined to form the final fracture. Clear evidence of backfilling (crack "healing") caused by an interdendritic liquid at the end of solidification was found in alloys R'142, R'125 and R'80 (Fig. 12a,b,c). No such evidence was found in any of the other alloys (Fig. 12d,e,f).

For each tested material, the fracture surface of the shortest pin that cracked 100% was evaluated by SEM. All fracture surfaces exhibited a dendritic "eggcrate" morphology that is characteristic of solidification cracking, Fig. 13. This fracture morphology is similar to the high and medium temperature region observed in Varestraint samples, Fig. 8. Alloys René 142 and René 125 had a mixed mode failure with both the dendritic morphology and small regions of flat fracture (Fig. 13b). The latter occurred due to bridging between solidification fracture surfaces due to tensile overload. This flat fracture bridging was not observed in the remainder of the examined samples, which exhibited dendritic fracture morphology, Fig. 13d to 13h.

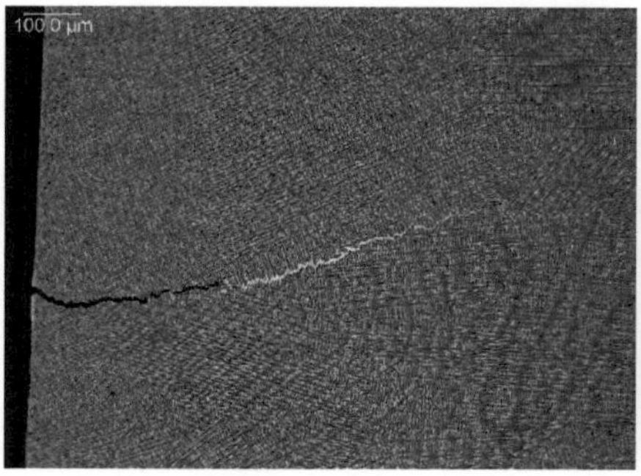

**a)** Crack initiation, propagation, and backfilling in René 142

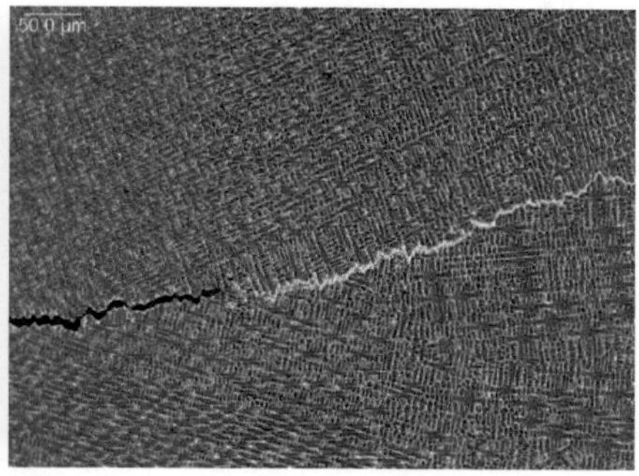

**b)** Crack initiation, propagation along the SGBs, and backfilling in René 142

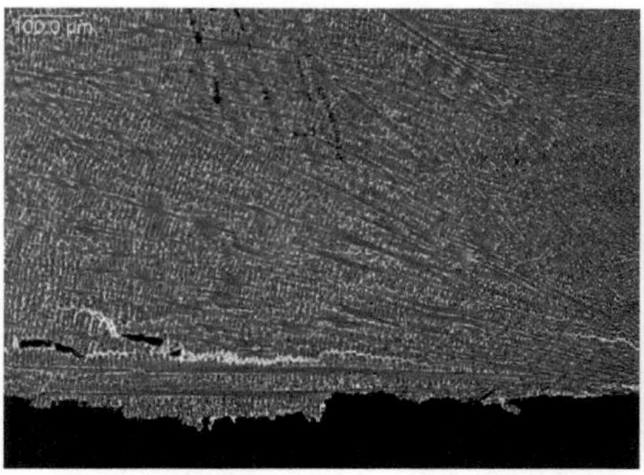

**c)** Multiple cracking, propagation along the SSGBs, and backfilling in René 125

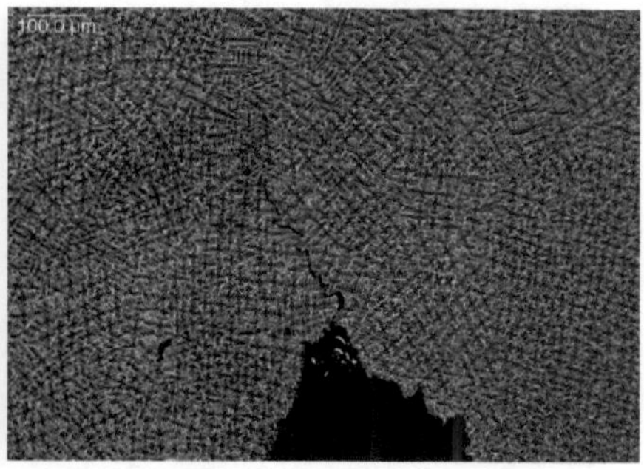

**d)** Cracking along the SSGBs in René 77

**e)** Multiple cracking and propagation along the SSGBs in alloy 718

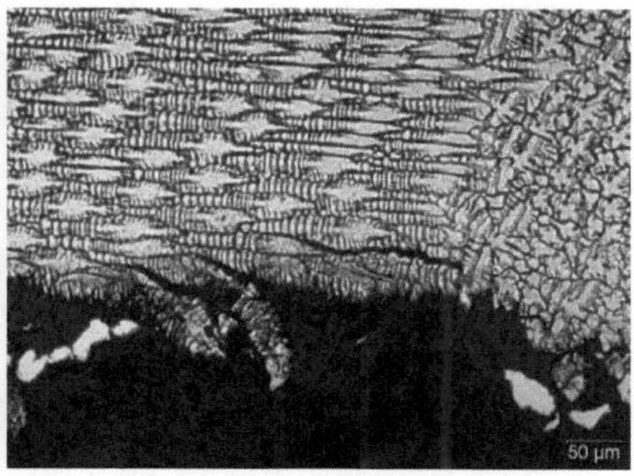


**f)** Multiple cracking and propagation along the SSGBs in alloy 600

**Fig. 12.** Solidification cracks in cast pins of the tested Ni-base superalloys

**a)** René 142 at 0.625 in.

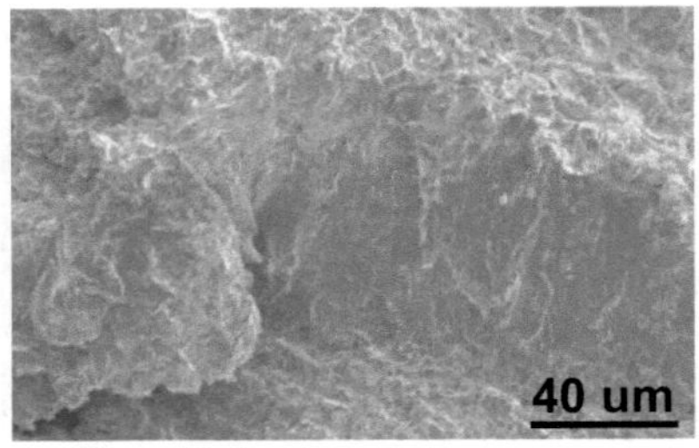


**b)** René 142 at 0.625 in.

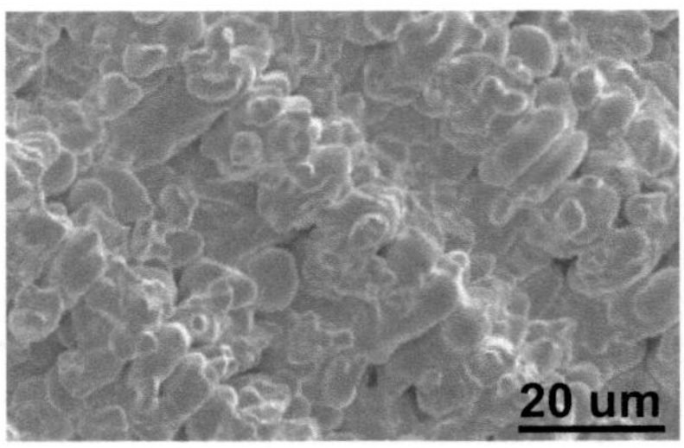


**c)** René 125 at 0.75 in.

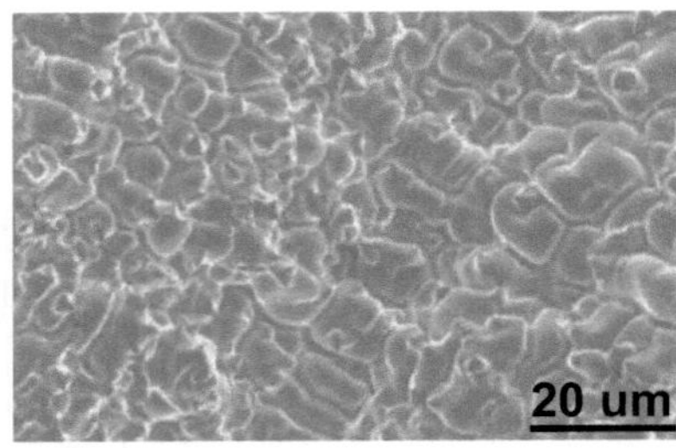


**d)** René 77 at 1.25 in.

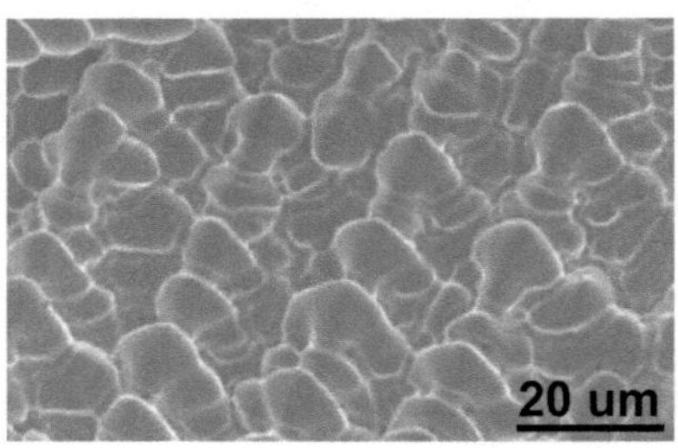


**e)** Inconel 718 at 1.75 in.

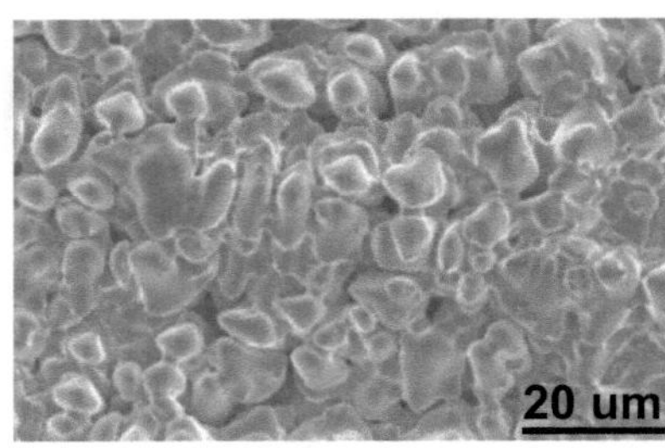


**f)** René 80 at 1.5 in.

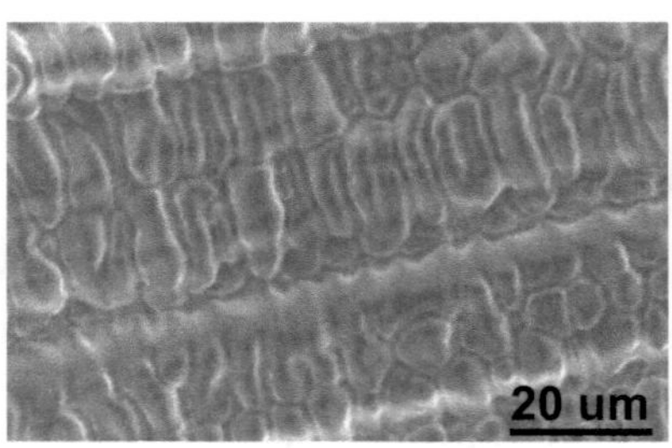


**g)** Inconel 600 at 2 in.

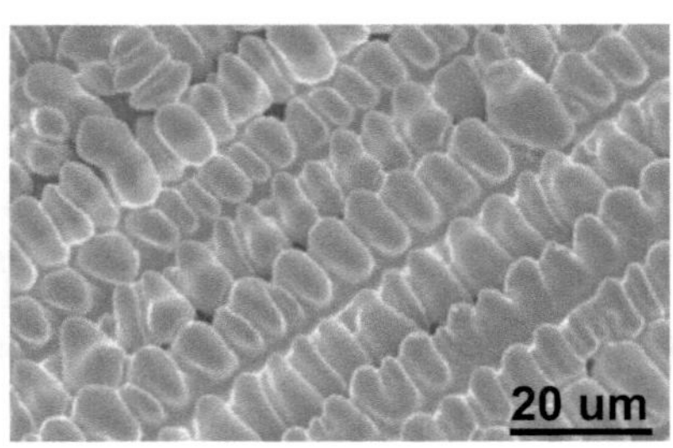


**h)** Waspalloy at 2 in.

**Fig. 13.** Solidification cracking fracture morphology in cast pins of the tested Ni-base superalloys

## Discussion

The second-generation cast pin tear test has been developed in an effort to create a robust weldability test that is able to rank the solidification cracking susceptibility of both "difficult-to-weld" alloys and "standard" alloys. The cast pin apparatus and testing procedure were designed to provide controllable and reproducible testing conditions, and to produce cast pins of good integrity. The cases of unacceptable shrinkage porosity in difficult-to-cast alloys were reduced to less than 10% of the tested samples. As a result, the reproducibility and reliability of the CPTT results were significantly improved.

The solidification conditions in the CPTT closely simulate the solidification conditions of low to medium heat input welds, such as those made by the GTAW process. These include the morphology of the solidification microstructure and the sequence of solidification, starting with a fine-grained "chill zone" at the mold wall, followed by cellular and columnar dendritic growth, and finishing with equiaxed dendritic growth at the center of the pin. The primary dendritic arm spacing was between 13 and 30 μm. Even higher weld heat input conditions can be simulated by using lower cooling capacity (lower conductivity) molds.

Metallographic and fractographic examination of cast pins have confirmed that the CPTT also simulates the weld solidification cracking in actual welds relative to the nature of crack nucleation and propagation, crack morphology, and crack healing. For example, cracking was primarily along solidification grain boundaries, just as it is in actual welds in Ni-base alloys and stainless steels.

The reliability of the CPTT results has been confirmed by comparison to six externally restrained hot cracking tests. Very good correlation was found between the weldability rankings of four high alloy stainless steels generated by CPTT and by PVR test.

The solidification cracking susceptibility ranking of the seven tested Ni-base superalloys is in good correlation with the field weldability experience with these alloys. This indicates that the improved, second generation CPTT technique is a reliable tool for evaluating the weldability of materials with wide variations in solidification cracking susceptibility, particularly those with high to moderate susceptibility.

The solidification cracking behavior of the Ni-base superalloys is a function of their complicated alloying systems and of the solidification conditions provided by the CPTT. The combination of chemical composition and cooling rate determine the solidification morphology, tendency for formation of eutectic phases, the properties of the latter and their capability to filling interdendritic shrinkage, and the general liquid-solid and solid-state shrinkage behavior.

A series of experimental and modeling investigations will be conducted in the future, aimed at quantifying the parameters that control the solidification cracking phenomenon under the conditions of the CPTT. The mold and cast pin design facilitates the application of relatively simple finite element modeling to study the processes of heat transfer, solidification, and development of stresses during solidification. The technique of single sensor differential thermal analysis – SS-DTA [6] will be utilized in conjunction with the CPTT to determine the solidification temperature ranges and the possible liquid-state precipitation and eutectic reactions. Such combined experimental- modeling approaches will allow the solidification morphology and solidification cracking behavior to be related to quantifiable parameters as cooling rate, solidification temperature range, liquid- and solid-state volume shrinkage, and stresses associated with solidification.

## Conclusions

1. A second-generation cast pin tear test (CPTT) was developed that is capable of testing and ranking the susceptibility of Ni-base superalloys to solidification cracking.
2. The improved design of the CPTT apparatus and the new CPTT procedure provide repeatable and controllable testing conditions, produce cast pins with acceptable integrity, yield reproducible and reliable test results, and are highly sensitive for ranking the weldability of difficult-to-weld alloys.
3. The CPTT has provided very good correlation to a variety of externally restrained solidification cracking tests in ranking the weldability of four high-alloy, fully austenitic stainless steels. The best correlation was obtained with the PVR test, which similar to the CPTT, restrains the samples in the direction of solidification.
4. A ranking of solidification cracking susceptibility has been generated that includes seven Ni-base high performance alloys. The weldability of these alloys as ranked by the CPTT is in good correlation with their field weldability.

5. The CPTT closely resembles the solidification conditions in actual fusion welds. Based on measurements of the primary dendrite arm spacing in the cast pins, the solidification conditions approximate those in low to medium heat input welds. These conditions can potentially be extended to simulate higher heat input welding by altering the mold cooling capacity.
6. The CPTT has a tremendous application potential in testing the weldability of a wide range of non-ferrous alloys and stainless steels. In combination with numerical modeling and a technique for single sensor differential thermal analysis, it will provide an efficient and inexpensive tool for alloy and consumable development.

## References

1. Hull F. C., Cast-Pin Tear Test for Susceptibility to Hot Cracking, Welding Journal, Vol. 38 (4), 1959, pp. 176s–181s.
2. Ryan D. P., Development of Modified Cast-Pin Tear Test to Evaluate the Solidification Cracking Susceptibility of Fully Austenitic Materials, MS Thesis, The Ohio State University, 2003.
3. Alexandrov B., Nissley N., Norton S., and Lippold J., Development of a Weldability Test for High Performance Base and Filler Materials, Report No. MR0606, Edison Welding Institute, July 2006.
4. Wilken K., Investigation to compare Hot Cracking Tests – Externally Loaded Specimen, IIW Doc. IX-1945-99 (II-C-168-99).
5. Finton T., Lippold J., and Bowers R., Comparison of Weld Hot Cracking Tests, Summary of an IIW Round Robin Study, IIW Doc. IX-H-459-99 (II-C-175-99).
6. Alexandrov B. T. and Lippold J. C., Relationship Between the Solidification Temperature Range and Weld Solidification Cracking Susceptibility of Stainless Steels and Ni-base Alloys, IIW Doc. IX-2163-05.

# SAW Cold Wire Technology – Economic Alternative for Joining Hot Crack Sensitive Nickel-Base Alloys

U. Reisgen, U. Dilthey, I. Aretov

RWTH Aachen University, Welding and Joining Institute, Aachen, Germany

## Abstract

The construction of industrial plants and gas turbines makes increasingly higher demands to the used materials. Consequently, austenitic CrNi-steels and nickel-based alloys are increasingly applied in power station construction, crude oil and petrochemistry and, moreover, in industrial furnace and turbine construction. Besides the good mechanical properties of these materials also a good process ability is required. This applies particularly to welding. Due to their low heat input, TIG and MIG (pulsed) welding are the most frequently used welding methods in the case of heat-resistant nickel-base super alloys and also in the case of stainless CrNi-steels. Due to the high heat input, submerged-arc welding is, as a rule, not applied for the welding of these steels. Submerged arc welding could, however, be used as efficient alternative if process modifications are available, which ensure higher hot cracking resistance by reduced heat input. Apart from the high weld quality which is a result of submerged-arc welding, the economic efficiency which is particularly marked by the high deposition rates is also interesting.

Submerged-arc welding knows different possibilities of reducing the energy to a lowest possible quantity. The easiest way to obtain this objective is a higher welding speed which reduces the energy-per-unit-length. A more sophisticated - but also more effective - approach is the direct withdrawal of heat from the weld pool by addition of filler material, for example, in the form of cold wire. The heat in the weld pool is not completely entering the base metal but is partly used for melting the cold wire.

The tests which were performed within the scope of this project were made on the nickel-based alloys 617 and 625. It has been the aim of the project to make a statement about the welding suitability and the economic efficiency of the SA cold wire technology; to this effect, the welding parameters and also the optimal electrode and cold wire parameters had to be determined. The specific reduction of the energy-per-unit-length is used to achieve a maximum deposition rate. Comparative tests were carried out using the conventional SA method and the SA thin-wire method.

One main point had been the examination of the cold wire influence on the weld geometry and on the chemical composition of the weld metal and thus also on the hot-cracking tendency. The hot-crack tests were carried out by means of the MVT test. Moreover, the influence of the modified SA cold wire technology on the distortion had been examined. Through the optimisation of the cold wire addition a high process reliability had been obtained

The transfer of the achieved results to industrial practice has been successful.

## Introduction

The main problems which occur in fusion welding of fully austenitic materials are caused by hot cracking. The reasons for the development of solidification cracks and melt cracks lie in the most complex interaction of several factors which are summarised into three groups. These factors are of metallurgical, mechanical and thermal nature. They are reflecting the properties of the welded joint. For a complete evaluation of the hot cracking susceptibility of a weld, all three factors with their respective degree of efficiency must be considered.

If a welding method is subject to modification, changes of several influential factors are often occurring at the same time. It has been the aim of this research work to examine the extent of the influence and of the factors' interactions.

## Testing Equipment

Conventional SA equipment has been used for the weld examinations. The equipment had been modified with an additional cold wire feeding unit, which works without applying any additional electrical energy to the cold wire. In order to adjust the feed speeds for electrode and cold wire independently of each other, two separate feeding units have been used. This allowed varying the feeding quantity of the cold wire without changing the cold wire diameter which resulted in a higher flexibility of the process.

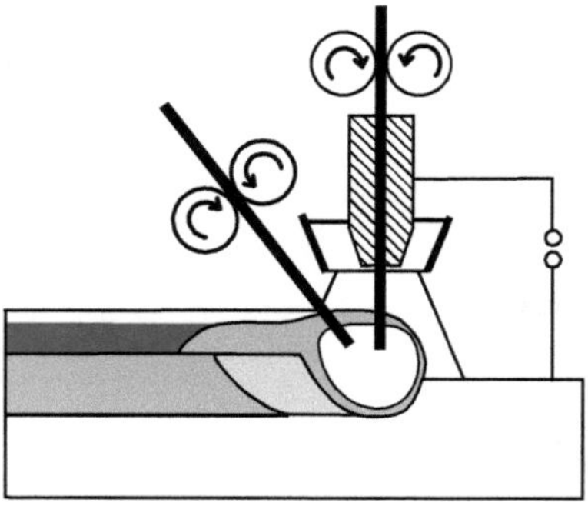

**Fig. 1.** Schematic representation of the modified SA cold wire method

The schematic view of the applied process modification is shown in Fig. 1. The cold wire is fed into the molten pool behind the electrode, passing the liquid slag and is then molten in the pool. The decision for a trailing arrangement of the cold wire feeding was based on the results from the preliminary tests. These tests showed that this arrangement allowed obtaining a wider seam with a better transfer to the base metal. The reinforcement of the weld is thus reduced and the ratio "weld width/weld height" is improved.

Another advantage of this arrangement is the possibility of melting the cold wire in the molten pool at significantly lower temperatures than would be the case if the drop transfer occurred during the melting of the cold wire in the arc region. If a leading arrangement is used, melting occurs only in the arc region which, again, results in the increased alloy burn-off. Using the trailing arrangement it is, thereby, possible to pass alloying elements into the molten pool which would otherwise have been burned off.

In order to ensure the melting of the cold wire in the molten pool and not in the arc region, the exact positioning of the cold wire is necessary to obtain a reproducible welding process. For the exact positioning of the cold wire, the following parameters must be considered:

- angle of slope of the cold wire
- distance between electrode and cold wire
- wire stick-out.

The parameters are depicted in Fig. 2.

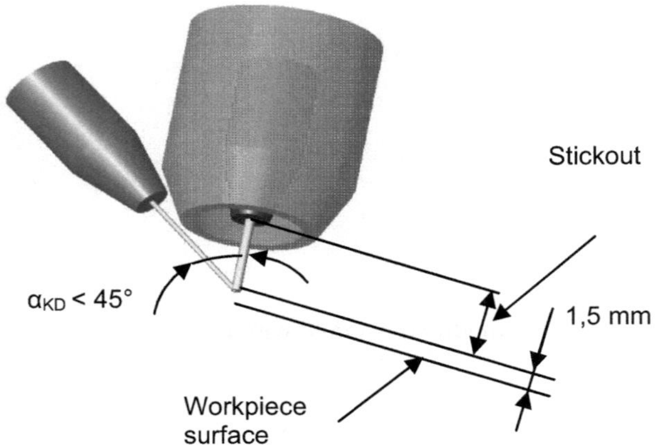


**Fig. 2.** Positioning of the cold wire feeding

## Influence of Cold Wire Addition on the Welded Joint

The following tests show the influence which cold wire addition exerts on the welded joint. In particular, the weld geometry, the structure, the chemical composition and the connection between cold wire addition and hot-cracking susceptibility have been considered.

### Weld Geometry

The first tests dealt with the determination of the influence which cold wire exerts on the weld geometry. For that, welds with different quantities of cold wire were carried out. The parameter "melting ratio" has been introduced for these tests. It specifies the ratio between the mass of the cold wire which has been added during welding and the mass of the electrode. With the increasing quantity of cold wire, weld reinforcement and weld width were also increasing, while the penetration depth was reduced (Fig. 3). The smaller penetration area resulted in a lower degree of dilution with the base metal (Fig. 4).

Caused by cold wire addition, the weld shows a tendency to reinforcement of the weld beads. In order to compensate the reinforcement, the

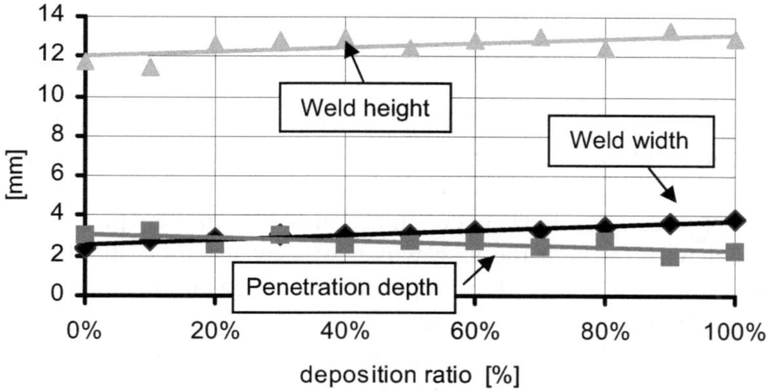


**Fig. 3.** Influence of cold wire addition on welds reinforcement, weld width and penetration depth

influence of the welding voltage on the weld geometry has been investigated more closely. It was determined that a voltage increase allowed the partial compensation of the developing weld reinforcement (Fig. 5). It must, however, be considered that the increase of the welding voltage has its process-technical limitations. The limited electrical and thermal load capacities of the slag and the dynamics of the welding process bring about process instabilities if the voltage is too high.

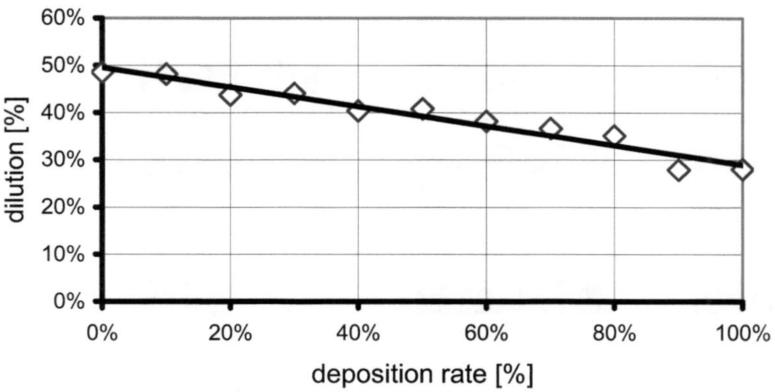


**Fig. 4.** Influence of cold wire addition on the dilution with the base metal

The cooling speed during crystallisation has great influence on the formation of hot cracks. Cold wire addition results in faster cooling of the molten metal; the crystallisation processes in the molten pool are greatly influenced.

These tests were performed in order to determine a potential influence on the temperature gradient in the molten pool in a solid state, provided that the deposition ratio had been varied.

For that, thermocouples have been dipped into the molten pool, right behind the cold wire through the slag which was by the time still liquid. The thermocouples allowed the reproducible measurement of the temperature gradient only within a temperature range of below 1000°C. It has been observed that the solidification interval for alloy 617 is between 1380°C and 1330°C. Because of the homogenisation of the temperature fields in the component it is not possible to establish an influence of the cold wire addition on the temperature profile below 1000°C. For further examinations it would be advisable to select a measuring method which allows the measurement of temperature ranges above the melting temperature of the filler materials.

The macro-sections (Fig. 5) do show differences with regard to the weld geometry. It is observed that – with the increase of the cold wire addition - the penetration area is getting smaller and the reinforcement is increasing (Fig. 6). This causes the shortening of the fusion line, the heat dissipation

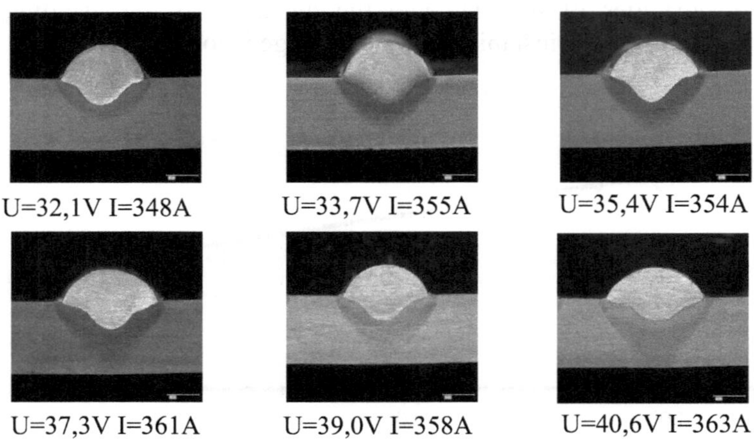


**Fig. 5.** Micro-sections of surface welds with increasing welding voltage and a cold wire addition of 50%. Used wire electrode: UTP UP 6170 Co Mod., weld flux: UTP UP FX 6170 Co, base material: austenitic CrNi-steel, welding speed: 50 cm/min

into the base metal is reduced. Particularly in SA welding, the heat dissipation into the base material is of special importance and the alteration of the fusion line may have large effects on the thermal balance of the molten pool and thus also on the solidification conditions.

Figure 6 shows that, up to the increase of the cold wire addition to 50%, also an increase of the weld cross-section takes place. This means that an increased quantity of metal is molten without any variation of the heat input. The heat in the molten pool is more effectively used through the cold wire addition than this is the case with standard SA welding methods. A conclusion is that the average temperature of the molten pool which has been produced with the SA cold wire method is lower than that of the molten pool produced with SA standard methods. This also brings about a changed curve of the molten pool isotherms and thus a lower temperature gradient during a SA cold wire process which again affects the crystallisation processes and the welded structure.

If the deposition rate is increased further, the excessive heat of the molten pool is used for melting the increasing quantities of cold wire addition. The heat of the molten pool is no longer sufficient for the melting of equal quantities of base material. The penetration area is reduced and the weld reinforcement area is, at the same time, increased.

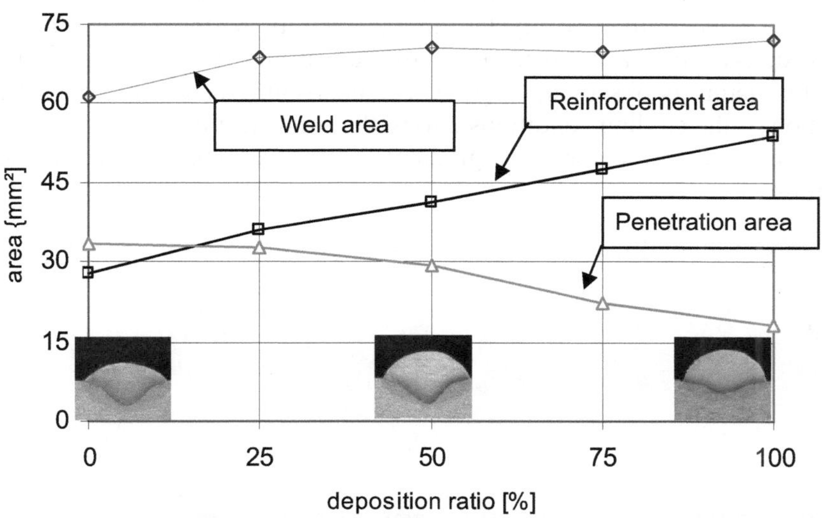


**Fig. 6.** Influence of the deposition ratio on the weld area, the penetration area and the reinforcement area

The fusion line is shortened. The heat transport from the liquid molten pool into the solid base material is carried out via the fusion line. That implies that: the longer the fusion line, the higher the excessive heat of the molten pool (deposition ratio – 0%), and, the shorter the fusion line, the colder the molten pool (deposition ratio – 100%). The increase of the cold wire quantities is accompanied by the displacement of the weld cross-sectional area in the direction of the workpiece surface.

If the deposition ratio is too high (above 100%), strong fluctuations of the weld geometry occur. This is characterised by changes of the weld width and the weld height and also by non-parallel weld flanks. The reason for this is that the cooling of the molten pool has been too strong. These fluctuations show that the welding process is no longer running stable and that it will be collapsing if the quantity of the cold wire addition is further increased.

### Structure

The examinations of the structure have been carried out by means of light and scanning electron microscopy. The macrographs of the surfacing beads showed mainly columnar crystals in the weld metal

As from a 200x magnification and higher, a finer distribution of the structure in the individual crystals in the vicinity of the fusion line is observed. With increasing distance to the fusion line, a coarsening of the crystal structure is observed, (Fig. 7). The coarsening is a result from the alteration of the cooling conditions during the solidification. It results from the decrease of the temperature gradient with increasing distance to the fusion line.

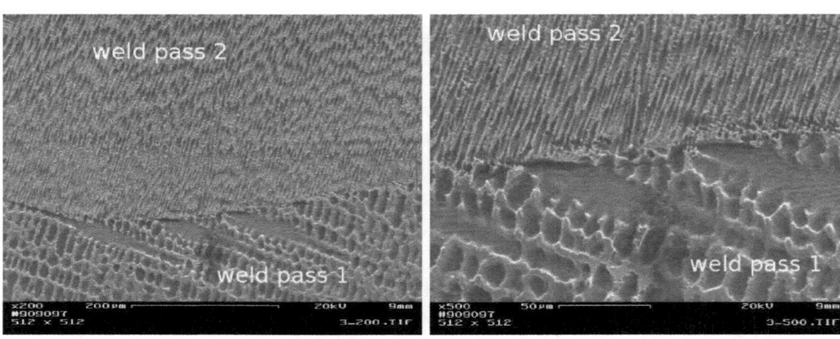


**Fig. 7.** Fusion line between two weld beads in multiple layer welding, using the SA Cold wire method, SEM micrographs

The separation processes are therefore, between high- and low-melting phases, facilitated in the direction of the weld surface. For this reason, dendrites with larger trunks and arms are developing with increasing distance from the fusion line while the regions with the low-melting phases between the dendrites are increasing .

In multiple-layer welds, the repetition of the crystal structure is observed in the individual beads. The structure is finer in the vicinity of the fusion line and coarser in the upper region. The grain boundaries are in the vicinity of the fusion line (top weld bead) clearly narrower than in the upper region of the weld bead (bottom weld bead).

Because of the melting down of these crystals with different structures (Fig. 8) at the fusion line, inheritance of the formed and deep-melting regions at the grain boundaries between lower and upper beads does not occur. This may restrict the crack propagation through several weld beads.

In contrast to the fusion line between two weld beads in multiple-layer welding, the transition between base metal and weld metal is frequently marked by epitaxial crystal growth. The relatively large grain boundary of the base material which is liquified at first during heating up, is contributing towards that. The grain boundary restricts the incompletely molten grains which act as nucleators for the weld metal crystals. Due to the restricted solubility of this eutectic fraction it is further distributing between the weld metal crystals (Fig. 9).

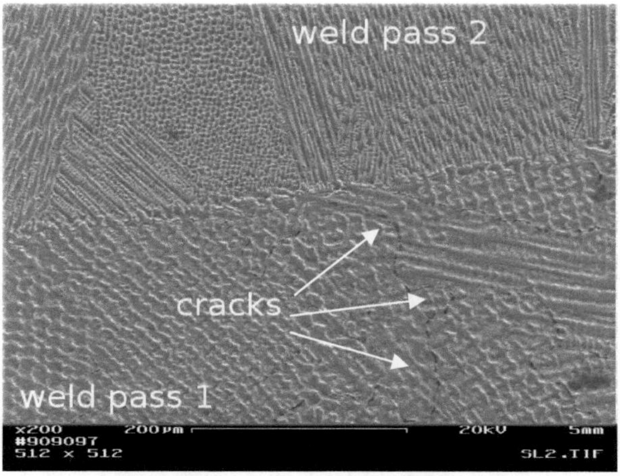


**Fig. 8.** Grain boundaries in the bottom and the top weld bead in multiple-layer welding using the SA cold wire method. SEM micrographs

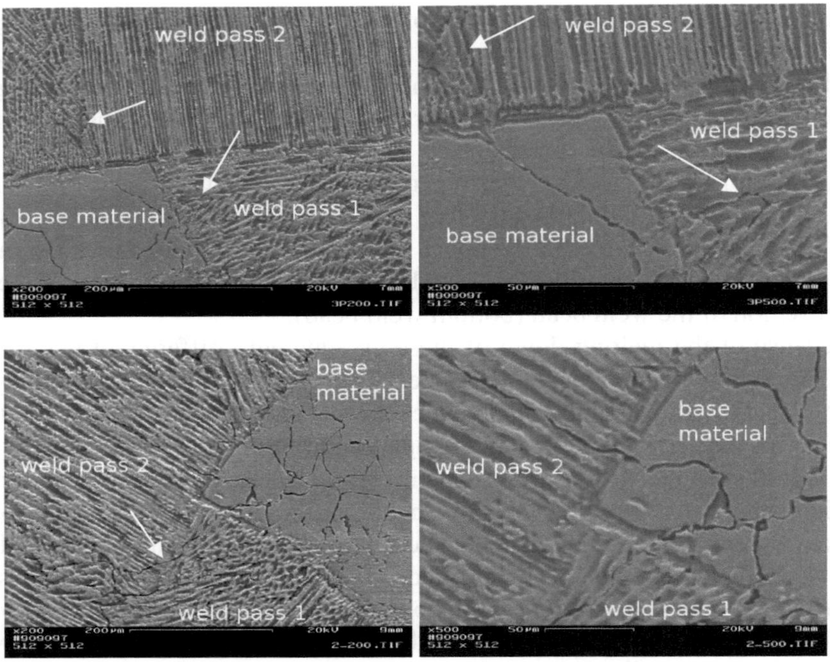


**Fig. 9.** Inheritance of the grain boundary from the base material to the weld metal, SEM-micrographs

If the critical tensile stresses are exceeded still in the high-temperature range, widening of the grain boundaries may occur (Fig. 10). At that, the width and the cross-linking of the grain boundaries, the temperature dependence and the size of the change of volume during cooling and also the high-temperature strength of the low-melting regions are decisive factors.

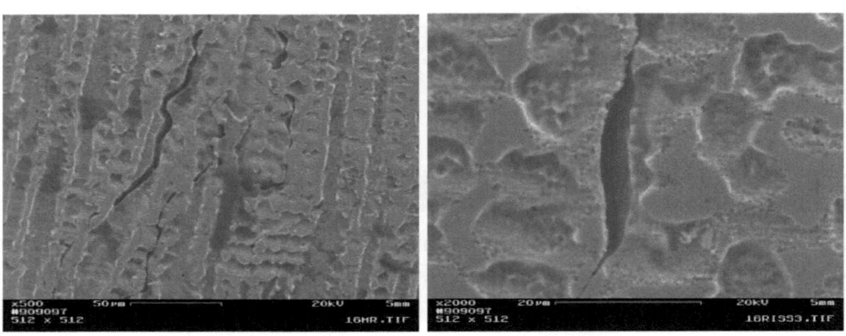

**Fig. 10.** Widening of the grain boundaries (formation of micro-cracks) in the weld metal, SEM micrographs

For a better clarification of these problems, further research work is required.

Figure 11 depicts two macro-sections. The macro-section (top) has been taken from a welding with additional cold wire feeding; the macro-section (bottom) has been taken from a welding without additional cold wire feeding. Both welds have been carried out with approximately equal welding parameters. The left and the right side show a magnification of the structure from the centre regions of the weld.

Especially the centre regions of the welds depicted in the photographs show clearly that the crystals (from the SA cold wire process) are directed and almost vertically positioned into one direction.

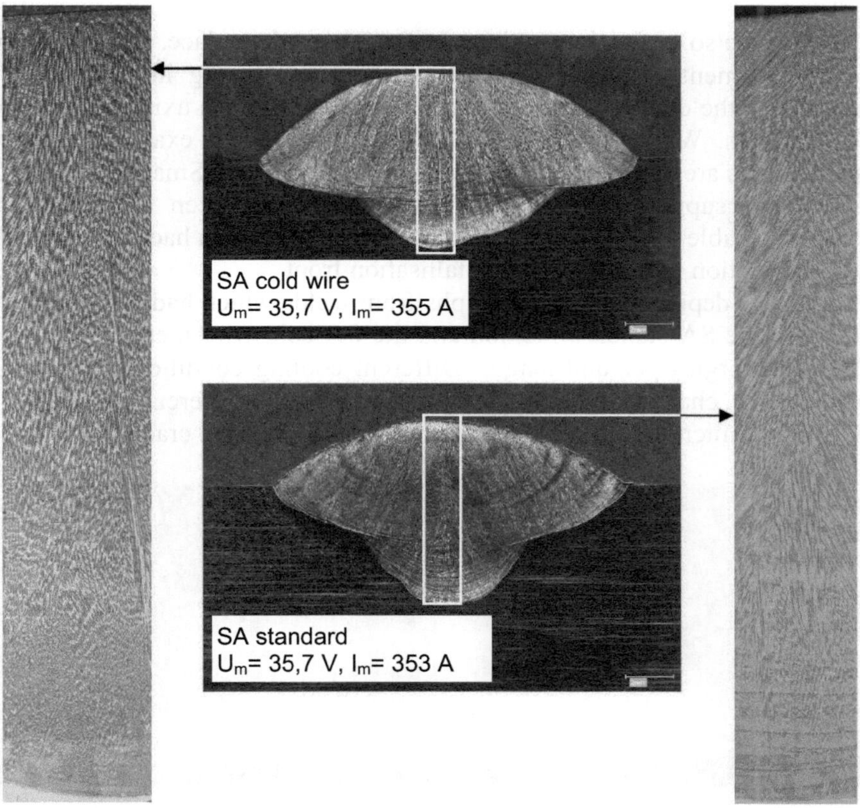


**Fig. 11.** Macro-sections with detailed photographs of a weld, Cold wire deposition ratio: 50% (*top*) , without cold wire addition (*bottom*), welding speed: 50 cm/min

The deviation from the growth direction is lesser than in conventional SA welding.

The angle where the crystals which are growing from both sides are meeting in the centre of the weld is clearly smaller than the angle which develops when using SA standard methods (Figs. 11 and 12).

The crystal growth direction in a SA standard process which is marked by a larger angle of contingence and larger growth angles is frequently enclosing deep-melting phases in the weld centre. Hot crack formation is supported by this (Fig. 12).

As far as SA cold wire methods are concerned, these eutectics are, through the smaller angle of contingence of the solidification fronts in the weld centre, transported to the surface of the weld and are there absorbed by the slag.

During the solidification, a change of volume takes place. This leads to the development of shrinking stresses which are acting in the counter direction of the dendritic growth and which are largest in axial position to the dendrites. With a larger angle of contingence, for example, higher tensile forces are acting on the centre region of the weld. Smaller angles of contingence support the inflow of molten metal between the dendrites which is capable to fill the micro segregation zones which had been caused by solidification in front of the crystallisation front.

Figure 13 depicts three micrographs from welds which had been carried out using the SA standard method and the SA cold wire method and two different energies per unit-length. Different cooling conditions and, as a consequence, changed solidification processes lead to different distribution and sizes of micro defects (micro-shrinkages and/or micro cracks).

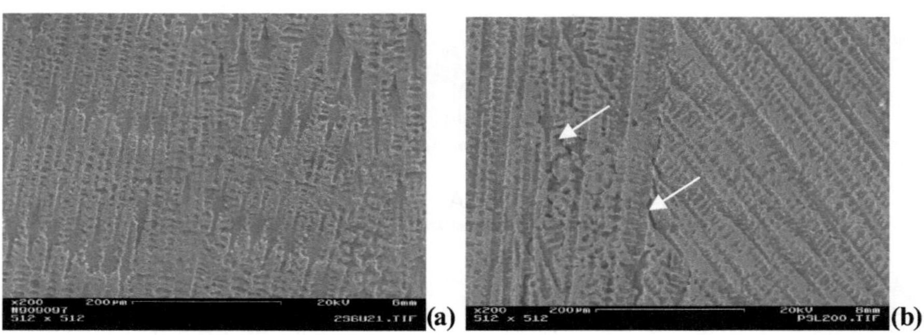

**Fig. 12.** Centre region of the weld bead in the SA cold wire method (**a**), contacting columnar crystals in the centre region of the weld bead in SA standard welding (**b**), SEM micrographs

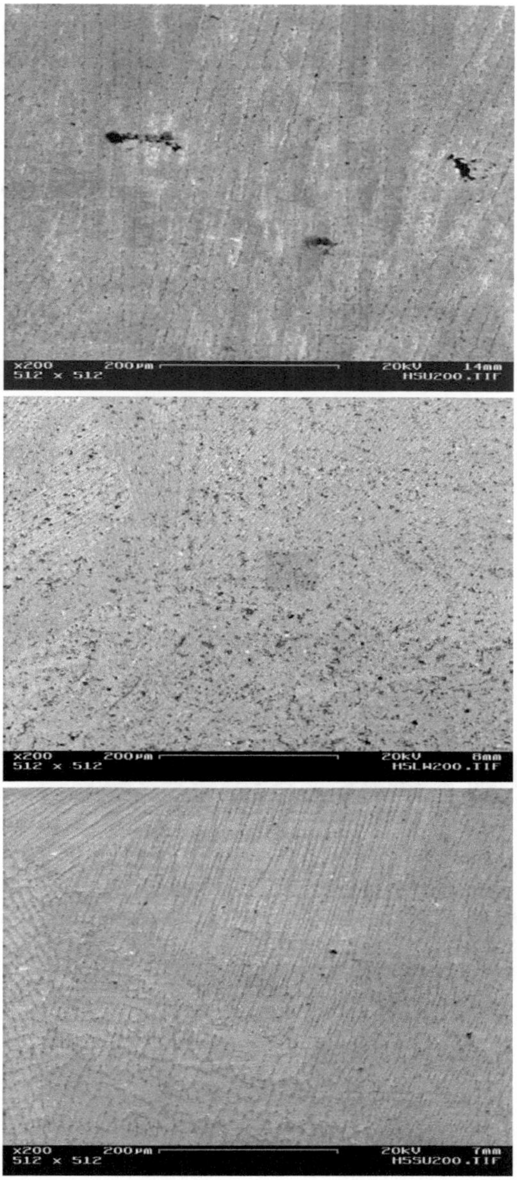


(a)

(b)

(c)

**Fig. 13.** SA standard method, lower region, transverse section (**a**), SA cold wire method, vs = 50 cm/min, lower region, transverse section (**b**), SA cold wire method, vs = 70 cm/min, lower region, transverse section (**c**)

While using the SA Cold wire method with a welding speed of 50 cm/min, the appearance of micro defects decreased. Their distribution on the dendrite boundaries is more homogenous. Micrographs show that these phenomena are less frequent in SA cold wire welds which had been produced with a welding speed of 70 cm/min.

Hardness measurements where the regions HAZ and weld metal in the root and flank region have been examined were carried out on these welds (Fig. 14). When considering these macro-sections it is noticeable that the hardness traverse in the base material is subject to higher fluctuation in the event of welding with cold wire than it is during a conventional SA weld. A possible reason for this might be the higher cooling speed which is a result from cold wire addition.

The additional cold wire feeding resulted also in the hardness increase in the lower region of the weld metal structure, Fig. 15. The influence of the cold wire addition is hardly noticed in the upper region of the molten pool since the cooling conditions of both methods are differing only slightly. The reason is that the melting of the cold wire takes place in the molten pool.

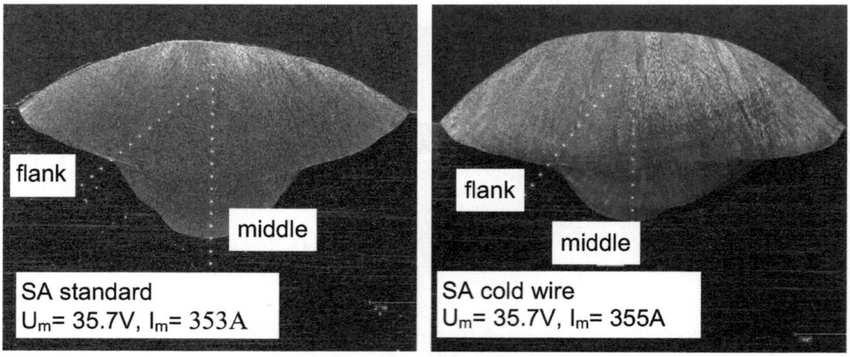


**Fig. 14.** SA standard method (*left*) and SA cold wire method (*right*)

## Chemical Composition

Alloy 617 is a nickel-cobalt-molybdenum alloy with excellent strength and creep properties through mixed crystal hardening of up to 1100°C. Due to the balanced chemical composition, the alloy possesses a very good resistance against high-temperature corrosion in the form of oxidation and carburization. In several tests, a connection between the aluminium content and the creep rupture strength values has been established. The aluminium content of between 1% and 1,2% resulted, in the case of alloy 617, in a clear improvement of the creep rupture strength [7].

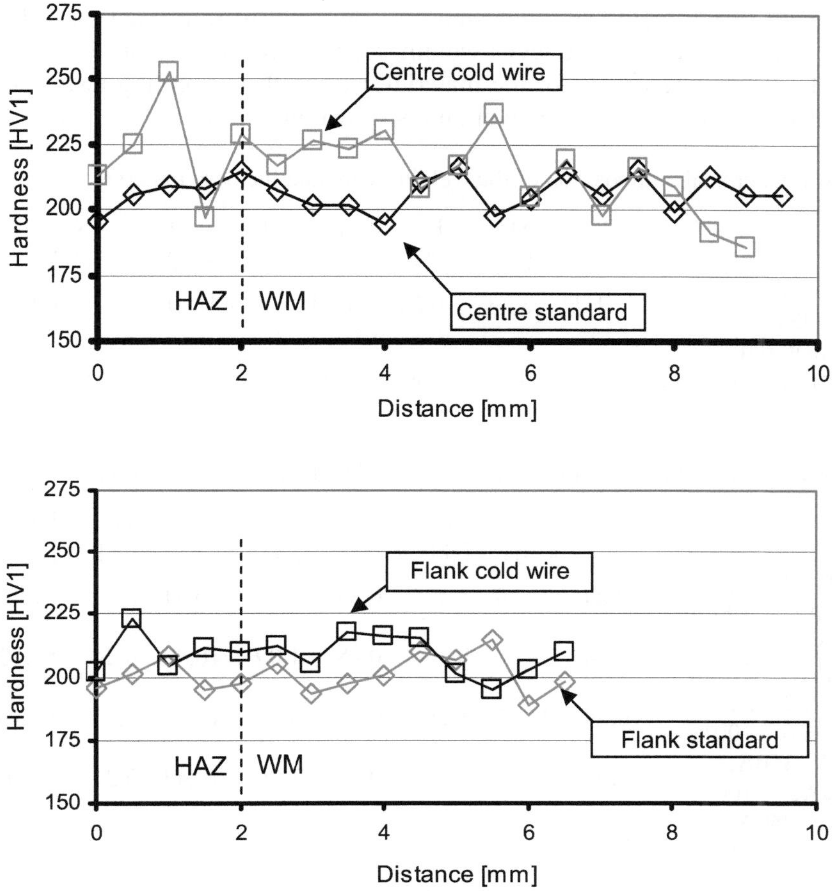


**Fig. 15.** Comparison of the hardness traverse of a weld made with the SA standard method and of a weld made with the SA cold wire method and a deposition ratio of 50% in weld centre and flank (welding speed 50 cm/min)

In this connection, modified welding filler materials with higher aluminium content were developed.

In standard submerged-arc welding, it is not possible to avoid the relatively high burn-off of aluminium which leads to low and unsufficient aluminium contents in the weld metal. It has been tried to find a solution for this problem by using the SA cold wire technology.

For the welding tests, a modified wire of the type UTP UP 6170 Co Mod. has been used. The chemical composition of this wire is specified in Table 1. For testing the influence of the cold wire addition on the Al content in the pure weld metal, welds on the austenitic CrNi steel were carried out. The specimens for the multiple-layer welding tests were built up in 5 layers. The chemical composition of the material used in these welding tests was each determined from the top layer and is shown in Table 2.

**Table 1.** Chemical composition of the welding wire and the base material

| | C [%] | Cr [%] | Mo [%] | Fe [%] | Al [%] | Co [%] | Ti [%] | Ni [%] |
|---|---|---|---|---|---|---|---|---|
| Wire analysis acc. to ISF | 0,05 | 21,77 | 8,38 | 0,24 | 1,23 | 10,44 | 0,31 | 57,00 |
| Base material analysis acc. to ISF | 0,05 | 18,47 | 0,27 | <69,6 | 0,01 | 0,14 | 0,01 | 8,51 |

**Table 2.** Chemical composition of the pure weld metal (built up in 5 layers)

| | C [%] | Cr [%] | Mo [%] | Fe [%] | Al [%] | Co [%] | Ti [%] | Ni [%] |
|---|---|---|---|---|---|---|---|---|
| 30V, 250A, $A_{s,KD}/A_{s,E}$ = 100% | 0,04 | 21,89 | 8,17 | 0,33 | 1,01 | 10,51 | 0,22 | 57,40 |
| 30V, 350A, $A_{S,KD}/A_{S,E}$ =100% | 0,04 | 21,78 | 8,14 | 0,48 | 1,09 | 10,49 | 0,24 | 57,40 |
| 35V, 250A, $A_{S,KD}/A_{S,E}$ =100% | 0,04 | 21,87 | 8,25 | 0,36 | 0,99 | 10,47 | 0,21 | 57,40 |
| 35V, 350A, $A_{S,KD}/A_{S,E}$ =100% | 0,04 | 21,90 | 8,15 | 0,66 | 1,00 | 10,44 | 0,21 | 57,20 |
| 35V, 250A, $A_{S,KD}/A_{S,E}$ =0% | 0,04 | 21,63 | 8,06 | 1,47 | 0,77 | 10,36 | 0,17 | 57,00 |

Four welding tests with a cold wire addition of 100% of the electrode mass have been carried out and the welding parameters voltage and current have been varied. The table shows that the Al burn-off remains constant with approximately 1%. The fifth test was carried out without additional cold wire feeding. The chemical analysis shows a clear decrease of the Al content down to 0.77%.

In comparison to the conventional SA welding, the tests with cold wire addition show a clear increase of the Al content in the weld metal. The reason is that through the trailing arrangement of cold wire addition the melting of the cold wire takes place at distinctly lower temperatures and is thus significantly reducing the aluminium burn-off.

### Joint Welding

The wire diameters 1.2 mm, 1.6 mm and 2 mm were examined. For all welds, electrodes and cold wire with equal diameters have been applied. During the tests it has been established that with the increase of the current, the melting potential of the molten pool was also increased.

The application of the electrode diameter of 1,2 mm resulted in a slight increase of heat input only. This was the reason why only a restricted, additional quantity of cold wire was fed to the process. Also, very precise positioning of the cold wire was necessary. Therefore, the process stability was not always ensured. With the application of the electrode diameters 1.6 mm and 2.0 mm, a very high process stability was achieved.

Figure 16 shows an example of a joint on alloy 617, plate thickness: 16 mm, and an electrode diameter of 2 mm. For the weld preparation, a V-type weld with a preparation angle of 60° had been chosen. Both plates were firmly clamped during welding. The root pass was welded manually using the TIG method. The 2 filler passes and 2 final weld passes were welded with the SA method, with a cold wire addition of 47% (Table 3). The macro-section in the picture shows good overlap and reliable fusion to the weld flanks. With those final weld passes, a good weld interface with the base material was obtained. A distinct increase of the deposition rate was achieved.

It has been observed that, also during welding in the groove, most crystals of each individual bead are continuing uninterruptedly from the fusion line to the weld surface. This facilitates the transport of the low-melting phases in the slag which again results in a positive influence on the hot cracking resistance [5].

The bead reinforcement and the increased deposition rate which are typical of the SA cold wire method have also been observed during welding tests made in the groove.

The slightly reinforced beads in the groove lead to the reduction of the tensile stress peaks which may result in longitudinal solidification cracks in the centre region of the weld bead [5].

Figure 17a shows a horizontal projection of the joint weld surface where a dye penetration test had been carried out. Surface cracks have not been observed. The results from the radiography examinations which are depicted in Fig. 17b do not show any slag inclusions, cracks or other discontinuities in the joint weld.

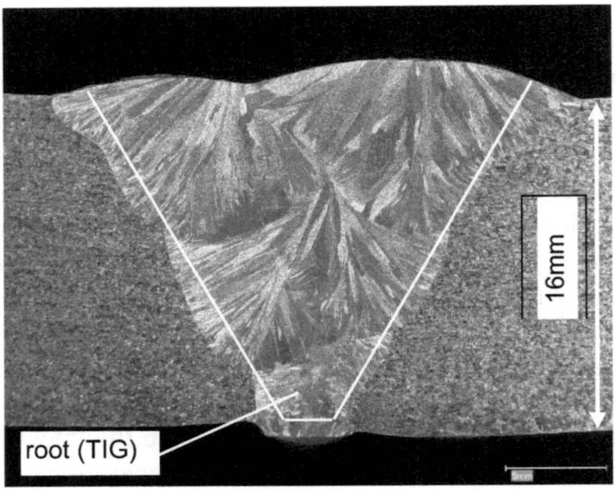


**Fig. 16.** Transverse section of a joint weld (Alloy 617) with a TIG welded root pass

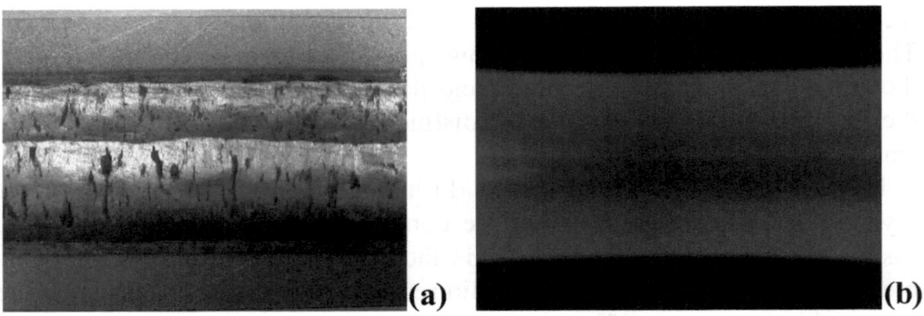


**Fig. 17.** Weld surface (**a**) and radiograph (**b**) of the weld

The next step was the examination of the cold wire influence on the angular distortion. For that purpose, joint welds with and without additional cold wire feeding have been carried out where one plate had been fixed and the other plate was allowed to shrink without restrictions (Fig. 18).

Through the cold wire addition which was caused by the higher deposition rate, the number of necessary weld beads was reduced. Moreover, the angular distortion was clearly reduced if welding was performed with cold wire addition

For a comparison of the supporting effect of the individual passes in welding, the welding speed in welding with cold wire addition had been increased in order to enable welding with the same weld build-up as in conventional SA welding. In this case also a clear decrease of the angular distortion was the result (Fig. 18). The exact influence of heat input and weld build-up is still subject to more detailed examinations.

**Table 3.** Welding parameters of the joint weld

| Layer | | 1 | 2 | 3 | 4 |
|---|---|---|---|---|---|
| I | [A] | 358 | 359 | 358 | 359 |
| U | [V] | 35,0 | 34,9 | 35,0 | 34,9 |
| $v_{EL}$ | [m/min] | 4,19 | 4,19 | 4,19 | 4,19 |
| $v_{KD}$ | [m/min] | 1,95 | 1,95 | 1,95 | 1,95 |
| $v_s$ | [cm/min] | 50 | 50 | 50 | 50 |
| $E_s$ | [kJ/cm] | 18,80 | 18,79 | 18,80 | 18,79 |
| $A_{s,KD}/A_{s,El}$ | [%] | 47% | 47% | 47% | 47% |
| $L_{ab}$ | [kg/h] | 10,49 | 10,49 | 10,49 | 10,49 |

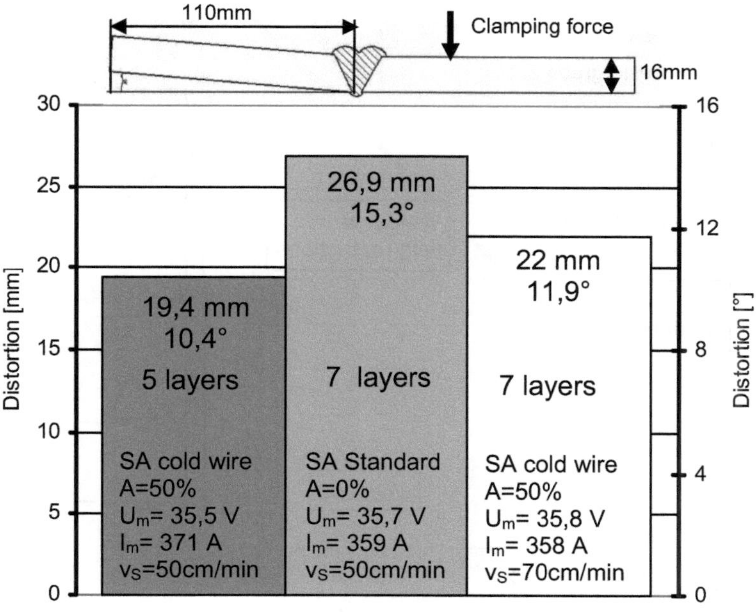


**Fig. 18.** Comparison of the distortion values using the SA standard method and the SA cold wire method

## Hot Cracking Tests using MVT-Tests

For a closer examination of the hot cracking tendency, MVT tests were made on SA welds. This is a sensitive and material-oriented testing method where the specimen is loaded externally. In the MVT test, the specimen is bent around a die with a defined radius and a defined speed. During the bending; the SA weld bead is re-molten by a TIG arc.

The three regions which are depicted in Fig. 19 (weldable, restrictedly weldable and not weldable) have been determined empirically through a large number of tests at the BAM, Berlin and have been stored in an expert data base. These three regions serve orientation purposes only since material-specific discrepancies may occur.

The diagram shows the results from the longitudinally welded SA specimens with the dimensions 100 mm × 40 mm × 10 mm. For the SA welding tests, constant welding current and constant welding voltage have been applied. The energy-per-unit length had been realised through the variation of the welding speed, its length is 50 cm/min and 70 cm/min, respectively.

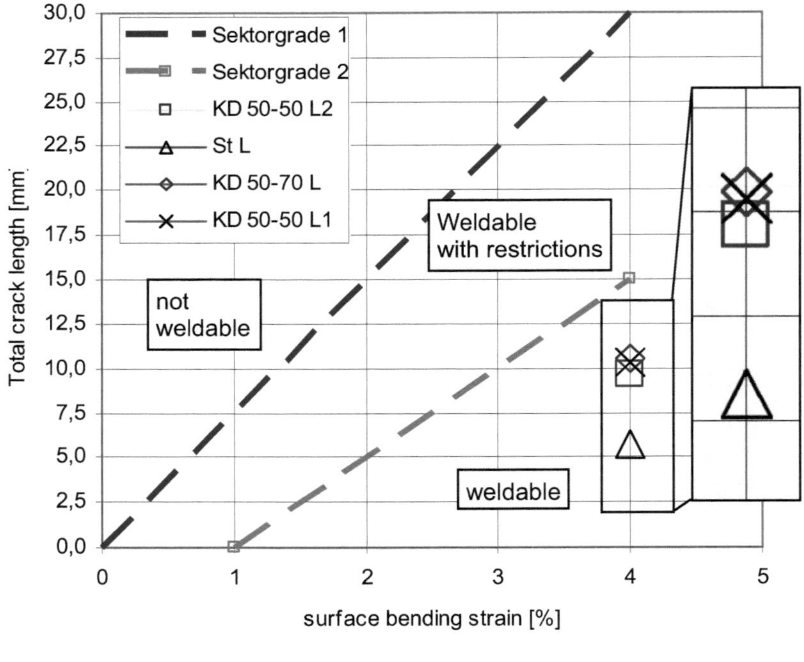


**Fig. 19.** Total crack length after application of the MVT test

The MVT tests have been performed at the BAM, Berlin. The Varestraint mode has been applied. All specimens have been tested with a travelling speed of 2.0 mm/s, 4% - strain and energy per unit length of 7.5 kJ/cm. The entire length of the hot cracks which have been produced during the test will be evaluated.

It is, however, important to mention that all results from the SA cold wire welded specimens are all within a close range and are forming a group which is close to the restrictedly weldable area. The diagram shows that the results do not show a clear difference with regard to the changed energy-per-unit length caused by the increase of the welding speed from 50 to 70 cm/min.

As already mentioned before, the addition of the cold wire is the dominating factor with regard to the deposition of the micro-alloying elements into the weld. This had been confirmed through the chemical analysis of the SA weld metal by means of MVT tests where a similar arrangement with regard to the Al content has been determined (Table 4). The low Al content of the weld made with the conventional SA method is correlating with a somewhat lower hot cracking susceptibility.

**Table 4.** Chemical composition of the SA weld metal of the MVT specimens

| | C [%] | Cr [%] | Mo [%] | Fe [%] | Al [%] | Co [%] | Ti [%] | Ni [%] |
|---|---|---|---|---|---|---|---|---|
| 36V, 367A, $A_{S,KD}/A_{S,E}$ =0%, $v_s$=50 cm/min | 0,02 | 21,52 | 8,35 | 0,56 | 0,90 | 10,65 | 0,29 | 57,30 |
| 36V, 367A, $A_{S,KD}/A_{S,E}$ =50%, $v_s$=50 cm/min | 0,03 | 21,29 | 8,59 | 0,57 | 0,97 | 10,63 | 0,31 | 57,20 |
| 36V, 368A, $A_{S,KD}/A_{S,E}$ =50%, $v_s$=50 cm/min | 0,04 | 21,87 | 8,37 | 0,45 | 1,01 | 10,67 | 0,27 | 56,90 |
| 36V, 352A, $A_{S,KD}/A_{S,E}$ =50%, $v_s$=70 cm/min | 0,04 | 21,30 | 8,58 | 0,57 | 1,01 | 10,68 | 0,31 | 57,10 |

Tests, carried out by other institutions, confirmed that the high Al content was responsible for the increase of the strength and the reduction of the ductility and the impact strength (internal research by company UTP). It is likely that the hot cracking susceptibility is also changed on a material level. This must be examined in more detail.

In the MVT test the remelting by a TIG arc changes the structure and has influence also on the effects of the cold wire.

For a closer examination of the process-technical influence of the cold wire, a PVR test would be reasonable. For this test, the specimens should be welded with the SA process. Here, not only the material side is considered, but also the process aspects and their influence on the hot cracking susceptibility.

## Industrial Implementation

Finally, a pipe weld had been carried out at the company Essener Hochdruck Rohrleitungsbau, using the SA cold wire method.

For this, two tube sections made of the material Alloy 617 with a wall thickness of 78 mm have been welded. A common circumferential weld preparation with a weld angle of approximately 24° has been chosen. The root pass was welded using the TIG method. The groove was then filled by manual arc welding up to a width of approximately 20 mm. The remaining passes were welded with the SA method and additional cold wire feeding.

The applied weld current was 290–310 A and the voltage was approximately 32–33 V. The welding speed was 45 cm/min. The cold wire was added with a ratio of 50% of the electrode feed speed.

## Conclusion

SA welding with cold wire addition exerts a considerable influence on the molten pool geometry. Among the influenced factors are the width, the depth, the length and also the shape of the molten pool.

By the addition of cold wire into the liquid molten pool, the cooling conditions and the solidification behaviour of the molten metal are strongly influenced. One reason is that the heat potential of the molten pool is better exploited. A constant quantity of heat input allows the melting of larger quantities of filler material; the cross-section of the weld may also be increased. Examinations of the structure and the hardness measurement confirm these observations.

The trailing cold wire feeding arrangement allows the reduction of the burn-off of some alloying elements which has a direct influence on the mechanical properties.

A connection between the alloying elements in the weld metal and the MVT results was also established.For the completion of the examinations about the influence of cold wire addition on the hot-cracking susceptibility with regard to material and process-related aspects, it is inevitable to carry out further tests.

The tests which have been carried out at the ISF demonstrated that the addition of cold wire resulted in a distinct reduction of distortion, compared with the conventional SA method.

The practicability of the process modification was demonstrated through the industrial implementation of the test results. It has been possible to realise the increase of the deposition rate by approx. 50%, referring to the deposition rate of the conventional SA welding methods.

## References


1. Schulze G (2004) Die Metallurgie des Schweißens. Spinger, Berlin
2. Böllinghaus T, Herold H (2005) Hot Cracking Phenomena in Welds. Springer, Berlin
3. Radaj D (1988) Wärmewirkungen des Schweißens. Springer, Berlin
4. Corporate Publication (1996) Erstarrungsformen von Rissen und Brüchen metallischer Werkstoffe.Verlag Stahleisen GmbH, Düsseldorf
5. Schuster J (2004) DVS-Bericht 233, Heißrisse in Schweißverbindungen. DVS, Düsseldorf
6. Karlsson L, Arcini H (2003) Synergic Cold Wire (SCWTM) Submerged Arc Welding of Highly Stainless Steels. Stainless Steel World 2003, KCI Publishing BV
7. Corporate Publication (2005) Werkstoffdatenblatt Nr. 4119 Nicrofer 5520 Co, ThyssenKrupp VDM
8. Corporate Publication (2000) Werkstoffdatenblatt Nr. 4118 Nicrofer 6020 hMo, ThyssenKrupp VDM
9. ISF-Aachen (2007) Verbesserung der Heißriss-Sicherheit beim UP-Schweißen von Nickelbasislegierungen unter dem Aspekt gesteigerter Wirtschaftlichkeit Forschungsvorhaben AiF-Nr.: 13.864N, DVS-Nr.: 1.049

# Part III
# Solidification Cracking of Aluminium Alloys

# Hot Tearing During Laser Butt Welding of 6xxx Aluminium Alloys: Process Optimisation and 2D/3D Characterisation of Hot Tears

D. Fabrègue[1], A. Deschamps[2], M. Suéry[2], H. Proudhon[1]

[1] INSA-Lyon, CNRS, MATEIS, Villeurbanne, France
[2] INP-Grenoble, CNRS, SIMAP, Saint-Martin d'Hères, France

## Abstract

The influence of process parameters on the occurrence of hot tearing during laser butt welding of 6xxx aluminium alloys with the presence of a filler wire is investigated. Lowering the welding speed and the solidification rate leads to welds with fewer defects. The chemical composition of the weld pool is shown to have a strong influence on hot tearing as well as the fastening device. The number of hot tears drops drastically when high silicon content is attained in the weld pool and when a vacuum fastening device is used. Hot tears are then characterised. They are shown to be intergranular, coming from fracture of liquid films without plasticity of the surrounding grains. The hot tear exhibits a round shape around the equiaxed grains at the centre of the fusion zone and a planar part between the columnar grains of the weld bead. By using three dimensional X-Ray tomography, the exact shape of the hot tears has been visualised.

## Introduction

The use of laser welding as a joining method is in rapid development in the industry, particularly in the aerospace sector, owing to its potential for weight savings and gains in productivity as compared to riveting. However, further development of this joining technique requires both an increase in the weld performance and service life (high joint coefficient, high strength, low distortions and residual stresses, good fatigue and stress corrosion properties, ...) [1], and an increase in the robustness of the welding process, especially in terms of defect control [2]. In order to obtain a fusion zone free of defects such as porosity or hot tears, the welding parameters need to be optimized. Among the important process parameters one finds the laser power, the welding speed and quantity of filler wire, which have a major role in controlling the microstructure and mechanical properties of the fusion zone and the adjacent material, namely the Heat-Affected Zone.

Very few studies have been carried out, which concentrate on the microstructure of laser welds, contrary to more classical welding processes such as arc welding [3, 4, 5]. Usually these studies deal with high strength alloys for the aerospace industry such as the 2xxx or the 7xxx series [6]. Most of the published work on laser welding is dedicated to parameter optimization, especially to decrease porosity levels in the weld [7], but literature on hot tearing during laser welding is quite limited. The present paper is concerned firstly with the study of the influence of process parameters (welding speed, quantity of filler wire, fastening device, …) on the hot tearing phenomenon in laser welds of 6xxx Al alloys series and secondly with the observation of the hot tears. The detailed microstructure of such welds has been already described in a former paper [8]. This study will concentrate on the observation of hot tears firstly thanks to usual micrography in different planes of the laser weld. Characterization by Synchrotron X-Ray tomography has been conducted to obtain 3D information on the defect.

## Experimental Procedure

The base material used in this work is AA6056, the chemical composition of which is given in Table 1. The material in the shape of 1.6 mm thick sheets was butt welded in the T4 state (namely after a solution treatment, water quench and several months of natural ageing). Welding was carried out along the rolling direction, on 400 mm × 150 mm samples, after degreasing. Butt joints were produced with a Nd:YAG laser operating at 3 kW as shown in Fig. 1 [8]. The focal point of the laser beam was set at the surface of the sheets.

**Table 1**. Composition of 6056 and AS12 alloys (weight percent)

| Alloy | Mg | Si | Cu | Mn | Fe |
|---|---|---|---|---|---|
| 6056 | 0.86 | 0.92 | 0.87 | 0.55 | 0.19 |
| AS12 (4047) | 0.1 | 12 | 0.3 | 0.15 | 0.8 |

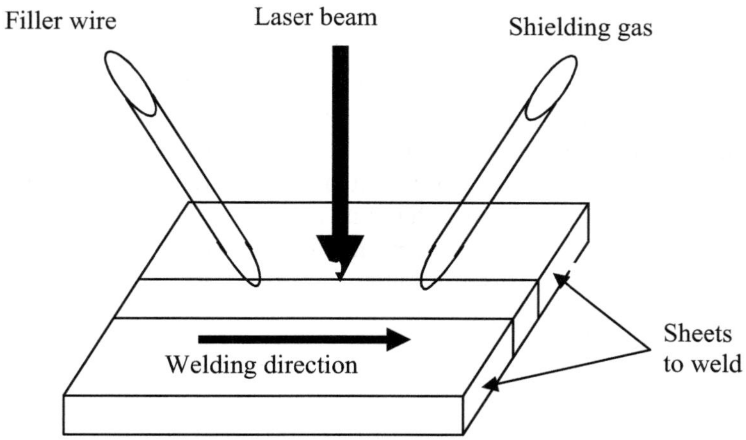


**Fig. 1.** Welding set up [8]

A 1 mm diameter filler wire of AA4047 (12%Si, see Table 1 for exact composition) was employed and helium was used as shielding gas at a flow rate of 20 $l.min^{-1}$. This particular filler wire has been chosen owing to its ability to prevent the formation of hot tearing [9]. For the different welding experiments, the welding speed was varied from 4 $m.min^{-1}$ to 6 $m.min^{-1}$ and the filler wire speed from 0 to 6 $m.min^{-1}$. The observation of hot tears has been carried out at a welding speed of 6 $m.min^{-1}$ and a filler wire speed of 1.5 $m.min^{-1}$. This choice of speed parameters ensures that numerous hot tears are obtained in the weld. Two different fastening systems have been investigated: a mechanical one using screws and a vacuum one using low pressure to maintain sheets on the welding table. In order to characterize the influence of the cooling rate, inserts have been used to change the thermal contact between the sheets and the steel welding table. Decreasing the cooling rate has been realized by inserting cork whereas brass has been used to increase it. One configuration with brass below and on top of the sheet has also been used (sandwich configuration). The number of hot tears has been characterized by using X-ray radiography. Optical observations have been carried out in the rolling plane. Samples for optical observations were polished and etched for 20 s with a Flick solution (10 ml HF, 20 ml $HNO_3$ and 95 ml water). Specimens for EBSD characterization were obtained after mechanical polishing by electropolishing with a solution of 330 ml nitric acid and

660 ml methanol. Samples for X-Ray tomography have been machined to obtain a stick of about 10 cm long, 2 mm wide and 2 mm thickness. It is constituted by the weld line and some base material on the sides. X-ray tomography has been carried out at the ESRF (European Synchrotron Radiation Facility, Grenoble) on the ID19 beam line. The specimen was imaged at 70 keV using 900 projections at different angular positions. The 3D volume was then reconstructed via a filtered back projection algorithm and further analyzed with the commercial software VGStudioMax (http://www.volumegraphics.com/products/vgstudiomax/). More details about this technique can be found in [10].

## Results

### Influence of Welding Parameters on the Occurrence of Hot Tearing

This first part is dedicated to the study of the number of hot tears on a given length of weld (40 cm) as a function of the process parameters. Figure 2 shows the influence of the welding speed on the occurrence of hot tearing for different filler wire speeds. The number of hot tears increases when the welding speed increases whatever the filler wire speed. These results are in agreement with those obtained by Cicala [11] on the same alloy but for different welding and filler wire speeds. This increase of the sensitivity to hot tearing can be due either to a higher instability of the weld pool (leading to the closure of the keyhole) or/and to the higher solidification rate when the welding speed increases.

The influence of the solidification rate on hot tearing has been widely studied over the past. It has been shown in particular that the presence of a second beam to decrease the cooling rate decreases the number of hot tears [7]. Early in the 70's, some models, predicting the sensitivity to hot tearing, considered that decreasing the solidification rate lead to less defects [12, 13]. To study the influence of the cooling rate different inserts have been used between the sheet to weld and the welding table (Fig. 3). These experiments confirm that the lower the cooling rate, the lower the number of hot tears. This could partly explain the higher number of hot tears observed in Fig. 2 when the welding speed was increased to 6 $m.min^{-1}$.

Figure 2 also shows that the quantity of filler wire has an influence on the occurrence of hot tearing: for higher filler wire speeds, the number of hot tears drops whatever the welding speed. The filler wire speed changes the chemical composition of the weld pool and thus of the weld nugget. In our case, the silicon concentration will increase as the quantity of filler wire increases. The silicon concentration has been calculated using optical observation and dilution consideration (for more details see [14]). Figure 4 shows the number of hot tears as a function of the silicon composition. The number of hot tears is high for low silicon content and then drops when the composition in silicon increases. This is in accordance with the literature [15]: for most of the alloying elements (Mg, Si, Cu), the hot tearing sensitivity during casting decreases with increasing alloying element concentration. This study shows that in the case of laser welding, the Si composition acts in the same way. Further experiments are required to verify and explain the minimum of hot tears observed for 2% of Si.

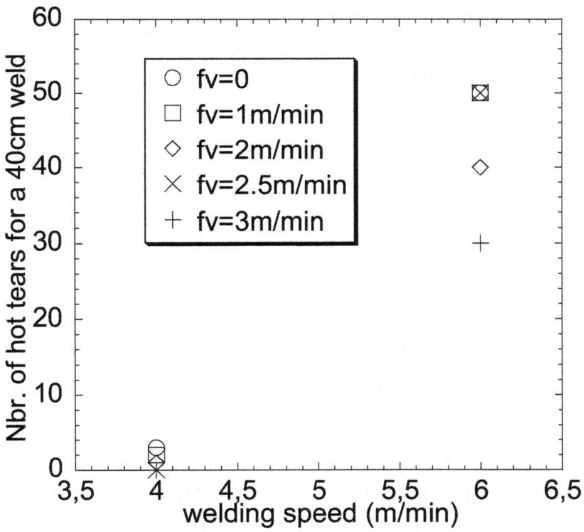


**Fig. 2.** Influence of the welding speed on the number of hot tears in a 40 cm long weld for various filler wire speeds (fv=filler velocity)

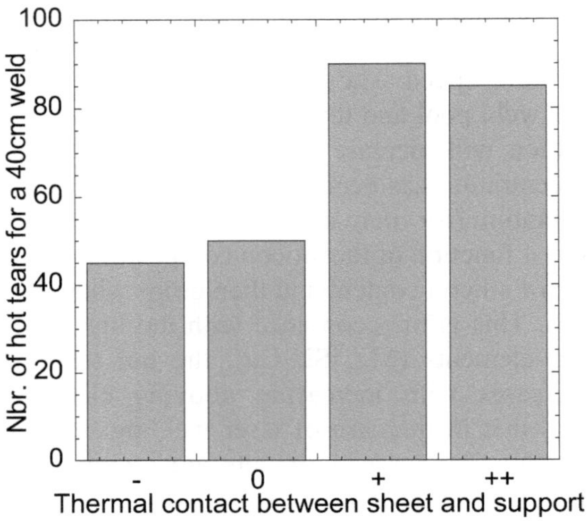


**Fig. 3.** Influence of the thermal contact between the sheets to weld and the welding table on the number of hot tears. The welding speed is 6 m/min and the filler wire speed is 1.5 m/min. The sign – corresponds to cork insert whereas the signs + and ++ correspond to an insert of a single sheet of brass and the sandwich configuration respectively

At last, the sensitivity to hot tearing could be influenced by the fastening system. Figure 5 presents the results obtained with the two types of fastening devices used in this study. The welding speed has been chosen in order to observe the maximum of hot tears with the classical fastening device (mechanical one). Figure 5 underlines the large influence of the fastening system on the occurrence of hot tearing during laser welding. FE calculations would be necessary to understand the difference on stresses/strains involved by the different fastening systems but these results definitely show that the vacuum fastening system must be preferably employed to obtain defect free welds. The fastening system can then be considered as a parameter of prime importance.

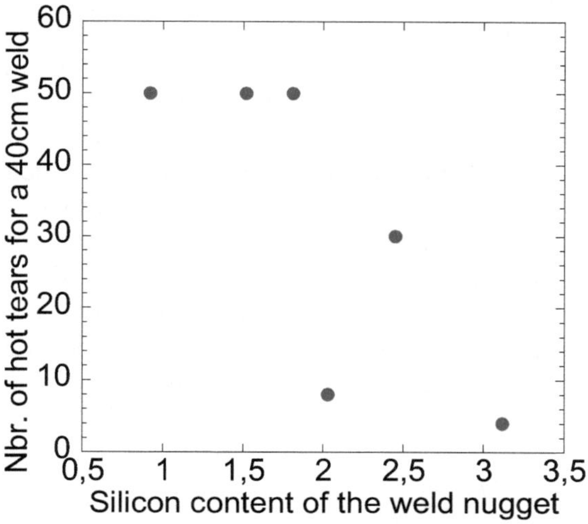


**Fig. 4.** Influence of the silicon content on the number of hot tears in a 40 cm long weld for a welding speed of 6 m/min

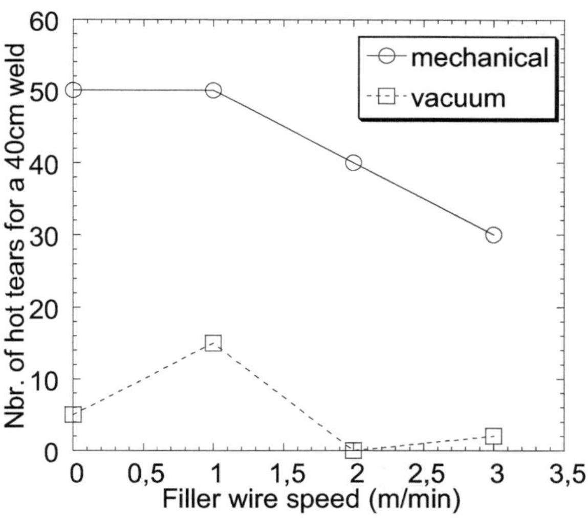


**Fig. 5.** Influence of the fastening device on the number of hot tears for various filler wire speeds. Welding speed is equal to 6 m/min

## Characterisation of Hot Tearing Phenomenon

Figure 6 shows the EBSD cartography of the fusion zone perpendicular to the welding direction. A hot tear is observed in black. On this figure the grain colour is randomly chosen and does not correspond to a preferential crystallographic direction. Indeed, no determined texture has been observed in the fusion zone. However, the EBSD picture confirms that the phenomenon of hot tearing is intergranular as already mentioned in the literature [16].

Figure 7 shows the fracture surface characteristic of the phenomenon of hot tearing in laser welding. The surface exhibits the dendrites which originate from the solidification process. Moreover, the dendritic surface is smooth indicating no plasticity. This leads to the conclusion that hot tearing is the result of the fracture of liquid films as already observed for hot tearing during casting [9]. Solidification stresses/strains pull apart the dendrites until a cavitation pressure is attained in the liquid and the fracture takes place [17]. Note that the secondary arm spacing of the dendrites observed in Fig. 7 is about 5 μm. Using the equation [11]:

$$t_f = M\left(\frac{\lambda_2}{5.5}\right)^3$$

with $\lambda_2$ the secondary arm spacing, and M a coefficient dependant of the material, allows calculating the solidification time $t_f$ and thus the cooling rate. With M equal to about $4.10^{-8}$ $m^3.s^{-1}$ as in [18] and considering the solidification interval of this alloy, the cooling rate is about 400 $K.s^{-1}$. This high cooling rate is characteristic of the laser welding process.

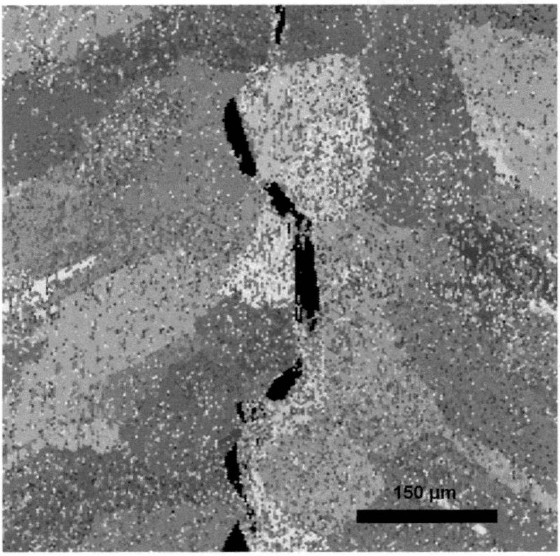


**Fig. 6.** EBSD map of the fusion zone perpendicular to the welding direction. Black zones correspond to the hot tear

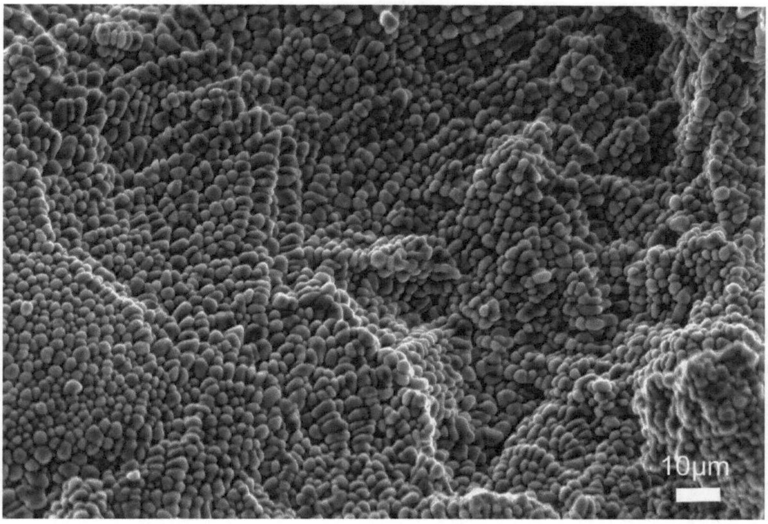


**Fig. 7.** SEM observation of the hot tearing surface

Hot tearing takes place at the very end of the solidification process [17], usually for solid fractions higher than 0.9. During laser welding, solidification occurs first at the fusion line because the main phenomenon responsible for cooling is conduction in the solid [19] (heat extraction is mainly assures by this means). Figure 8 shows the micrograph in the rolling plane of a laser weld containing hot tears. From this figure, a hot tear can be seen. The crack exhibits a quite planar shape when it goes between the columnar grains and a round shape around the equiaxed grains of the centre of the fusion zone. The hot tear results from the strains applied on the mushy zone. These stresses come from firstly the solidification shrinkage which is quite high in aluminium alloys (except for the 4xxx series) and secondly from thermal contraction of the solid skeleton. Moreover external loads (for example coming from the fastening) can apply on the mush zone.

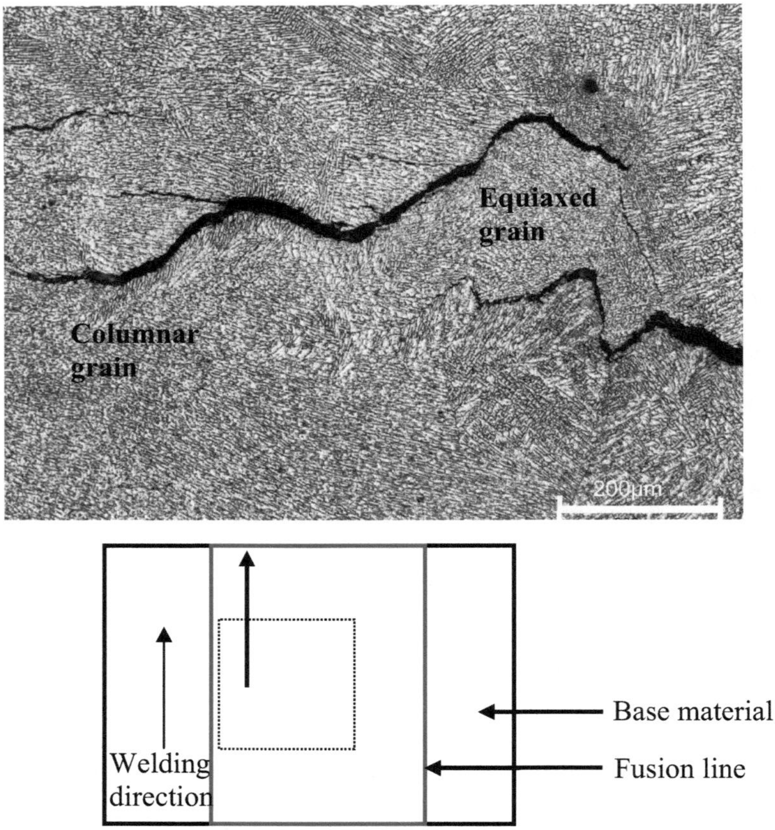


**Fig. 8.** Optical micrograph of the fusion zone seen from above

Observation of the crack by X-ray tomography gives a more general picture of the exact shape of the hot tear and of its way of propagation. Figure 9 shows a 2D image where the hot tear is clearly visible. The extraction of the 3D shape of the crack can be done by a grey level segmentation process. The result of this operation is shown in Fig. 10. The surface of the weld bead is represented here in grey. As it can be seen two hot tears (highlighted by two different colours) can be observed in this weld. The 3D tomography permits firstly to conclude that the two parts of the red tear are actually only one defect. This could not have been seen on a 2D picture. Moreover, Fig. 10 shows perfectly the linear part of the tears following the basaltic grains of the fusion zone: the surface is quite planar. Figure 11 exhibits the same hot tears seen from above. The outer surface of the fusion line is still outlined in grey. This view permits one to observe more accurately the centre zone of the hot tears. The deviation around the equiaxed grains of the centre, as guessed from the 2D micrograph (Fig. 8), is clearly visible here. The planar surface of the borders becomes more round as the crack goes around the grains.

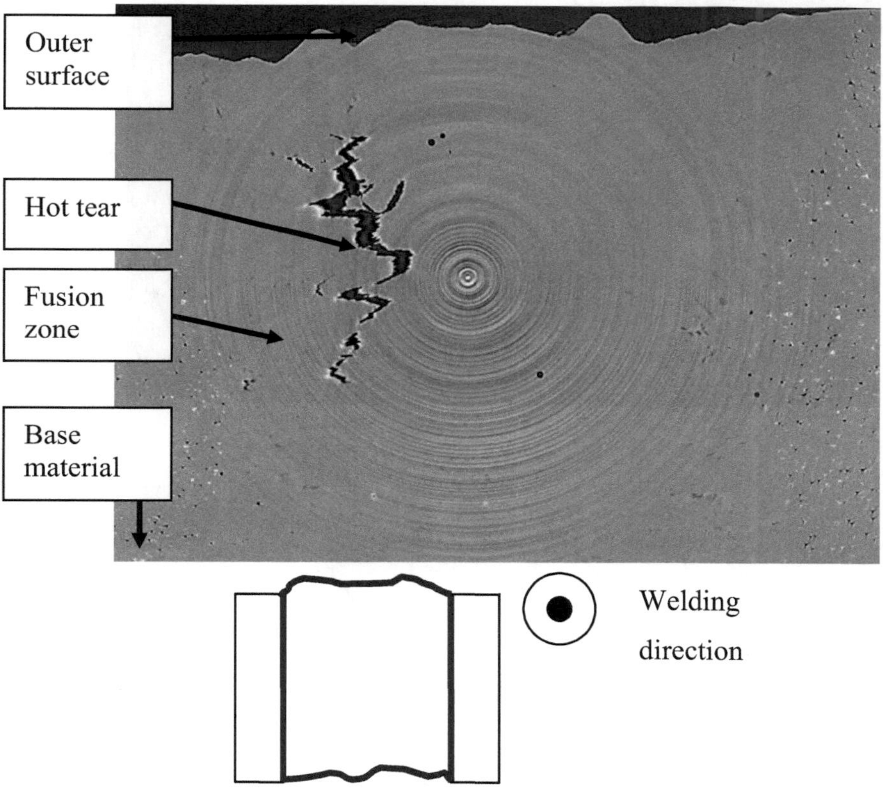


**Fig. 9.** X-ray picture showing a hot tear. The rings are artefact

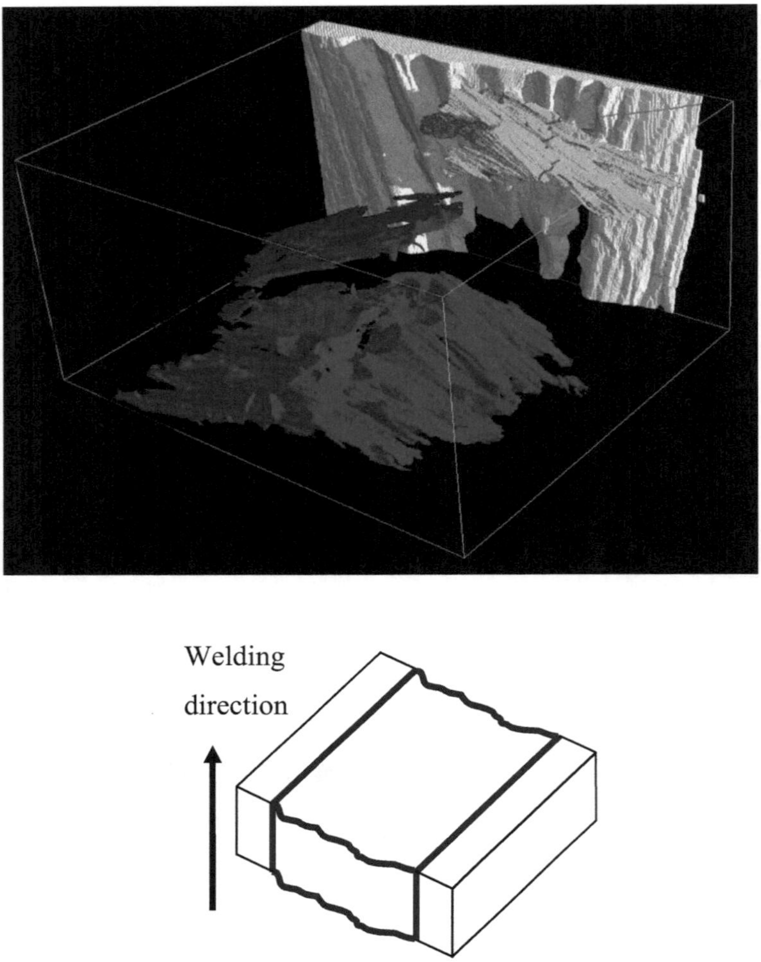


**Fig. 10.** 3D X-ray tomography of the fusion zone containing two hot tears. The grey surface represents the outer surface of the fusion line

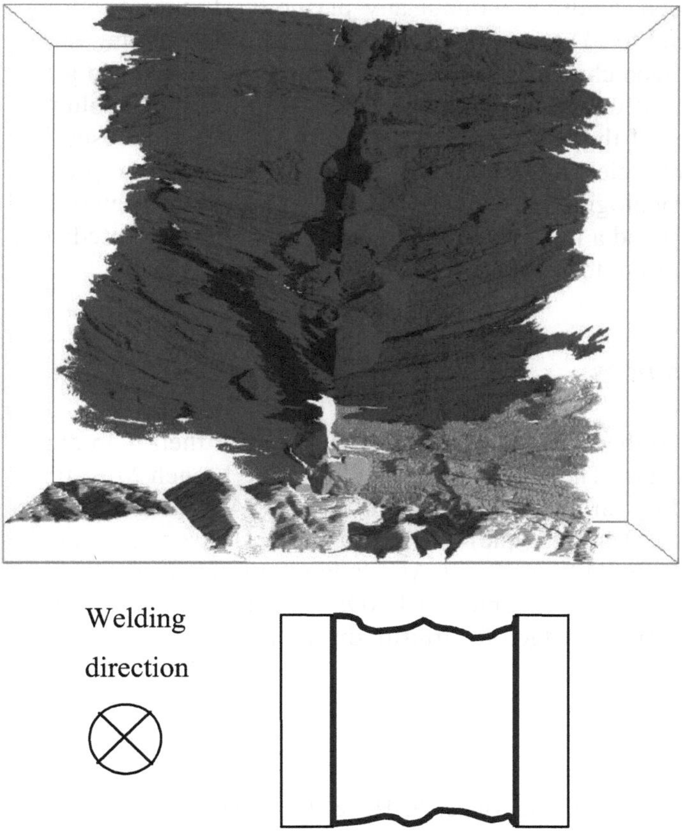


**Fig. 11.** 3D X-ray tomography of the fusion zone seen from upward containing two hot tears. The grey surface represents the outer surface of the weld bead

## Conclusions

This paper firstly allows a better understanding of the key process parameters controlling the quality of laser welds in term of hot tearing. Welds with fewer defects can be achieved by decreasing the welding speed and the solidification rate and by increasing the silicon content in the weld pool. The vacuum fastening device improves significantly the quality of the welds compared to the classical mechanical fastening device.

Secondly this study permits a better understanding of the hot tearing phenomenon. Optical microscopy, SEM and EBSD clearly show the intergranular character of this defect. Hot tearing occurs by fracture of the liquid films between dendrites without any plasticity. Thanks to various 2D micrographs and to 3D tomography, propagation of hot tearing during laser welding has been characterised: it partly goes around equiaxed grains located in the centre of the fusion line and partly follow the columnar grains on the sides of the weld bead. All these results could be used in a more general finite element simulation of the laser welding process. Coupled to the rheological behaviour of the mushy zone (determined elsewhere [20, 21]) and a hot tearing criterion, it would permit a prediction of the occurrence of hot tears.

## Acknowledgements

The authors want to thank L. Salvo and J.-J. Blandin for their help during the ESRF experiments. The authors are grateful to the French Ministry of Industry for funding in the framework of the ASA (Allègement des Structures dans l'Aéronautique) RNMP project, carried out in collaboration with EADS and ALCAN, which are also thanked for providing the materials. L. Kirschner of EADS-CCR and J.C. Ehrström of ALCAN CRV are acknowledged for fruitful discussions.

## References

1. Shinzel C, Hohenbeger B, Dausinger F, Hugel H (2000). Laser welding of aluminium – extended processing potential by different wire positions. Proceedings of SPIE, Eds. C. Xiangli, F. Tomoo, A. Matsunawa, Osaka, Vol. 3888, 380.
2. Seto N, Katayama S, Matsunawa A (2000). High-speed simultaneous observation of plasma and keyhole behavior during high power $CO_2$ laser welding: effect of shielding gas on porosity formation. Journal of Laser Applications 12(6): 245–250.
3. Kuk JM, Jang KC, Lee DG, Kim IS (2004). Effects of temperature and shielding gas mixture on fatigue life of 5083 aluminum alloy. Journal of Materials Processing Technology 155–156: 1408–1414.
4. Menzemer C, Lam PC, Srivatsan TS, Wittel CF (1999). An investigation of fusion zone microstructures of welded aluminium alloy joints. Materials Letters 41: 192–197.
5. Hou KH, Baeslack WA (1996). Characterization of the heat-affected zone in gas tungsten arc welded aluminium alloy 2195-T8. Journal of Materials Science Letter 15: 239.
6. Liu C, Northwood DO, Bhole SD (2004). Tensile fracture behaviour in CO2 laser beam welds of 7075-T6 aluminium alloy. Materials and Design 25: 573–577.

7. Haboudou A, Peyre P, Vannes AB, Peix G (2003). Reduction of porosity content generated during Nd:YAG laser welding of A356 and AA5083 aluminium alloys. Materials Science and Engineering A 363: 40–52.
8. Fabrègue D, Deschamps A (2002). Microstructural study of laser welds Al6056-AS12 in relation with hot tearing. Materials Science Forum, Proceedings of the 8th Int. Conf. on Aluminium Alloys, Eds. P.J. Gregson, S. Harris, Cambridge 396–402, 1567–1572.
9. Instone S, StJohn D, Grandfield J (2000). New apparatus for characterising tensile strength development and hot cracking in the mushy zone. International Journal of Cast Metals Research 12: 441–456.
10. Proudhon H, Buffière JY, Fouvry S (2007). Three-dimensional study of a fretting crack using synchrotron X-ray micro-tomography. Engineering Fracture Mechanics 74(5): 782–793.
11. Cicala E (2203). Etude de la fissuration à chaud des alliages d'aluminium 6056T4 soudés par lasers Nd :YAG continu. Post doctoral research report, LTM, Université de Bourgogne.
12. Clyne TW, Davies GJ (1979). Comparison between experimental data and theoretical predictions relating to dependence of solidification cracking on composition. In: Solidification and Castings of Metals, The Metals Soc., 275–278.
13. Feurer U (1976). Mathematisches Modell der Warmrißneigung von binären Aluminiumlegierungen. Giesserei-Forschung 28: 75–80.
14. Fabrègue D (2004). Microstructure et fissuration à chaud lors du soudage Laser d'alliages d'aluminium 6056. PhD thesis, INPG.
15. Zao H, Debroy T (2001). Weld metal composition change during conduction mode laser welding of aluminium alloy 5182. Metallurgical and Materials Transactions B 32. 163–172.
16. Spittle JA, Brown SGR, James JD, Evans RW (1996). Mechanical properties of partially molten aluminium alloys. Proceedings of the 9th Symposium on Physical Simulation of Casting, Hot Rolling and Welding, ed. Suzaki, Sakai and Matsuda, 81–91.
17. Rappaz M, Drezet JM, Gremaud M (1999). A new hot tearing criterion. Metallurgical and Materials Transactions A 30(2): 449–456.
18. Kurz W, Fisher DJ (1989). Fundamentals of Solidification. Third Edition, Trans Tech Publications, Ltd., Aedermannsdorf.
19. El-Ahmar W, Julien JF, Gilles P, Taheri S, Boitout F (2006). La robustesse de la simulation numérique du soudage TIG, application sur des structures en acier 316L. Proceeding of the Matériaux 2006 conference.
20. Fabrègue D, Deschamps A, Suéry M, Drezet JM (2006), Non-isothermal tensile tests during solidification of Al-Mg-Si-Cu alloys: Mechanical properties in relation to the phenomenon of hot tearing. Acta Materialia 54(19): 5209–5220.
21. Fabrègue D, Deschamps A, Suéry M, Poole WJ (2006). Rheological behavior of Al-Mg-Si-Cu alloys in the mushy state obtained by partial remelting and partial solidification at high cooling rate. Metallurgical and Materials Transactions A 37(5): 1459–1467.

# The Integral Approach – a Tailored Method to Optimize Structural Behavior and Weldability

H. Gruss[1], A. Pshennikov[2], H. Herold[3]

[1] MBDA/LFK, Unterschleißheim, Germany
[2] Otto-von-Guericke University, Institute for Machine Design, Magdeburg, Germany
[3] Otto-von-Guericke University, Institute for Materials and Joining Technology, Magdeburg, Germany

## Abstract

The structural design at current welded skin-stringer-joints is geared to a riveted structure. Consequently, the geometrical discontinuities between skin and stringer are responsible for the stress concentration in the weld seam at stringer ends.

Many projects related to welded fuselage structures investigate problems concerning either the weldability or the structural behaviour under flight phase conditions, but not the structural behaviour in terms of the flight phase and the welding process in its complexity.

Hence, it is important to focus on the development of a complex approach to improve the weldability and to work over the designing for welding of the current welded structure. The Integral Approach enables to combine the structural behavior under welding design criterions with an optimization of weldability. The stringer geometry is modified by the implementation of slots in the stringer web.

From the technical welding experiences points to the tendency that the weld seam cracks may be reduced by the so called clamping effect. A tailored welding strategy in combination with slots leads to desired residual compressive stresses in several particular stringer parts. These stresses prevent the volume expansion due to the heat input and the hot crack resistance grows up.

On the other hand during the flight phase the stress concentrations at the bearing weld seam ends decrease due to the implementation of slots. The influences of the same geometrical parameters and welding directions on the structural behavior are shown for the flight phase as well as the weldability.

## Introduction

Ten years ago laser beam welding was establish in the commercial aircraft industry (Fig. 1). At the present time skin and stiffener in form of stringers are welded in the lower shells of some sections at Airbus types A318, A340 and A380 (Fig. 2). Against the riveted structures the welded ones have a completely different fracture mechanical behavior. At riveted joints the stringer is able to take over loads when at the same time the skin collapses. By contrast skin and stringer fail simultaneously at a welded structure. The different behavior bases on the differential and integral design.

New aircraft aluminum alloys are designed after criteria like weight, residual strengths, damage tolerance or corrosion but not under considerations concerning weldability. New materials like AA2098 or 7xxx-alloys contain elements like Li or Zn. Both are characterized by a high vapor pressure. This circumstance provokes instabilities of capillary during the welding. Experiments have shown that these base alloys have a lower solidification cracking resistance compared to the applied alloys in fuselage like AA6013 or AA6056 [1].

Solidification cracks result from transverse shrinkage. They extend some millimeters and could be visible on the weld seam surface (Fig. 3) or limit on the inner weld seam. Hilbinger argues that the reason for the appearance of hot cracks lies in the magnitude of its dimension [2]. That is that microscopic cracks based on process instabilities and macroscopic ones have a global cause like process design and position of weld seam. Solidification cracks belong to the second group.

The Integral Approach enables the combination of the structural behavior under welding design criterions with an optimization of weldability. The objective of this approach is to improve the performance of the welded structure and the transition to the upper shell. There are higher tensile stresses in different to the lower one. A structural compressive stress is somehow preferred for reduced susceptibility to hot cracking because compressive stresses work against the thermal expansion due to the local heat input and heat conduction in the structure.

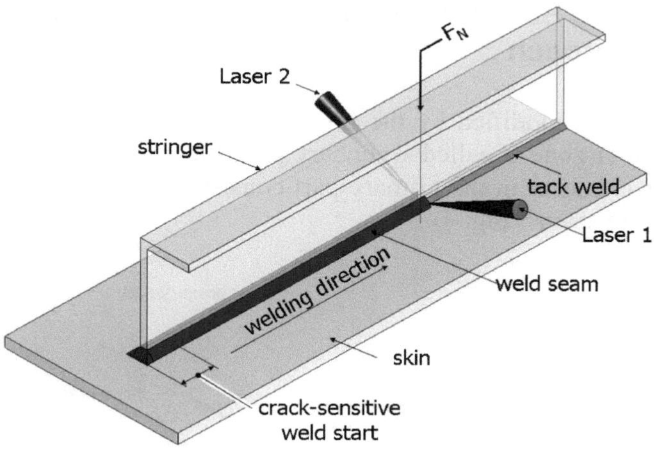


**Fig. 1.** Joining of stringer and skin by both sides simultaneous laser beam welding

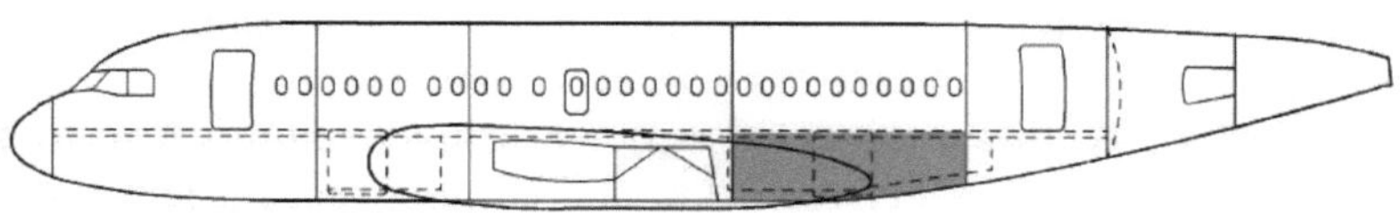

**Fig. 2.** skin and stringer are welded in the lower shell of the section no. 16 by A318 (*gray*-coloured)

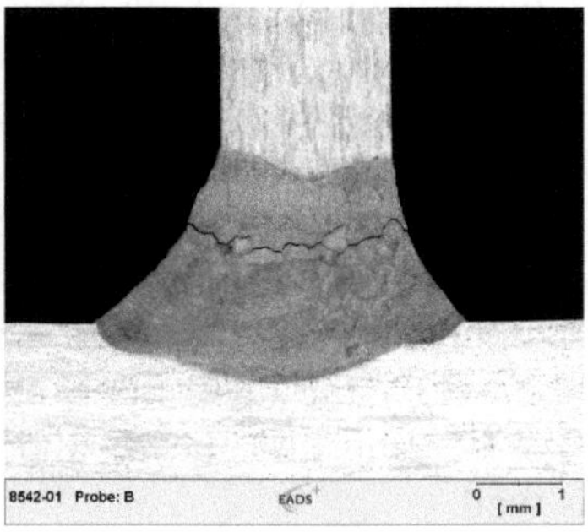


**Fig. 3.** Position of a solidification crack at a welded T-joint configuration. Base metal: AA6056, filler: AlSi7

## Interaction Between Structural Behavior and Weldability using Integral Approach

The stringer geometry is modified by the implementation of slots in the stringer web. The outer zone is called secondary part; the inner one is the primary part. The weld seam in the primary part consists of the inner and the outer primary weld seams (Fig. 4).

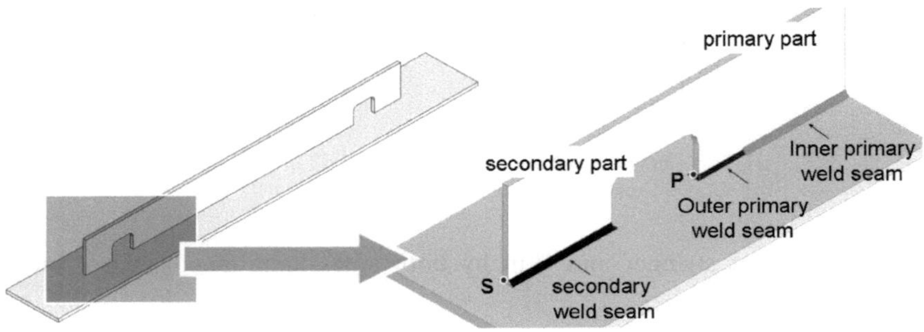


**Fig. 4.** Modification of the stringer with the Integral Approach by designed two slots, each at the stringer end. It is only shown one end to simplify matters

The secondary parts, that are located at outer stringer regions lead to a decrease of the discontinuous between skin and stringer (eccentricity) and therefore to a higher stiffness at the ends of the outer primary parts. In this way the bending stresses that result by the loads during the flight phase are reduced locally. The notch stresses that depend on the bending stresses decrease as well.

From the technical welding point of view the tendency that the weld seam cracks is reduced by the so called clamping effect. After welding of the inner primary and the secondary parts both areas which surround the weld seams shrink. The still not-welded outer primary part is pulled down. In these areas residual compressive stresses are accumulated. With the welding of the outer primary part the compressive stresses are induced as well, but both compressive stresses have an opposite direction to each other. Therefore the material expansion is impeded by the heat input, resulting in a lower volume shrinkage due to the cooling and the resulted tensile stresses on the mushy zone are lower as well.

This work is focussed on the structural behaviour during the welding process, less the material behaviour. The next sections prove the influence of slot on the structural behavior as well as on the hot cracking behavior.

## Slot Influence on Structural Behavior

The fuselage consists of several sections which each other are joined with stringer couplings. Each coupling is composed of one short stringer and a circumferential butt strap. At all Airbus types couplings are located below a frame (Fig. 5). The cabin pressure and the self-weight of the fuselage lead to tensile stresses in the structure.

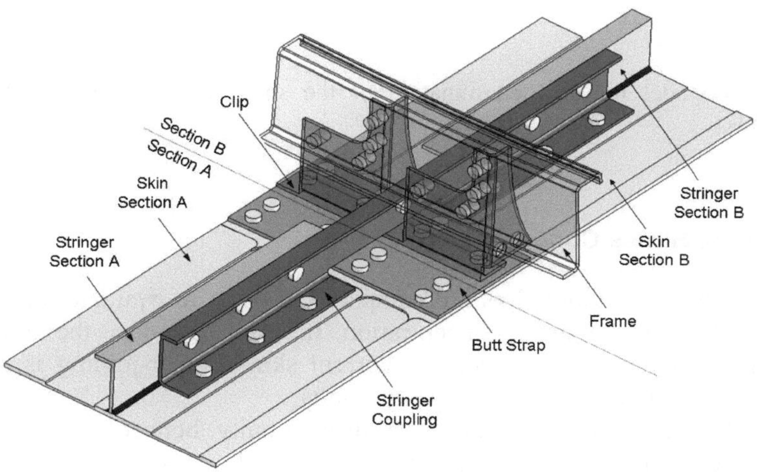


**Fig. 5.** The stringer coupling for the connection of several fuselage sections

### Analysis of Stress Concentration

An optimized current skin-stringer-configuration (without slots) is the determination base for the structural analysis. The stress concentration factor (notch factor) at the stringer end results by the ratio between the absolute stress concentration and the average longitudinal tensile stress in the skin (Fig. 6). The stress concentration at the stringer ends is a combined effect of skin bulging at stiffness discontinues and eccentricity by the middle surfaces of skin and stringer.

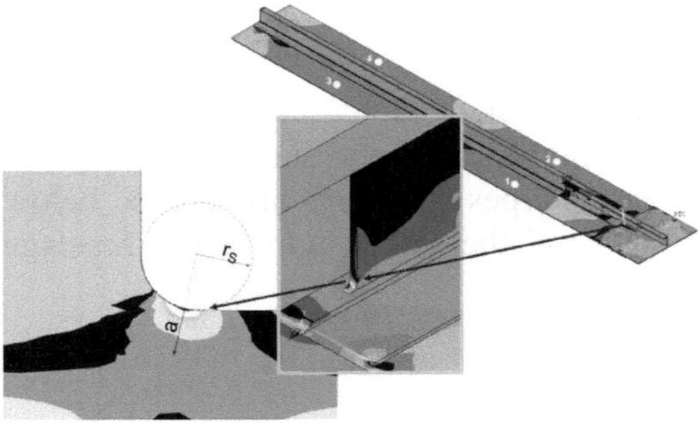

Fig. 6. FEA-Model for the determination of the stress concentrations at the stringer ends

## Evaluation of Stress Concentration

Stringers with and without slots are compared in order to evaluate the influence of slots on the stress concentration. In order to reduce the stress concentration at the stringer ends the current skin-stringer-joining is revised by a socket below the stringer (Fig. 7 left).

The socket leads to a reduction of discontinuity between skin and stringer. The reference parameters are $d_{STR}$ = 15, $l_{FH}$=1.4, $l_{FW}$=6, $l_{GD}$=0, $l_{GW}$=0. These ones also apply for the structure with slot. Figure 7 right shows the structural model for a stringer with slot. The slot relevant reference parameters in mm are $d_{Str}$ =15, $l_{SLOT}$ =15, $h_{SLOT}$ =15, $l_S$ =30.

The characteristic lines in Fig. 8 show that the maximum stress concentrations in the primary weld seam at stringers with slots are 20% lower than at stringers without slots. But the stresses in the outer regions of the secondary weld seams are increased about 11%. This fact is based on the about $l_s$ longer lever of the stringer coupling.

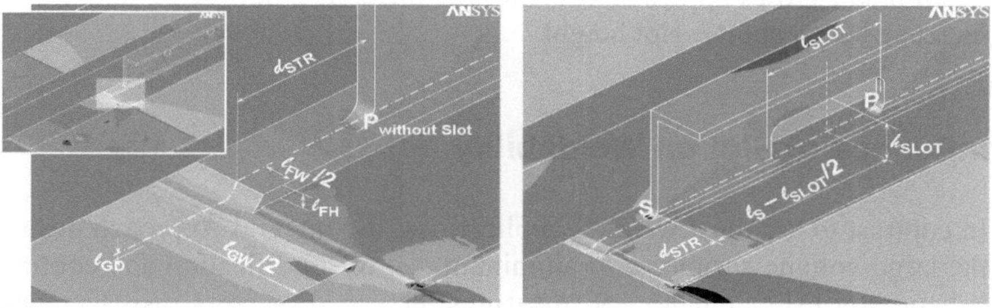


**Fig. 7.** Structural models for the calculation of stress concentration. (*left*) for the revised current skin-stringer-design without slot. (*right*) for a stringer with slot

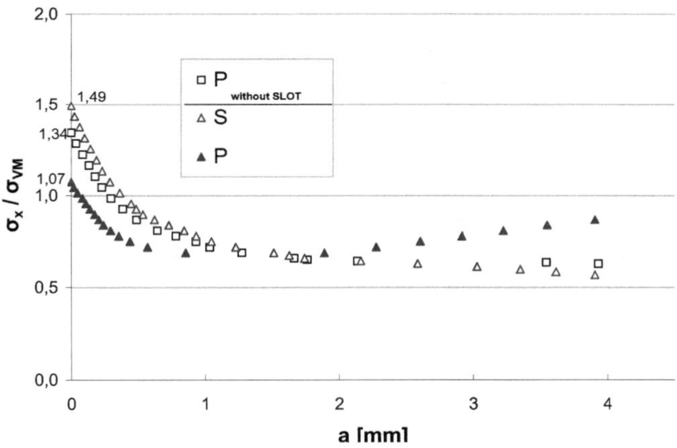


**Fig. 8.** The comparison of the stress concentrations for stringers with and without slot at the start-end-positions in primary part (P) and secondary part (S) (Fig. 4). The geometrical parameter for the stress concentration at the stringer-edge a is shown in Fig. 6

The edge distance ($d_{STR}$) has not any significant influence on the maximum stresses in the bearing primary weld seam. But stresses in the secondary weld seam (S) grow up about 65% at an edge distance $d_{STR}$ (see Fig. 7) by 15 mm compared to 5 mm. The longer lever is the reason for the increasing of stresses again.

The secondary part length ($l_S$) and the slot length ($l_{SLOT}$) have not any influence on the stress concentration at the primary (p) and secondary (S) weld seam. A modifying of the slot height ($h_{SLOT}$) has a significant influence on the stress concentration in P and S. In case of reduction from 15 to 10 mm the stress notch factor in P decreases roughly 11% but the stress notch factor grows up 6% in S. Hence the difference between P and S does increase with a shorter slot height.

## Influence of Slot on Weldability

In contrast to the most welding applications, the Integral Approach desires the large volume shrinkage of aluminum due to solidification and cooling. Pshennikov [3] calculates and measures the speed of cross displacements $v_q$ as well as the changing of cross displacements during the time of existing Brittleness temperature range (BTR) $\Delta u_{YBTR}$ as a criterion to assess the risk of hot cracking at the weld seam ends. He developed effective methods to avoid hot cracks at unsteady stages of seam ends by design and assembly for welding as well as considering the welding sequence for assemble the run-out-plates. An important effect to avoid hot cracking at the seam end is the induction of compressive stresses by joining of the two run-out-plate-halves. In shipbuilding is used the one-sided submerged arc welding of large steel panels.

Further the run-out plates are provided with slots and are able to shrink in consideration of an adapted welding and tacking strategy as well as a sufficient cooling-off time before the real welding process starts. Both steel plates try to move diametrically opposed transverse to the welding line due to the induction of thermal stresses. But the shrinkage movement by the run-out plates causes a diametrical orientation relating to the divergent movement of steel plates. The material is cooling-off and the displacements within the BTR are reduced. The BTR is fixed by the alloy composition, but the displacements during the time of existing BTR are determined by the structure.

The fuselage design does not allow using any run-in or run-out plates. Following the shrinkage effect must appear on the outer stringer regions. Experimental investigations and finite element analysis have shown a lower hot crack resistance than at the ends of current stringers.

### The Active Principle

The skin is fixed on a vacuum clamping board. Against this background the focus lies on the residual stress in the stringer as well as the strains relating to hot cracking. The terms primary and secondary part refer to the stringer. By the use of a suitable welding regime it is welded the inner primary part (1) first and afterwards the secondary one (2). These welded sections shrink and pull down the still not-welded outer primary part (3) to skin. A higher hot cracking resistance in P is achieved (Fig. 9).

The residual tensile stresses (1 and 2) obtain its maximum within the weld seams and decrease with increasing stringer height. Thereby it is important that both residual tensile stress areas are joined by a so called "Bridge". This bridge ensures that after the first two welding steps the outer primary zone is pulled down to the stringer whereby there are occured residual compressive stresses in this one (Fig. 10).

At the third welding step the base metal is heated due to the thermal stress and it tries to expand. The residual compressive stresses around P and the surrounding cold areas constrain the volume expansion. Depending on the residual compressive stresses in the outer primary part it results lower tensile stresses within the BTR. At this principle it is spoken about a clamping effect, because the third welding section is enclosed by two tensile stress areas via a bridge.

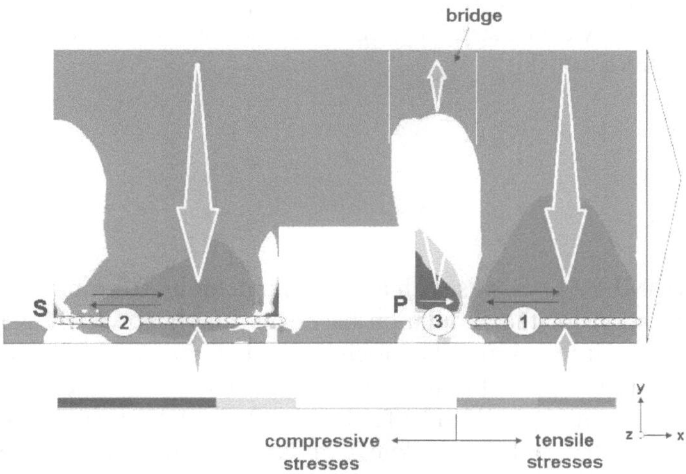


**Fig. 9.** Active principle bases on tensile and compressive stresses

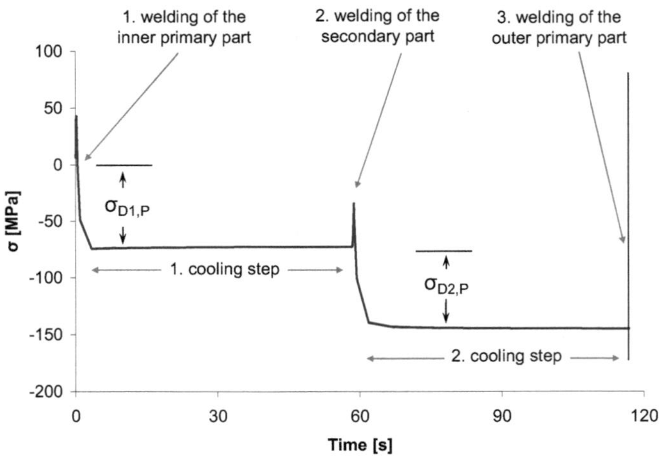


**Fig. 10.** The resulted stresses (y-direction) at the start-end-position of the primary part (P) after the first two welding steps. The characteristic compressive stresses lines refer to the reference model

## Determinations

There are used two Nd:YAG-Lasers with a focus diameter by 0.3 mm. The heat source's calibration for the simulation bases on a laser power by 2.3 kW, on a welding velocity by 4 m/s. The focus point is $f_L$=0. Shielding gas is He with a flow rate by 30 l/min (Fig. 11).The used base metal is AA6056 and the filler metal is AlSi7 (Table 1), 82% dilution rate. The stringer thickness is 2 mm and the thickness of skin is 3 mm. Before the welding process the stringer is tacked on the skin.

**Table 1.** Chemical composition of AA6056 and AlSi7 [wt%]

| Alloy | Al | Si | Fe | Cu | Mn | Mg | Cr | Zn |
|---|---|---|---|---|---|---|---|---|
| AA6056 | 96.98 | 0.89 | 0.09 | 0.60 | 0.59 | 0.71 | 0.012 | 0.13 |
| AlSi7 | 92 | 6.8 | 0.6 | – | – | 0.6 | – | – |

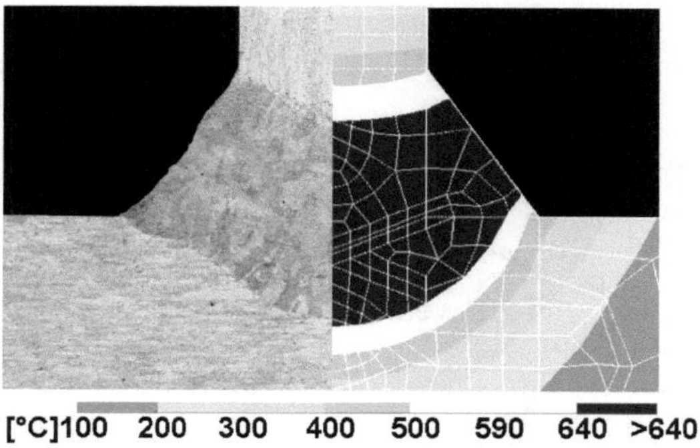


**Fig. 11.** Comparison of the cross section and the simulated geometry of welding seam

## Influencing Parameters

In Fig. 12 are shown several influencing parameters on the hot cracking tendency, which are described either by geometrical parameters or by a vector. Five geometrical parameters arising from the stringer configuration, the tack welding regime as well as the order of welding steps ( $l_{SLOT}$, $h_{SLOT}$, $l_S$, d und e). The slot width and height are defined by the parameter $l_{SLOT}$ und $h_{SLOT}$. The parameter $l_S$ determine the secondary part length which is equivalent to the primary weld seam length.

The distance from the inner slot edge to the beginning of the tacking weld seam is defined by the parameter d. The parameter e arises from the distance from to inner slot edge and the start/end point of the inner primary welding seam. In order to the welding direction result six degrees of freedom for all three weld seams. The welding directions are determined by the variables A1 / A2 for the inner primary weld seam, B1 / B2 for the secondary weld seam and C1 / C2 for the outer primary weld seam.

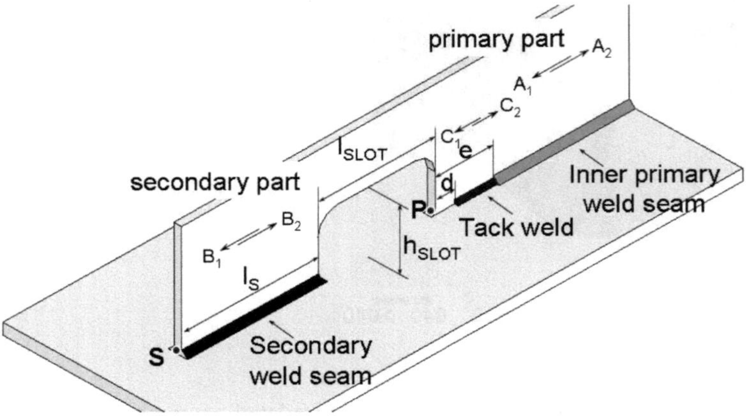


**Fig. 12.** Influencing parameters controlling the hot cracking behaviour at the position P (start-end-position in the primary part)

## Evaluation of Displacements within the BTR

### *The Reference Model*

The slot width ($l_{SLOT}$) and height ($h_{SLOT}$) are fixed on the minimal allowable dimension by 7 mm. The secondary part length is defined by 25 mm. The parameter e is determined by 10 mm (Table 2). The displacements at start-end-position of the primary part (P) are the reference value (R) for the following investigations (Fig. 13).

**Table 2.** Parameters for the reference model

| $l_{SLOT}$ [mm] | $h_{SLOT}$ [mm] | $l_S$ [mm] | d [mm] | e [mm] | welding directions | | |
|---|---|---|---|---|---|---|---|
| 7 | 7 | 25 | 0 | 10 | A2 | B2 | C2 |

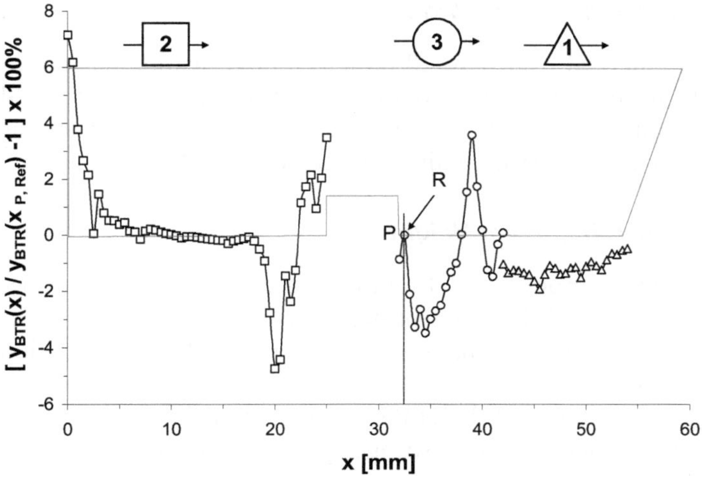


**Fig. 13.** Characteristic BTR-displacement lines calculated for the reference model. In order to compare the displacements with each other the graphical presentation of the BTR-displacements happens in relation to the displacements on position p at the reference model (100%)

Basically the BTR-displacements of the inner primary weld seam have a lower magnitude in relation to the displacements both of the secondary and of the outer primary weld seam. This circumstance is related to the higher structure stiffness. At the beginning of the first weld step (inner primary weld seam) the location where the beam is coupled into the base metal is surrounded by tack welds. In this way the stringer is fixed previous and behind the laser beam.

The outer edge of the secondary weld seam (2) represents a current stringer edge basically. The largest hot crack tendency is observed on the outer edge. The stringer is only fixed from one side. As soon as the stringer start is melted due to the heat input, the structure resistance is to low in order to counteract the resulted compressive stresses effectively. According to this the strains achieve their maximum values.

At the outer primary part position (P) the strains are lower due to the “bridge” in contrast to the secondary one. The bridge connects the primary and secondary parts and lead to the larger stiffness. Furthermore both tensile stress areas on the left and right of the slot cause a compressive stress area. This facts counteract to the compressive stresses that based on the heat input.

### Geometric Parameter: $h_{SLOT}$, Slot Width

The slot length is enlarged from 7 to 20 mm. The modification has no influence on the displacements of the inner primary and the secondary weld seams. At the beginning of the third weld step (P) the maximum strains grow up about 50% in comparison with the reference model.

### Geometric Parameter: $h_{SLOT}$, Slot Height

The enlarged slot height from 7 to 20 mm has an effect on the strain behavior at the secondary and the outer primary weld seams. The rising of strains about 90% in the outer primary weld seam results by the reduction of the web width above the slot (Fig. 14). That leads to a decrease of the mechanical section modulus und therefore to a reduction of residual compressive stresses in the outer primary stringer part, which counter the heat expansion (Fig. 15).

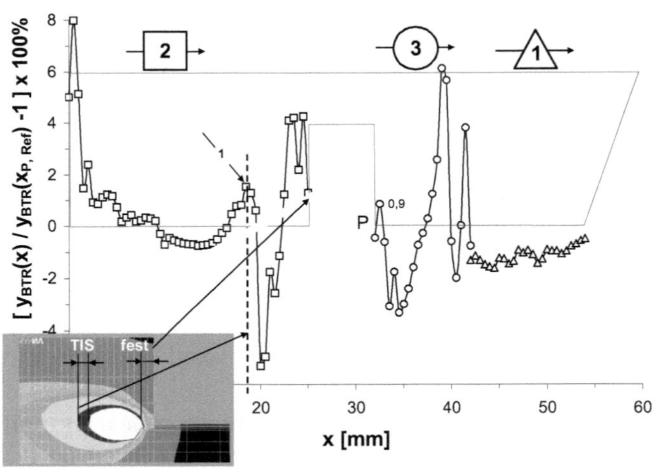


**Fig. 14.** Effect of slot height on displacements at P. The enlarged slot height from 7 to 20 mm causes higher displacements about 90%

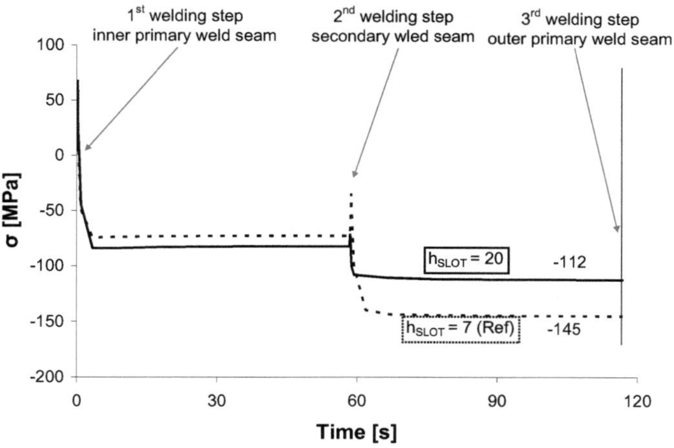


**Fig. 15.** Effect of slot height on residual stresses at the start-end-position of the primary part (P). The enlarged slot height leads to a lower stringer stiffness and therewith to a lower level of residual compressive stresses in P.

### *Geometrical Parameter: $l_s$, Length of Secondary Part*

An increase of the secondary part ($l_s$) from 25 to 40 mm leads to a not significant rising of the residual compressive stresses in the outer primary part. Not the secondary part length is the crucial factor but rather the increase of the residual tensile stresses which are caused by the volume shrinkage of the secondary part is important for the hot crack behavior at the beginning of the third weld step (Fig. 16).

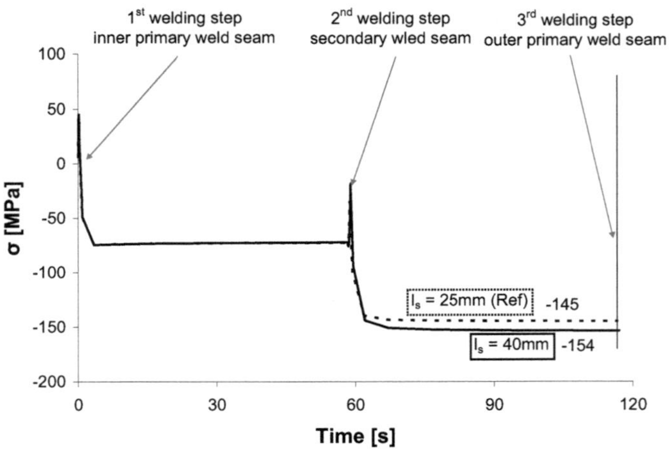


**Fig. 16.** Effect of the parameter $l_S$ on the residual stresses at the start-end-position of the primary part (P). The residual compressive stresses in P increase insufficiently, in spite of a extended secondary part length

#### *Geometrical Parameter: d, Length of Tack-Free Primary Part*

If the outer primary part is not tacked (d=10 mm), then this part is not able to develop residual compressive stresses in P. The resulting low stress level in the outer primary part (P) leads to higher displacements about 430%.

#### *Geometrical Parameter: e, Distance Between Inner Slot Edge and Start/End of the Inner Primary Part*

The shortening of the distance between the inner slot edge and the start/end of the inner primary part from 10 to 2 mm leads to a partial melting of the inner primary weld seam during the welding of the outer primary part. The tensile stresses which are assigned to the inner primary weld seam are released. Due to this the resulted tensile stress in the inner primary part and the compressive stress in the outer primary part at the P-position are decreased.

#### *Geometrical Parameters: $l_{SLOT}$, $h_{SLOT}$, e*

In this case it is investigated the combined effect of several parameters. The slot width $l_{SLOT}$ and height $h_{SLOT}$ are 40 mm and 20 mm respectively.

This configuration leads at the P-position to a reduced stiffness, so the secondary part is not able to act as a hold-down device.

Due to the enlargement of parameter e – it describes the distance from the inner slot edge to the start of the inner primary weld seam – from 10 to 20 mm the residual tensile stresses that bases on the inner primary weld seam has not any influence on the still not-welded outer primary part.

Both residual tensile stress areas interfere with each other only insufficient. The low residual compressive stresses at the P-position lead to an increase of strains about almost 500% compared to the reference model.

#### *Welding Direction: A1 / A2*

The compressive stresses at the P-position increase, if the welding direction of the inner primary weld seam will be changed. This circumstance bases on two events at the P-position. First, at the border between the inner and the outer primary part the molten pool is characterised by a steady state behaviour and therewith extended than at the opposite welding direction. Second, the heat concentration is full formed. In both cases a larger volume is either molten or heated. On cooling the material shrinks and resulting larger residual compressive stresses, which causes compression of the mushy zone within the BTR that means a hot cracking avoiding.

#### *Welding Direction: B1 / B2*

A changed welding direction at the secondary part results in lower strains on the outer secondary weld seam. In relation to the reference model the displacements on the outer secondary part decrease about 20%. Nevertheless this welding direction should be prevent in order to avoid hot cracking and cracks in the end crater pool in the extreme stressed stringer ends.

The reverse welding direction in the secondary part has not any bearing on the hot cracking behavior for the third weld step (outer primary weld seam).

#### *Welding Direction: C1 / C2*

The change of the welding direction at the inner primary part results to higher strains about 200%. Furthermore the end crater cracks are expected on the start-end-positions at the primary and secondary parts (P and S).

## Summary of Applied Integral Approach

The designed slots in the advanced stringer geometry reduce the stress concentrations and increase the hot crack resistance at the outer primary part (P) assumedly. In order to the conditions in the start-end-positions P and S during the flight phase and the welding, there are following conclusions for the structural design and the welding procedure:

1. The width and height of slots ($l_{Slot}$, $h_{Slot}$) should be chosen as small as possible in order to reduce the stress concentrations and to increase the superposition between both residual tensile stresses areas, which are generated by several welding steps of the inner primary and the secondary parts.
2. The main factor, limiting the residual comressive stresses at the start-end-position of the primary part (P) is not the length of the secondary part ($l_s$), but rather the heat input.
3. A short edge distance ($d_{STR}$) – the distance between the edges of the socket and the start-end-position of the secondary part (S) – reduces the stress concentration at S.
4. In order to avoid end crater cracks the ends of weld seams should be not located at any stringer edges but rather in regions within the stringer with lower stress levels during the flight phase.

A complex welding order results by the existence of two slots (Fig. 17). The welding direction has to be changed at the several welding steps. The stringer must be tacked on the skin before welding procedure can start.

The composition of 4 welding steps and their welding directions result by the welding analysis for the whole stringer geometry:

- Step 1: The first welding step starts in the middle of the stringer (inner primary part) and ends at the distance e in front of the slot at the stringer end.
- Step 2: The welding direction of the second step is diametrical to the first one and starts also in the middle of the inner primary stringer part. This welding steps stops at the distance e in front of the slot at the stringer start.
- Step 3: The welding direction changes again. Two weld seams are produced by this step. The first weld seam spans along the secondary part and starting on the start –end-position $S_{Start}$. The second one completes the outer primary part and starts on the slot at the stringer start $P_{Start}$.
- Step 4: The welding of the secondary slot needs the same welding step sequence than step 3, but starting at $S_{end}$ and $P_{end}$ in opposite direction.

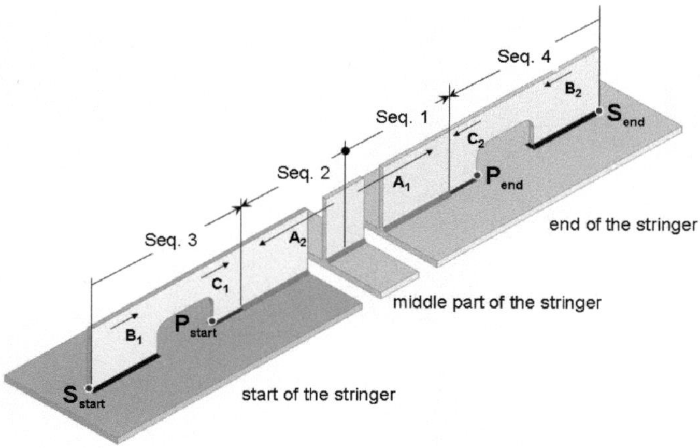


**Fig. 17.** Recommendation for the order of the several welding steps and their direction

## References


1. Herold H (2007) Rissminderung beim Schweißen von Al-Legierungen mittlerer und höherer Festigkeit (Hot crack reduction of middle and high strength aluminium alloys), AIF-Projekt, Nr. 13.983B / DVS 01.047, Otto-von-Guericke-Universität Magdeburg
2. Hilbinger R M (2001) Heißrissbildung beim Schweißen von Aluminium in Blechrandlage (Hot cracking initiation on welding of aluminium due to criticals plate-edge-positions), Doctor Thesis, Universität Bayreuth
3. Pshennikov A (2005) Entwicklung von Maßnahmen zur Heißrissvermeidung beim Einseitenschweißen langer Schweißnähte. (Development of methods for hot cracking prevention on one-side-welding of larger weld seams), Doctor Thesis, Otto-von-Guericke-Universität Magdeburg

# Weld Parameter and Minor Element Effects on Solidification Crack Initiation in Aluminium

N. Coniglio, C.E. Cross

Federal Institute for Materials Research and Testing, Berlin, Germany

## Abstract

Earlier work has established that a critical amount of 4043 filler is required to avoid solidification cracking when arc-welding 6060 aluminium, depending upon local strain conditions. For example, when the mushy zone behind the weld pool experiences a tensile strain from combined thermal and shrinkage stresses, the possibility exists for crack initiation. For a greater rate of strain, it has been determined that a greater 4043 dilution (i.e. higher weld metal silicon content) is required to avoid crack initiation. Making use of the Controlled Tensile Weldability (CTW) test and local strain extensometer measurements, a boundary has been established between crack and no-crack conditions for different local strain rates and filler dilutions, holding all other welding parameters constant. Using this established boundary as a line of reference, additional parameters have now been examined and their influence on cracking has been characterized. These parameter influences have included studies of weld travel speed, weld pool contaminants (Fe, O, and H), and grain refiner additions ($TiAl_3$ + Boron). Each parameter has been independently varied and its effect on cracking susceptibility quantified in terms of a critical strain rate required to initiate cracking for a given 4043 filler dilution.

## Introduction

Al-Mg-Si extrusion alloys are widely used in fabricated structures because of their good corrosion resistance, moderate strength achieved through heat treatment, and good weldability when using an appropriate filler metal. When welded autogenously, however, these alloys have been found to be highly susceptible to solidification cracking [1, 2]. Typical welding filler metals, such as Alloy 4043 (Al-5%Si) or Alloy 5356 (Al-5%Mg), shift the weld pool composition to a hybrid alloy mixture that is less crack sensitive [3]. Because the filler and base metal have different compositions, the amount of filler dilution will influence the composition of the weld pool, determining its metallurgical and mechanical properties and cracking susceptibility [2, 4].

Although it is well established that use of an appropriate filler metal improves weldability, the amount of filler dilution required to avoid cracking has remained an undefined quantity. Curiously, this has never been considered an important issue to industry, perhaps because standard welding practice routinely results in sufficient dilution to avoid cracking. This would particularly be true for gas-metal arc welding, where high filler dilution (30–60%) is common. This explanation has been further substantiated by recent work examining the solidification cracking of Alloy 6060, showing that only a small increase in weld metal silicon content (above base metal levels) is required to avoid cracking [5].

Research work spanning fifty years has lead to the general belief that tensile stress and strain at the trailing end of the weld pool are instrumental in initiating solidification cracks [6-14]. Strain accumulated as a result of solidification shrinkage and thermal contraction serves to pull weld metal grains apart in the mushy zone, resulting in failure of grain boundary liquid films. Prokovorov [7, 8, 9] and Senda [14] have established ductility limits for cracking; i.e. maximum strain that can be accumulated over the solidification range (Brittle Temperature Range- BTR). Strain rate was realized as being important by these researchers, but only in so far as it serves to determine how much strain can be accumulated during the time of solidification. Accordingly, numerous weldability tests have been developed specifically to measure critical strain rate [5, 15-17]. Meanwhile, developments arising from solidification cracking models applied to castings, have suggested that strain rate may play a more direct role in the actual fracture mechanism by controlling the pressure drop in the interdendritic liquid [13, 18, 19].

Building upon the concept that Si improves weldability and that weldability can be defined by a critical strain rate, studies were initiated to examine this interrelationship. Specifically, the critical strain required for cracking an aluminum 6060 weldment was measured for different 4043 filler dilutions [5]. This was done using a Controlled Tensile Weldability (CTW) test, where a controlled transverse strain was applied during welding. A crack-no crack boundary was defined for variable conditions showing that higher local strain rates require higher 4043 dilutions to avoid cracking. The objective of the present study was to characterize the behaviour of this crack-no crack boundary when varying metallurgical factors (e.g. grain size, iron content), welding parameters (torch travel speed), and external contamination (hydrogen and oxygen contents).

## Background

Each of the factors examined in this study could conceivably have an effect on weld solidification cracking susceptibility. Details regarding these possible interactions are discussed below.

### Role of Silicon

Jennings et al. measured the influence of composition on cracking susceptibility of high purity Al-Mg-Si ternary alloys by the means of a ring casting test [20]. They found a ridge of high cracking susceptibility along the Al-$Mg_2Si$ quasi-binary line (Mg/Si = 1.73, weight ratio). A peak in cracking susceptibility is observed at 0.5 wt.%Si and 0.2 wt.%Mg, as shown in Fig. 1. Alloy 6060 sits close to this peak, and is shifted to less crack sensitive regions when diluted with Alloy 4043, according to the compositions given in Table 1.

**Table 1.** Measured chemical analysis for aluminium alloy 6060 and 4043 mixtures (wet chemical analysis): (a) 6060-T4, (b) 6060-T4 + 10%4043, (c) 6060-T4 + 20%4043, and (d) 4043

| | Composition (wt.%) | | | | | | | | | | |
|---|---|---|---|---|---|---|---|---|---|---|---|
| | Si | Mg | Fe | Mn | Cu | Cr | Ni | Zn | Ti | Ga | Zr |
| (a) | 0.42 | 0.59 | 0.19 | 0.02 | 0.01 | 0.004 | 0.004 | 0.009 | 0.02 | 0.01 | 0.001 |
| (b) | 0.90 | 0.53 | 0.19 | 0.02 | 0.04 | 0.00 | 0.00 | 0.02 | 0.04 | 0.01 | 0.00 |
| (c) | 1.39 | 0.47 | 0.19 | 0.02 | 0.07 | 0.00 | 0.00 | 0.03 | 0.06 | 0.01 | 0.00 |
| (d) | 5.25 | 0.001 | 0.21 | 0.02 | 0.3* | --- | --- | 0.1* | 0.2* | --- | --- |

*handbook values
--- no value specified

Considering its relationship to the quasi-binary line of the Al-Mg-Si ternary system, it is expected that when welding with 4043 filler, and hence increasing the silicon content of the 6060 weld metal, the microstructure should consist of increasing amounts of silicon phase in addition to the quasi-binary eutectic phase $Mg_2Si$ [21]. However, due to the presence of the impurity iron (0.2 wt.%Fe, Table 1), the phase reactions that actually occur are considerably more complex as indicated in Table 2. Even at low concentrations, iron has a strong influence on solidification structure, due to its strong tendency to partition (equilibrium partition ratio: $k$=0.03) and form compounds with silicon and aluminum.

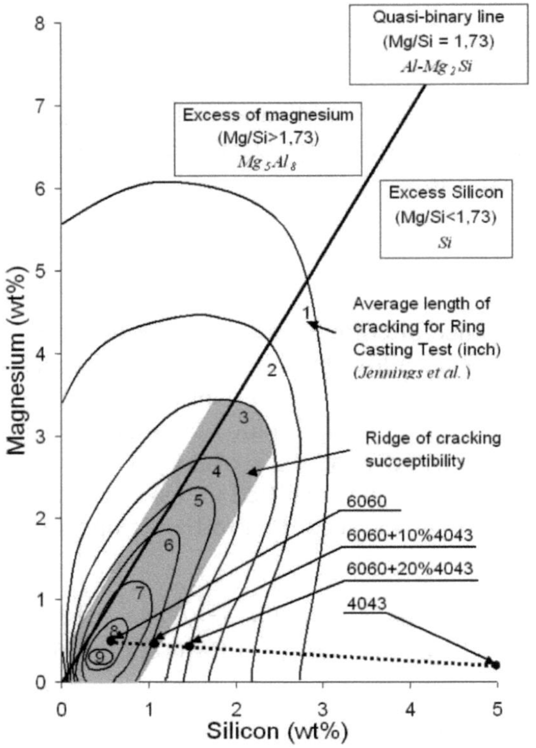


**Fig. 1.** Quasi-binary line superimposed on ring casting data of Jennings et al. [20] showing solidification cracking susceptibility for Al-Mg-Si ternary alloy system

Based on thermal analysis, microstructure observations and available literature, small increases in Si content have been found to have a major effect on the solidification path of 6060 castings [22]. Increased Si from 0.42 wt.% to 1.39 wt.% causes the solidus temperature to drop from 577°C to 509°C, increases the quantity of interdendritic constituent from 2% to 14%, and results in different phase formation. Binary β ($Al_5FeSi$) and $Mg_2Si$ phases are replaced with ternary β ($Al_5FeSi$), π ($FeMg_3Si_6Al_8$), and a low melting quaternary eutectic. Higher Si content has also been observed to result in grain refinement and reduced solidification shrinkage. Whether this trend persists at higher cooling rates (e.g. as experienced in welding) is a subject of ongoing work.

**Table 2.** Phase reactions in the Al-Mg-Si-Fe quaternary system for composition range corresponding to aluminum 6060 base metal at low and high 4043 dilutions

| No. | Phase Reactions | Temperature (°C) | |
|---|---|---|---|
| | | 6060 | 6060+20%4043 |
| 1 | $L+TiAl_3 \rightarrow \alpha(Al)$ | 666 | 666 |
| 2 | $L \rightarrow \alpha(Al)$ | 653-650 | 650-647 |
| 3 | Coherency Point | 642-637 | 642-639 |
| 4 | $L \rightarrow \alpha(Al)+\beta$ | 609-597 | ----- |
| 5 | $L \rightarrow \alpha(Al)+Mg_2Si$ | 597-589 | ----- |
| 6 | $L \rightarrow \alpha(Al)+Si$ | 587-578 | 584-577 |
| 7 | $L \rightarrow \alpha(Al)+Si+\beta$ | ----- | 564-556 |
| 8 | $L+\beta \rightarrow \alpha(Al)+Si+\pi$ | ----- | 539-534 |
| 9 | $L \rightarrow Mg_2Si+Si+\alpha(Al)+\pi$ | ----- | 524-509 |

## Role of Iron

Iron, even when present at low impurity levels, plays an important role in determining solidification microstructure as noted above. Although normally found at around 0.20 wt.% (0.30 wt.% max.) in most wrought aluminium alloys, iron has a 0.80 wt.% max. limit in the 4043 filler alloy [23]. It is normally considered an undesirable element due to its resulting in reduced mechanical properties and corrosion resistance, where both problems are tied largely to the formation of coarse, iron-bearing intermetallic phases that are cathodic relative to the aluminium matrix [24].

During solidification of castings, high temperature iron bearing intermetallic phases are believed to block interdendritic channels, leading to poor feeding of shrinkage and porosity formation [25-27]. Singer et al. measured the cracking susceptibility of high purity Al-Fe-Si alloys by means of restrained welds [28]. Plotting the mean crack length as a function of iron and silicon content (Fig. 2), they observed no cracking for the condition Fe/Si>1. Cracking susceptibility increased when increasing silicon content or decreasing iron content.

Lu and Dahle [25] studied the influence of iron content on the solidification path and cracking susceptibility of cast Al-7Si-0.4Mg alloys. Cast samples were quenched at around 590°C to observe the microstructure. Increasing the iron content from 0.3 to 0.7 wt.% favours the binary β-phase (forming at a temperature over 600°C) over the ternary β-phase (forming at temperatures lower than 580°C). Moreover, an increase in iron content resulted in an increase in porosity, but decreased cracking susceptibility. It is believed that, because these coarse binary β needles form at high temperature, they block the liquid feeding and promote porosity [25] and serve as effective pore nucleation sites [27]. The improvement in cracking resistance may also be due to an increase in high temperature strength with high iron content [26], or because β-phase needles act like solid bridges spanning the α-dendrites [25].

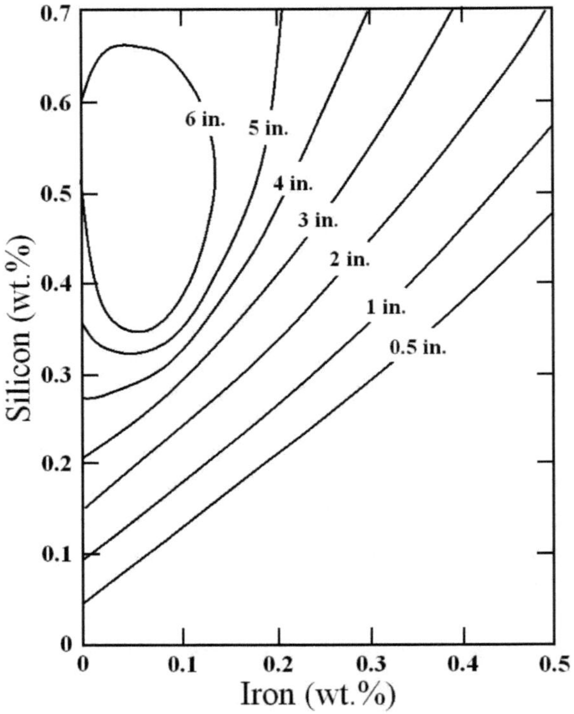


**Fig. 2.** Mean crack length (in inches) on restrained welds of Al-Fe-Si ternary alloys as a function of iron and silicon content [28]

## Grain Refinement

Grain size directly affects the strain distribution over the grain boundaries in the weld mushy zone. Refining the grain structure increases the number of grain boundaries, thereby reducing the strain seen by each boundary. Reduced cracking susceptibility of aluminium alloy 7108 has been related to grain refinement, adding grain refiners scandium and titanium-boron, i.e. TiBor [29]. Using the circular patch test, cracking was avoided when adding a minimum of 0.25 wt.%Sc or (0.02 wt.%Ti + 0.004 wt.%B), corresponding respectively to a grain size 80 μm and 180 μm. Note that the grain size in a 7108 weld without TiBor and Scandium was 290 μm.

## Travel Speed

Strain and stress fields around the weld pool have been experimentally observed and simulated [10, 11, 16, 30-34]. These fields are dependent on weld processing parameters as well as restraining conditions. The influence of torch travel speed on solidification cracking during welding has been extensively studied, but is not well understood [16, 30-34]. Some studies have observed an improvement in weldability when increasing the torch travel speed [30-32]. These measurements were made for the GTA process and travel speeds between 2.5 and 13 mm/s. Chihoski [30, 31] experimentally observed compressive and tensile cells around a moving weld pool, and postulated that no cracking will occur if the mushy zone is under compression. At high travel speeds, a compressive cell is located at the mushy zone, thereby avoiding cracking. With decreasing travel speed, this compressive cell diminishes and is replaced with a tensile cell that is highly susceptible to cracking. However, some simulations [33] and experimental observations [34] show an increase in cracking susceptibility with increasing travel speed. But in these situations, the welding speeds are high; between 50 and 100 mm/s for laser welding [34] and between 16 and 25 mm/s for GTA welding [33].

Weld travel speed also influences the weld pool shape, which is known to have a strong effect on weldability [35]. Increasing weld travel speed modifies the weld pool shape from round to teardrop, which favours stray grain formation and long continuous centreline grain boundaries. The orientation of these long grain boundaries are perpendicular to the thermal and mechanical strains, leading to a high cracking susceptibility. This counters Chihoski´s observations where high torch travel speed should improve weldability. Therefore, both weld pool shape and local strain cells must be considered when determining solidification cracking susceptibility.

## Oxygen and Hydrogen Contamination

Oxygen and hydrogen contaminations are typically unavoidable, because these elements are always present in the vicinity of the molten weld pool. For example, humid air is often aspirated into the shielding gas. Concerning oxygen, molten aluminium has a strong tendency to form aluminium oxides. These oxides have a density close to that of molten aluminium, and hence tend not to separate if mixed. Entrapped oxide films may influence the cracking susceptibility, providing sites for crack nucleation due to decohesion [36]. Such oxide films may also impair the fluidity and feeding ability of the alloy [24], or affect the solidification path, serving as sites for phase nucleation [37]. Above 750°C, the $\gamma$-$Al_2O_3$ will be present, which acts as the preferred nuclei for $\beta$-$Al_5FeSi$ needles. The $\alpha$-$Al_2O_3$ oxide suppresses nucleation of the $\beta$-phase, which as noted before is the phase believed to promote porosity.

Hydrogen contamination can arise from moisture or hydro-carbons (e.g. machine oil). Hydrogen in molten aluminium often results in porosity, because of the sharp drop in solubility when going from liquid to solid. During solidification of pure aluminium, the solubility of hydrogen drops from 500 to 50 cubic millimetre per 100 grams of aluminium with 1 atm of hydrogen over the coupon [24]. The role of porosity on cracking is not well understood, and it could conceivably have either positive or negative effects. Dissolved hydrogen should make it easier to cavitate, which is one of the possible mechanisms for crack formation [36, 38, 39]. But pores also serve to feed shrinkage, reducing the interdendritic volume that needs to be fed, and hence lowering the interdendritic pressure drop [40].

## Critical Strain Rate – Dilution Map Concept

The influence of 4043 filler dilution on aluminium 6060-T4 weldability has been previously studied by means of the CTW test [5]. Compositions of both alloys are given in Table 1. The principle of the CTW test is to apply a controlled transverse strain rate during welding. The 4043 filler speed was varied to obtain 4043 dilutions in the weld from 0 to 16%. An extensometer placed underneath the weld joint measured local strains. Strain rates were afterwards calculated and the critical strain rate to initiate cracking was determined for different 4043 dilutions.

Cracking susceptibility, plotted as a function of local strain rate, 4043 dilution, and the corresponding silicon content, is shown in Fig. 3. Cracking has been found dependent on both local strain rate and 4043 dilution, as revealed by the crack-no crack boundary (shadow region) in Fig. 3. Higher local strain rates require higher 4043 dilution to avoid cracking.

The position of the crack-no crack boundary along the strain rate axis and its slope are significant. Its relative position along the x-axis represents the influence of thermo-mechanical factors. Alloys having this boundary occurring at higher strain rates are considered more weldable, since critical strain rate is representative of weldability. The slope of this boundary reveals the influence of thermo-metallurgical factors on weldability. The more vertical this boundary (i.e. higher slope), the less influence filler dilution has on weldability. Accordingly, an alloy is considered more weldable if less 4043 is needed to avoid cracking.

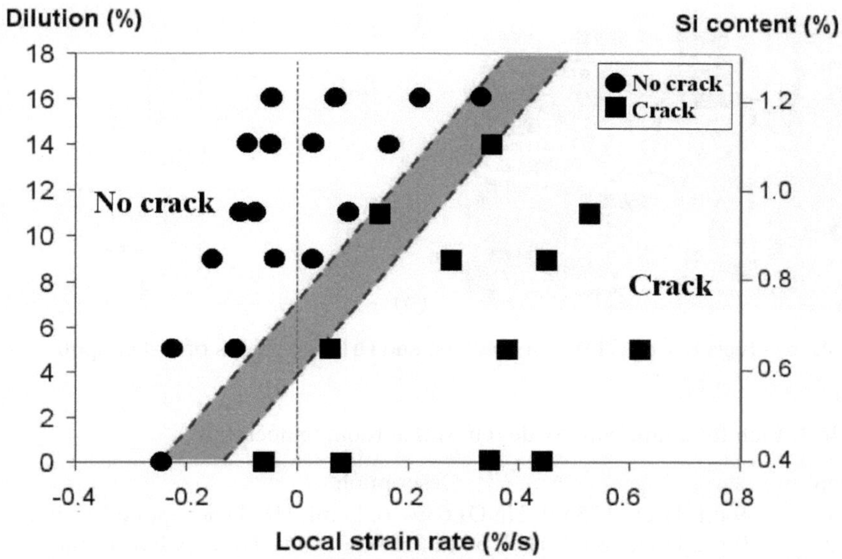


**Fig. 3.** Critical strain rate – dilution map for alloy 6060-T4, characterizing cracking susceptibility in terms of local strain rate and 4043 filler dilution using CTW test [5]

## Experimental Approach

### Controlled Tensile Weldability (CTW) Test

Solidification cracking susceptibility was studied by means of the CTW test (Fig. 4a), which consisted of applying a controlled transverse strain rate during welding, with strain applied normal to the welding direction and away from the weld. Experimental equipment consisted of a unidirectional tensile machine (load capacity 500 kN). The tensile transverse crosshead speed was varied from 0 to 0.083 mm/s by increased steps of 0.017 mm/s. This globally applied strain rate resulted in variations in local strain rate (i.e. region adjacent to weld pool), which was measured with an extensometer.

Weld coupons of 120 mm × 40 mm × 4 mm were cut from extruded bars of 6060-T6. These coupons were welded between two bigger aluminum plates (300 m × 150 mm × 8 mm) (Fig. 4b), which were clamped into the CTW machine.

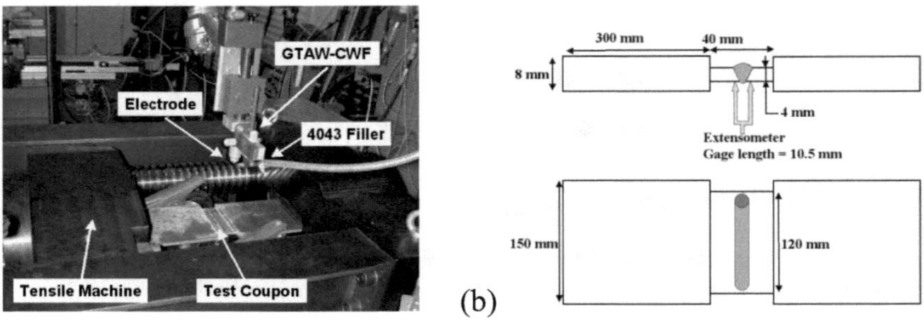


(a) (b)

**Fig. 4.** (**a**) Overview of CTW test machine and (**b**) dimensions of test coupon

**Table 3.** Etch for aluminium oxide removal at room temperature

| Etch | Description |
|---|---|
| E1 | 869 mL $H_2O$, 125 mL $HNO_3$ 65%, 6.25 mL HF 48% : applied 15 min |
| E2 | 100 mL $H_2O$, 66 mL HCl, 66 mL $HNO_3$, 16 mL HF : applied 1 min |

**Table 4.** Base welding parameters for GTAW-CWF process

| | |
|---|---|
| Current | : 110 A |
| Voltage | : 17.8 V |
| Arc Gap | : 2 mm |
| Electrode diameter | : 3.2 mm |
| Electrode Type | : Tungsten + 1%$LaO_2$ |
| Electrode Tip Angle | : 30° |
| Torch Gas | : Helium |
| Gas Flow Rate | : 0.33 L/s |
| Polarity | : DCEN |
| Torch Travel Speed | : 4 mm/s |
| Wire Diameter | : 0.8 mm |
| Wire Feed | : 0–41.7 mm/s |

### Welding Parameters

Welding was performed using the gas-tungsten arc, cold-wire feed process (GTAW-CWF). The base welding parameters are listed in Table 3. The arc voltage was kept constant using an arc voltage control system, maintaining a 2 mm arc gap corresponding to a 17.8 V arc voltage. The heat input was chosen to obtain a bead-on-plate, full penetration weld. Filler wire speed and tensile cross-head speed were experimental variables. Prior to the CTW test, the oxide layer on the test coupon was chemically removed (etch E1, Table 4), followed by degreasing with acetone.

### Test Sequence

The CTW test sequence is summarized in Table 5. A pre-load of 15 kN was applied prior to welding to compensate for thermal expansion of the weld coupon during welding, maintaining it in tensile conditions even at a 0 mm/s transverse cross-head speed. The arc was initiated by touch contact between the electrode and the weld coupon. In a 100 mm long weld, the transverse cross-head speed was applied 30 mm after the start of welding. Local strain was measured at weld mid-length. At the end of the weld, the arc was abruptly extinguished, providing information as to the shape of the weld pool.

**Table 5.** CTW test sequence

| Step Number | Distance from the Weld Start (mm) | Related Action |
|---|---|---|
| 1 | 0 | Start weld |
| 2 | 30 | Start cross-head travel |
| 3 | 50 | Electrode at the top of the extensometer |
| 4 | 90 | Stop cross-head travel |
| 5 | 100 | Stop weld |

### Local Strain Measurement

An extensometer was placed underneath the weld at mid-length of the weld (Fig. 4b) to measure local strain across the weld pool during the welding process. Local strain rates were calculated from these measured strains and the critical strain rate for cracking was determined. The extensometer output was recorded at a 100 Hz frequency during the whole test using *CATMAN 4.5* computer software for data acquisition and a *Spider 8* analog-to-digital converter. In addition, a few select strain rate measurements were made at the observed crack initiation site to correlate these measurements with those measured at weld mid-length. The gage length of the extensometer was 10.5 mm which, by definition, influences local strain measurements.

The strain and strain rate values were taken at the coherency temperature, which corresponds to the point during solidification where the secondary arms of two adjacent dendrites are beginning to be in contact. According to some theories [13, 18, 41–43], the coherency temperature corresponds to the critical temperature where interdendritic liquid feeding ability (healing) is sharply reduced due to the continuity of the solid phase. The calculated strain rate value was also taken at the coherency temperature, at the critical stage where cracking begins. The position of coherency relative to the torch along the centreline was determined by temperature and weld pool shape measurements.

The arc was suddenly extinguished at the end of each test, freezing in the weld pool shape, the boundary of which corresponds to the liquidus of the weld metal. The position of the liquidus relative to the torch along the weld centreline was measured. Temperature measurements were then performed to determine the position of coherency point relative to the liquidus along the weld centreline.

### *Temperature Measurement*

The cooling curve for the weld pool was monitored during solidification for all experimental conditions to determine the position of the coherency point relative to the liquidus. A hole, 0.6 mm diameter and 1 mm deep, was drilled from the bottom of the weld coupon, 35 mm from the start of the weld. A 0.5 mm outer diameter, sheathed, electrically grounded, nickel/chrome-nickel thermocouple was pre-placed inside the hole before welding. The thermocouple output was recorded at a 200 Hz frequency during solidification using *CATMAN 4.5* computer software for data acquisition and a *Spider 8* analog-to-digital converter.

## Material

### *Extrusions*

Extruded aluminum 6060-T6 bar was used as the base material, and was compared against 6060-T4 bar material studied previously [5]. Both materials came from different heats and had slightly different compositions.

Composition of the 4043 filler and both 6060 heats were measured with wet chemical analysis and spectrometry (Table 6). 6060-T4 and 6060-T6 extrusions had a different hardness (respectively 40 and 83 HV0.5) and varied, in particular, in alloying elements Mn, Cu, and Zn. The size of the weld coupons were 120 mm × 40 mm × 4 mm (Fig. 4b).

**Table 6.** Measured wet chemical analysis and emission spectrometry for aluminum alloys 6060, 4043, and controlled mixtures of 6060 with master alloys Al-10Fe and Al-5Ti-B (wt.%) : (a) 6060-T4, (b) 6060-T6, (c) 4043, (d) insert 6060+Tibor (6060+15%(Al-5Ti)), (e) insert 6060+Fe (6060+15%(Al-10Fe)), (f) weld pool 6060+insert (6060+Fe), and (g) weld pool 6060+ insert (6060+Tibor)

| | Composition (wt.%) | | | | | | | | | | |
|---|---|---|---|---|---|---|---|---|---|---|---|
| | Si | Mg | Fe | Mn | Cu | Cr | Ni | Zn | Ti | Ga | Zr |
| (a) | 0.42 | 0.59 | 0.19 | 0.02 | 0.01 | 0.004 | 0.004 | 0.009 | 0.02 | 0.01 | 0.001 |
| (b) | 0.51 | 0.51 | 0.21 | 0.04 | 0.03 | 0.003 | 0.003 | 0.04 | 0.02 | 0.01 | 0.001 |
| (c) | 5.25 | 0.001 | 0.21 | 0.02 | 0.3* | --- | --- | 0.1* | 0.2* | --- | --- |
| (d) | 0.43 | 0.43 | 0.18 | 0.03 | 0.02 | 0.003 | 0.003 | 0.04 | 0.77 | 0.01 | 0.001 |
| (e) | 0.43 | 0.43 | 1.68 | 0.03 | 0.02 | 0.003 | 0.003 | 0.04 | 0.02 | 0.01 | 0.001 |
| (f) | 0.50 | 0.50 | 0.43 | 0.04 | 0.03 | 0.003 | 0.003 | 0.04 | 0.02 | 0.01 | 0.001 |
| (g) | 0.50 | 0.50 | 0.21 | 0.04 | 0.03 | 0.003 | 0.003 | 0.04 | 0.14 | 0.01 | 0.001 |

*handbook

--- value not specified

### *Inserts*

The objective was to vary weld metal composition using inserts having a controlled composition (Table 6). Inserts were machined from cast ingots made from controlled mixtures of 6060-T6 with master alloys (e.g. Al-5Ti-B or Al-10Fe). These inserts were then pre-welded into 6060-T6 CTW coupons in preparation for CTW wedability testing.

Controlled mixtures of 6060-T6, 6060+15%(Al-5Ti-B), and 6060+15%(Al-10Fe) were cast. The composition of the resulting ingots, when using master alloys, differs from 6060 respectively for titanium (0.8 wt.% instead of 0.02 wt.%) and iron (1.7 wt.% instead of 0.2 wt.%). The casting mold is shown in Fig. 5, with inside dimensions 140 × 10 × 30 mm, and mold thickness 10 mm. All mold materials with direct exposure to molten aluminum were pre-coated with boron-nitride spray. Each cast heat weighed approximately 130 grams and was melted in a graphite crucible placed inside an electric furnace held at 800°C. Oxide dross was skimmed just prior to casting, and the melt was rigorously stirred in order to ensure thorough mixing with the master alloy. In the case of Al-5Ti-B, 6060-T6 was thoroughly melted before adding the TiBor master alloy, in order to reduce time exposure of $TiAl_3$ particles in the melt, and thus minimize particle dissolution and promote efficient grain refinement. Five inserts (2 mm × 2 mm × 140 mm) were machined from each cast ingot. For purposes of providing an experimental control, inserts of cast 6060 material were also prepared, left untreated with any master alloy.

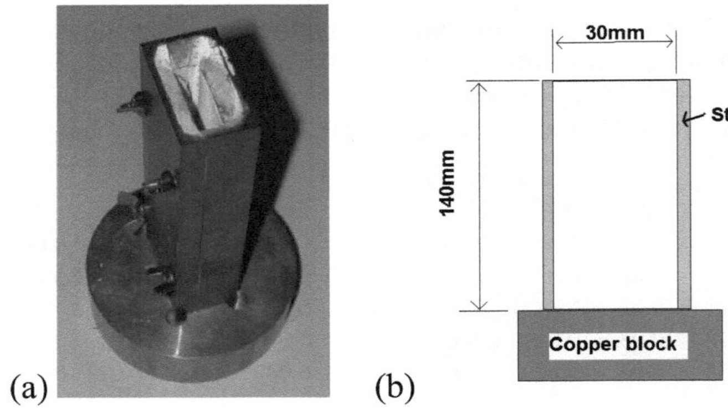


**Fig. 5.** Casting mold (**a**) photograph and (**b**) schematic, used to obtain ingots for weld insert production

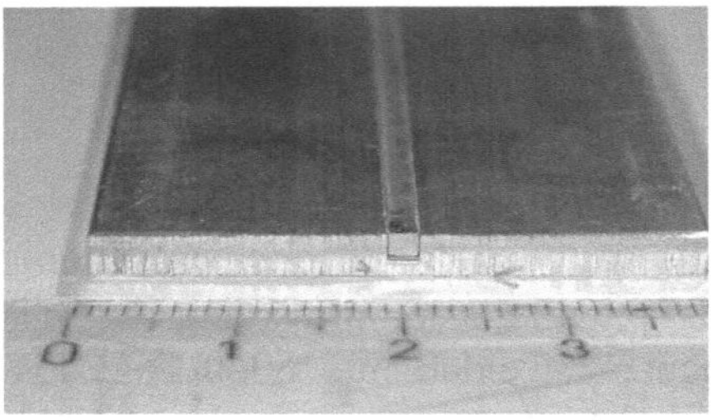

**Fig. 6.** Insert placed into groove of 6060-T6 coupon before pre-weld

Square grooves (2 × 2 mm) were machined along the centreline of 6060-T6 CTW test coupons (i.e. extrusions) to receive the inserts (Fig. 6). Prior to assembly, the oxide layer was chemically removed by etching both weld coupon and insert (etch E1, Table 4), followed by degreasing with acetone. The insert was tapped into the groove and pre-welded using the welding parameters listed in Table 7. Variable polarity current was used here to help to remove oxides at the weld surface. The pre-weld consisted of a bead-on plate, partial penetration weld (6 mm wide and 3 mm thick), which completely melted the insert.

### Shielding Gas

In the case of controlled oxygen and hydrogen contamination, helium flow rate was maintained constant (0.33 L/s) and mixed with gases from premixed bottles of either Ar+2%$H_2$ or Ar+1%$O_2$. Because of the effect of gas additions in changing the weld heat input, current was varied in order to maintain a constant weld pool size. Ar+2%$H_2$ flow rate was fixed at 0.27 L/s to obtain a sufficient amount of dissolved hydrogen in aluminum to promote limited interdendritic pore formation. Regarding oxygen contamination, Ar+1%$O_2$ flow rate was fixed at 0.03 L/s, corresponding to the maximum gas flow rate possible while maintaining arc stability. At a 0.08 L/s Ar+1%$O_2$ flow rate, the arc became highly erratic, likely due to the large quantity of aluminium oxides formed on the weld pool.

**Table 7.** Welding parameters for pre-welding inserts

| | |
|---|---|
| Positive Current /Duration | : +80 A /20% |
| Negative Current /Duration | : –220 A /80% |
| Frequency | : 50 Hz |
| Arc Gap | : 1.5 mm |
| Electrode diameter | : 3.2 mm |
| Electrode Type | : Tungsten + 2% $ThO_2$ |
| Electrode Tip Angle | : 30° |
| Torch Gas | : Argon |
| Gas Flow Rate | : 0.3 L/s |
| Polarity | : Variable polarity |
| Torch Travel Speed | : 4 mm/s |

### Critical Strain Rate - Dilution Maps

The crack- no crack boundary in a critical strain rate – dilution map was determined for each experimental condition by measuring cracking susceptibility at two levels of dilution: 0% and 15–20%. 4043 filler speed was chosen to reach a dilution between 15% and 20%. Dilution is an important concept in determining weld pool composition, approximated using Eq. (1):

$$\text{filler dilution} = \frac{B+C}{A+B+C} \times 100\ (\%), \tag{1}$$

where $A$ is the melted area of 6060 base metal, and $B+C$ is the difference between the total area of the weld metal and $A$ (Fig. 7). Calculated dilutions are summarized in the Appendix.

For both low and high dilution levels, the transverse cross-head speed was incremented in steps of 0.017 mm/s, noting the value where cracking occurred. Because of the fixed incremental step of transverse cross-head speed (0.017 mm/s), the exact location of the crack-no crack boundary lies somewhere within this fixed step. This corresponds to the difference between the highest measured strain rate without cracking and the lowest one to initiate a crack. Hence, the accuracy of testing is limited by the magnitude of this step.

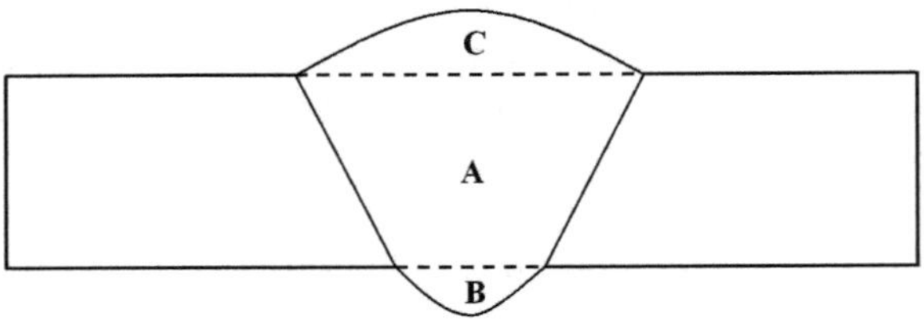


$$\text{dilution} = \frac{B+C}{A+B+C} \times 100\ (\%)$$

**Fig. 7.** Illustration for filler dilution calculation from weld metal cross-section

## Results and Discussion

### Characterization of Weld Pool Shape

Strain measurements were made at the point where the mushy zone (coherency point) passes between the extensometer. The weld pool temperature was recorded during solidification to know where the mushy zone and coherency was located relative to the torch position. Thermal analyses performed previously measured the liquidus at 660°C and coherency at 624°C for the 6060 alloy [5]. For all experimental conditions, it has been found that the weld temperature drops during solidification from 660°C to 624°C in approximately 0.2 s. Thus, knowing the travel speed, the distance from the fusion line to the coherency point can be calculated.

The weld pool shape and resulting grain structure was observed and characterized from the top surface. The application of the etch E2 (Table 4) on the top surface of the weld revealed the grain structure and weld pool shape, obtained by extinguishing suddenly the arc. Figure 8 illustrates the measured distances to characterize the weld pool shape. For all the welding conditions, the dimensional characteristics of the weld pool were: top width 8 mm, and distance behind electrode 6 mm, with a deviation of ± 1 mm. The measurements of the weld pool shape are summarized in the Appendix.

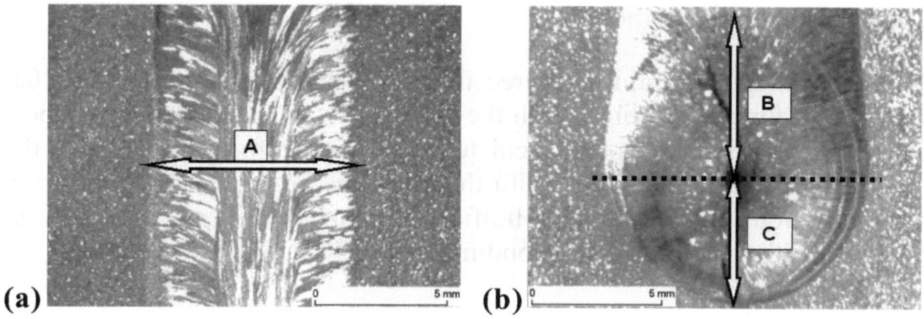


**Fig. 8.** Weld pool measurements : (**a**) top width (A) and (**b**) distance behind electrode (B) and in front of electrode (C)

Metallographic transverse cross-sections were cut from the mid-length of the weld. These sections were ground and polished to 1 μm and then chemically polished using a slightly basic solution of colloidal silicon dioxide. The cross-sections were observed with optical microscopy. Weld bead dimensions were measured in all experimental conditions (see Appendix) and remained almost constant. An example of weld cross-sections made on 6060-T6 coupons with 0% and 17% 4043 filler dilution coupons are shown in Fig. 9. Cross-sections for all experimental conditions revealed similar weld pool shape. More detailed weld metal dimensions for 0% 4043 dilution are: 8 mm overbead width, 6 mm underbead width, 4 mm centreline thickness, and 24–27 mm² cross sectional area. For 17% 4043 dilution these dimensions are: 8 mm overbead width, 6 mm underbead width, 5 mm centreline thickness, and 28–32 mm² cross sectional area.

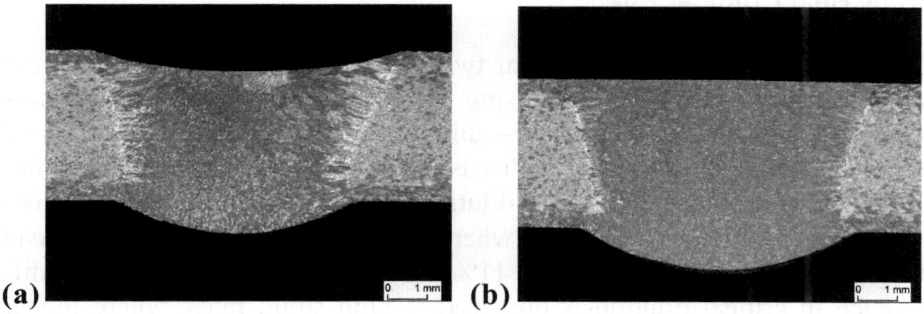


**Fig. 9.** Weld cross sections of 6060-T6 coupons made with (**a**) 0% and (**b**) 17% 4043 dilution

## Local Strain Measurement

An example of the strain measured in a CTW test is plotted in Fig. 10a, shown as a function of time, with the CTW test sequence (Table 5) superposed. Figure. 10b shows the calculated strain rate from the point (3) to the point (4), i.e. during cooling. With the objective being the study of strain rate conditions during weld solidification, time has been redefined in Fig. 10b, with 0 seconds corresponding to step 3 (Table 5).

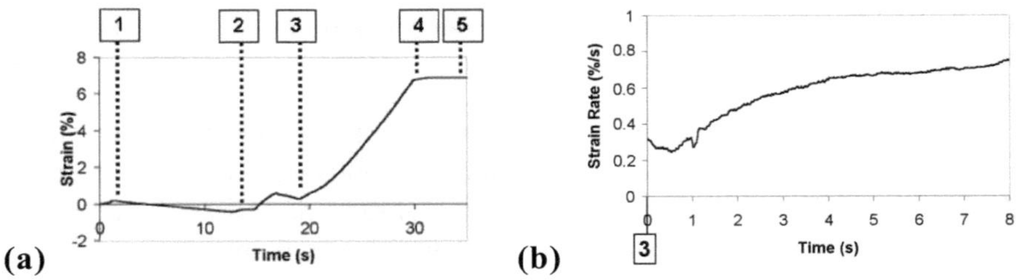


**Fig. 10.** (**a**) CTW test sequence superposed on the measured strain and (**b**) calculated strain rate, for the test conditions 6060-T4 with 9% 4043 dilution and 0.067 mm/s cross-head speed

When welds were made using base welding parameters on 6060-T6 coupons (Table 3), it was observed that cracking always initiated 25 mm after the weld start, which corresponds to the position of the weld pool mushy zone at the instant of transverse strain application (step 2 in Table 5).

## Alloy Heat / Temper Effect

Aluminium 6060 extrusions from two different heats and tempers were evaluated with the CTW test using the same base welding parameters (Table 3). A critical strain rate – dilution map comparing 6060-T4 and 6060-T6 is shown in Fig. 11 as discussed in the background. 6060-T6 has been studied at 0 and 17% 4043 dilution, whereas data for 6060-T4 exists from previous work (Fig. 3) [5] where CTW tests were run over a broad range of dilutions (0%, 5%, 9%, 11%, 14%, and 16% 4043). A large difference in critical dilution is observed at high strain rates, where higher 4043 dilution is required to avoid cracking in 6060-T6. For example, at a local strain rate of 0%/s, 7% 4043 dilution is required to avoid cracking in 6060-T4, while 17% dilution is required for 6060-T6. It is not known at this point whether this difference in weldability is due to differences in hardness (40 versus 83 HV0.5) or composition (Table 6). This demonstrates, however, how material with the same alloy designation can exhibit different weldability depending on specific conditions.

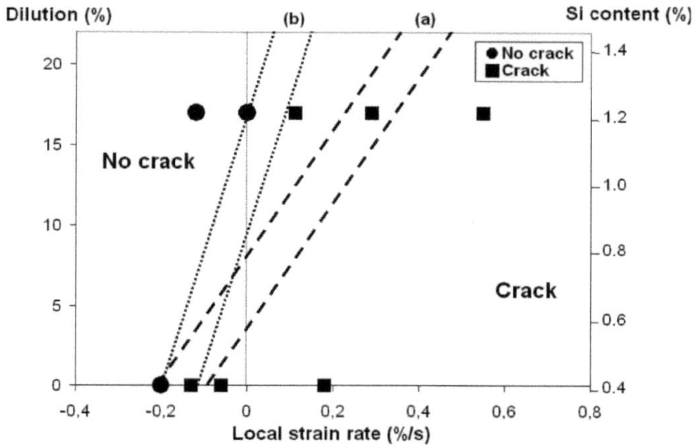


**Fig. 11.** Critical strain rate-dilution map: (**a**) 6060-T4, previous data from [5] and (**b**) 6060-T6, with data points shown for 0 and 17% 4043 dilution

## Travel Speed Effect

CTW tests were made on aluminium 6060-T6 using the base parameters (Table 3), but varying the current and weld travel speed. In addition, the welding current was varied so as to maintain a constant weld pool size. Weld travel speed was run at 2, 4, and 6 mm/s, and the current was varied respectively from 80A to 145A. Data for 4 mm/s was obtained from a previous test (Fig. 11). Dilutions studied were 0%, and either 18% or 16% 4043 respectively for 2 and 6 mm/s weld travel speeds. The crack – no crack boundaries for these conditions are compared against the boundary found using the base welding parameters (Table 3), i.e. 4 mm/s weld travel speed and 110A, as shown in Figs. 12 and 13. Increasing weld travel speed should improve weldability following the relationship of Chihosky [30, 31], and this appears to be the case, but only at high strain rates. Specifically, at 18% 4043 dilution, solidification cracking occurs for a +0.12%/s local strain rate at 2 and 4 mm/s weld travel speed, while no crack occurs at +0.25%/s strain rate for a 6 mm/s travel speed. On the other hand, 6060-T6 is less weldable at 0% 4043 dilution with increasing weld travel speed.

This is likely due to a change in weld metal grain structure, where stray centreline grains were observed to occur only at 4 and 6 mm/s weld travel speed.

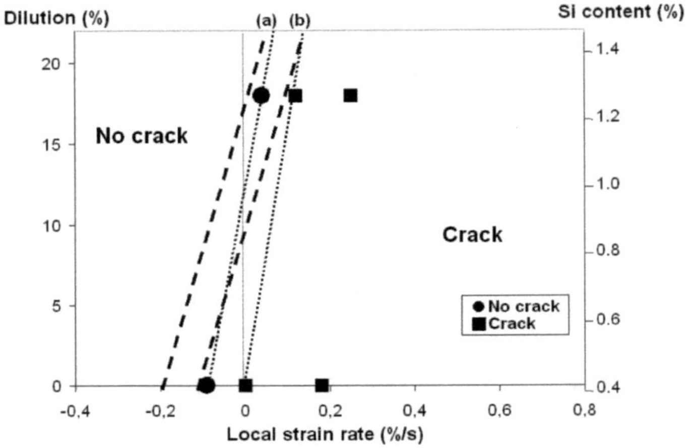


**Fig. 12.** Critical strain rate – dilution map comparing crack-no crack boundaries at weld travel speeds of (**a**) 4 mm/s and (**b**) 2 mm/s when welding 6060-T6. Data points are shown for 2 mm/s

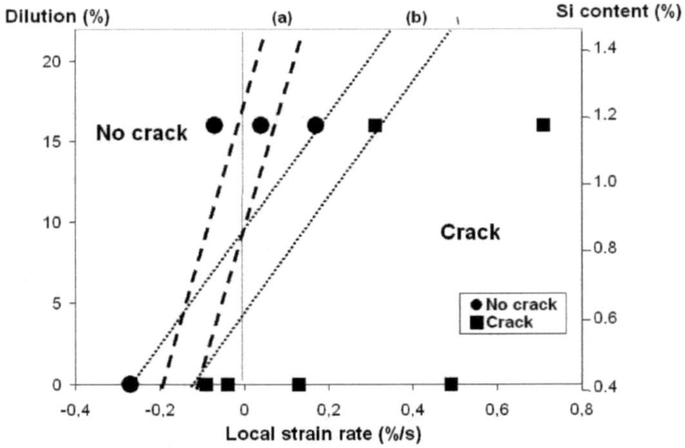


**Fig. 13.** Critical strain rate – dilution map comparing crack no crack boundaries at weld travel speeds of (**a**) 4 mm/s and (**b**) 6 mm/s when welding 6060-T6. Data points are shown for 6 mm/s

## Insert Control Test

Weld coupons with 6060 inserts containing no extra alloying elements were evaluated with CTW testing thus providing a control to determine the influence of the insert and corresponding weld coupon preparation on weldability. The base welding parameters were used (Table 3), but the current was decreased to 95A to maintain a constant weld pool size. Cracking susceptibility was measured for 0 and 17% 4043 dilution. Figure 14 compares the crack-no crack boundary boundaries for 6060-T6 with base welding parameters and 6060-T6 with 6060 inserts. A slight difference is observed which may come from the methodology to determine these boundaries (i.e. fixed increasing step of the cross-head speed). The use of inserts appears to have no significant influence on the crack-no crack boundary. Thus, weldability differences due to insert alloy additions (following sections) will be assumed to represent the effect of the alloy addition, and not the insert itself.

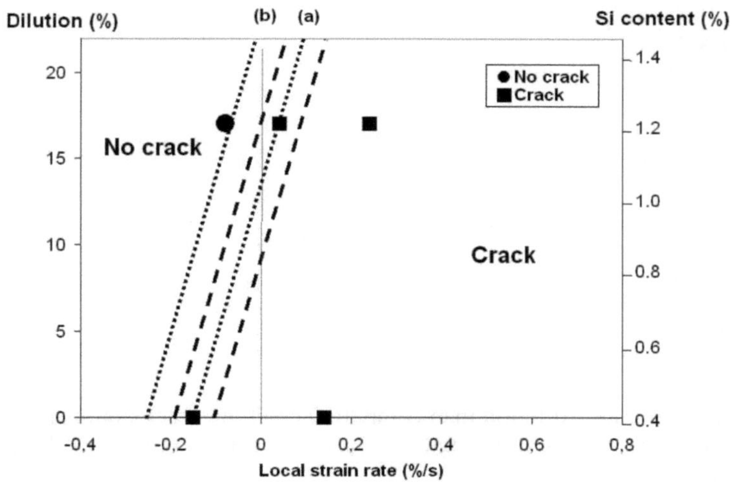


**Fig. 14.** Critical strain rate – dilution map comparing crack-no crack boundary for (**a**) 6060-T6 and (**b**) 6060-T6 with 6060 insert. Data points are shown for 6060 inserts

## Iron Impurity Effect

Inserts at high iron content were machined from a 6060+1.7%Fe ingot (Table 6). The influence of iron content on cracking susceptibility was first observed during the casting of ingots. While 6060 ingots cracked at the base of the pouring spout, there was no cracking in 6060+1.7%Fe ingots. This supports observations of Lu and Dahle [25], where high iron content prevented cracking in cast Al-Mg-Si alloys. The welding parameters were the same as those used for 6060 inserts. The 4043 dilutions studied included 0% and 20%. The weld pool composition resulting from using the high iron insert had a corresponding high iron content (i.e. increase from 0.2 to 0.4 wt.%Fe) as given in Table 6. The crack-no crack boundaries are compared in Fig. 15. Further investigations are needed to better define these boundaries, but it appears there is little effect of iron.

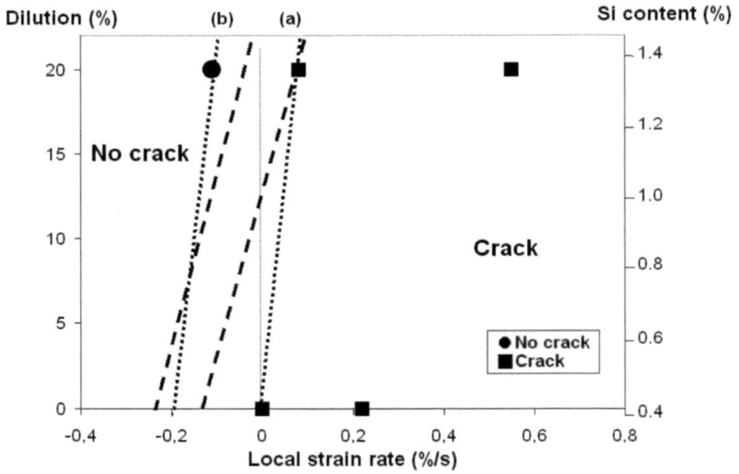


**Fig. 15.** Critical strain rate – dilution map comparing crack-no crack boundary for 6060-T6 with (**a**) 6060 insert and (**b**) 6060+1.7%Fe insert. Data points are shown for 6060+1.7%Fe insert

## Grain Refinement Effect

CTW tests were performed with inserts high in TiBor content machined from a 6060+0.8%Ti+0.16%B ingot (Table 6). A few scattered macropores were observed in the weld cross-sections. The influence of TiBor on cracking susceptibility was first observed in cast ingots, where 6060 ingots cracked at the base of the pouring spout, and 6060+0.8%Ti+0.16%B ingots did not crack. The welding parameters were the same as those used for 6060 inserts. The 4043 dilutions studied were 0% and 15%. The weld pool composition with the Ti rich insert differed by its higher titanium content (from 0.02 to 0.14 wt.%Ti) as given in Table 6. Crack-no crack boundaries are compared for both insert compositions in Fig. 16. The grain refiner addition significantly improved weldability, particularly at low 4043 dilutions. This is not surprising considering the potent effect of refinement reported for aluminium weldability improvement [29]. Moreover, it seems that TiBor improves weldability with more efficiency than 4043, the boundary being completely moved to higher strain rates for all dilutions. Finally, the boundary is almost vertical; revealing that 4043 dilution has little influence on the weldability of grain refined 6060 welds. In essence, the grain refinement resulted from TiBor exceeds any metallurgical effects resulting from 4043 filler dilution.

4043 filler is known to reduce the mechanical properties of 6xxx base metal, and so the improved weldability comes at a price. This suggests that it may prove feasible to develop a filler that is high in grain refiner, while low in Si. It is interesting to note that Ti content in 4043 filler tends to be very high (0.2 wt.%Ti) compared to the initial Ti content in 6060-T6 (0.02 wt.%Ti). Increasing 4043 filler dilution increases the Ti content in the weld metal, promoting grain refinement [5]. Thus, improved grain refinement should be achieved with increasing 4043 dilution, which may explain why the crack-no crack boundaries tend to be closer at high 4043 dilutions (Fig. 16). Future studies are required to determine the improvement in weldability when increasing the only Si content of the weld without increasing the Ti content.

## Oxygen Contamination Effect

CTW tests were performed using an oxygen containing shielding gas (flow rate: 0.33 L/s He + 0.03 L/s Ar + 1%$O_2$). This required a decrease in the welding current to 100A to keep a constant weld pool size. 4043 dilutions of 0% and 18% were studied. High quantities of oxides were observed at the weld surface. Crack-no crack boundaries are compared between normal and oxygen-containing gas in Fig. 17, where use of oxygen is found to slightly improve weldability. At 18% 4043 dilution and a local strain rate of +0.09%/s, a crack should occur when welding with helium, but does not initiate with oxygen added to the helium. This trend goes counter to the belief that oxygen forms oxides films that can nucleate cracks [24]. However, it may just be that a heavy oxide is formed at the pool surface and does not get mixed into the weld pool.

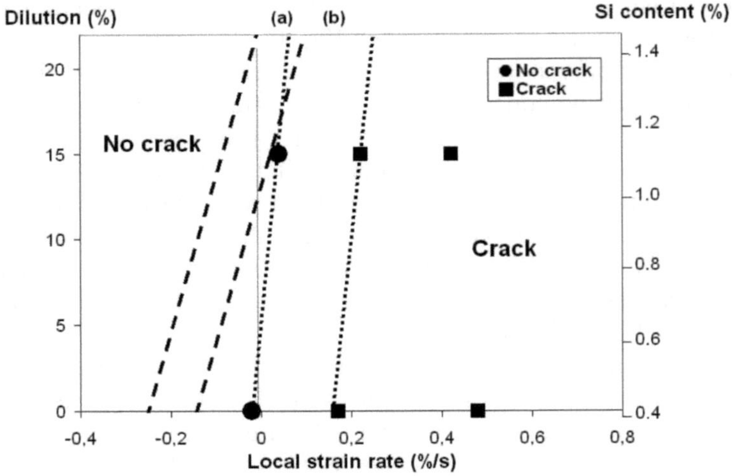


**Fig. 16.** Critical strain rate – dilution map comparing crack-no crack boundary for 6060-T6 with (**a**) 6060 insert and (**b**) 6060+0.8%Ti insert. Data points are shown for 6060+0.8%Ti insert

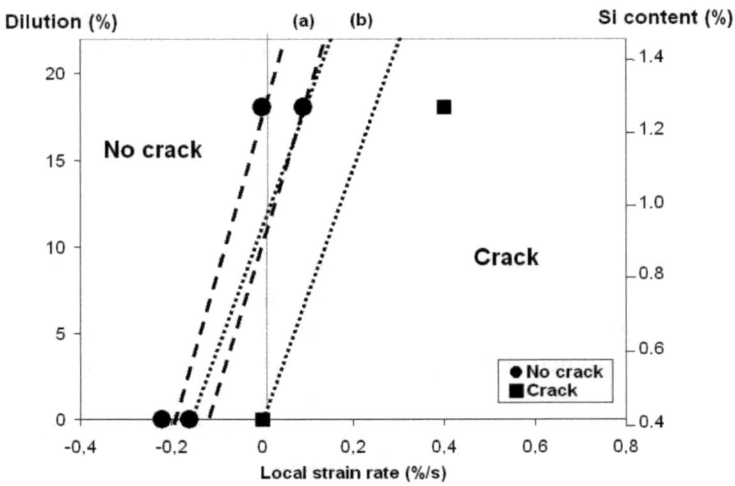


**Fig. 17.** Critical strain rate - dilution map comparing crack-no crack boundary for 6060-T6 welded with (**a**) 0.33 L/s He and (**b**) 0.33 L/s He + 0.03 L/s ($Ar+1\%O_2$). Data points are shown for He+($Ar+1\%O_2$) shielding gas

## Hydrogen Contamination Effect

CTW tests were performed using a hydrogen containing shielding gas (flow rate: 0.33 L/s He + 0.27 L/s Ar + 2%$H_2$). This required a decrease in the welding current to 105A to keep a constant weld pool size. Dilutions were studied at 0 and 18% 4043. Limited interdendritic pores were observed in the weld cross section (Fig. 18). Since no oxides were observed on the weld surface, it is assumed that hydrogen served as an effective reducing element. Crack-no crack boundaries are compared between welds made with helium and helium-argon-hydrogen in Fig. 19. Hydrogen is found to improve weldability, especially at high 4043 dilution. In fact, at 18% dilution and +0.15%/s local strain rate, 6060-T6 cracked when welded with helium, but did not crack when welded with helium-argon-hydrogen. This suggests that hydrogen pores may be feeding shrinkage and thus reducing the pressure drop. This leads to an interesting choice between having cracking or porosity.

## Sensitivity of CTW Weldability Test

One of the aims of the present work was to use the CTW test to evaluate how small variations in alloy composition may affect weldability. In pursuing this aim, the sensitivity and limitations of this new weldability test have also been established. Some of the inherent problems associated with this test are discussed below.

### *Position of Extensometer*

In executing the CTW test sequence (Table 5), the extensometer used to measure local weld strain was positioned at weld mid-length. However, ideally the best place to measure strain is at the crack initiation site, which was found to always occur at the point where the mushy zone sits when strain is first applied, 25 mm after the weld start. In order to determine how strain at these two locations may differ (i.e. 25 mm versus 50 mm), four tests were performed with the extensometer positioned 25 mm from the weld. These values were then compared against the standard mid-length measurements, as shown in Table 8. These values show good agreement (less than 0.05%/s difference), except for one condition at high cross-head speed where there was a large difference (0.29%/s). This suggests that the standard sequence should be modified for future tests, to relocate the extensometer to the 25 mm position.

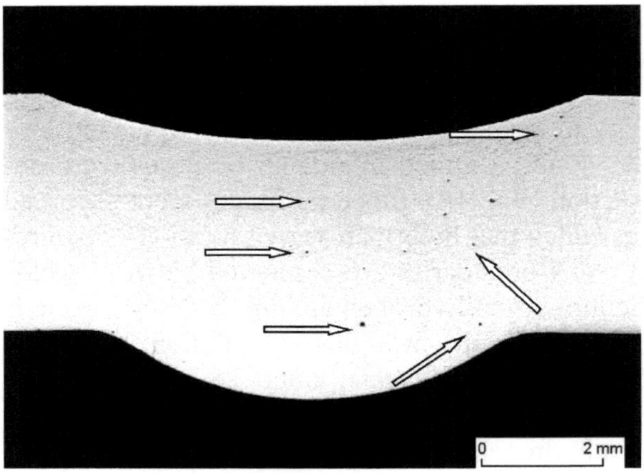


**Fig. 18.** Un-etched weld cross sections of 6060-T6 coupon made with He+Ar+$H_2$ and 0% 4043 filler dilution revealing interdendritic pores (*arrows*)

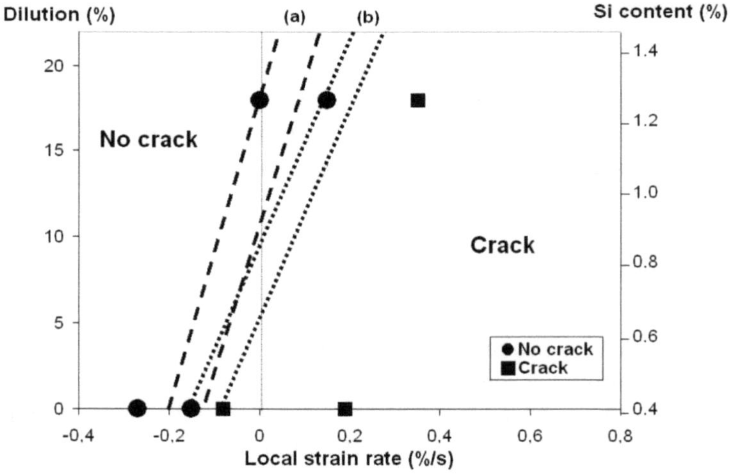


**Fig. 19.** Critical strain rate – dilution map comparing crack-no crack boundary for 6060-T6 welded with (**a**) 0.33 L/s He and (**b**) 0.33 L/s He + 0.27 L/s (Ar+2%$H_2$). Data points are shown for He+( Ar+2%$H_2$) shielding gas

**Table 8.** Comparison between strain rates based upon extensometer measurements made 25 mm from weld start and at weld mid-length

| Cracking susceptibility | Dilution (%) | Cross-head speed (mm/s) | Measured strain rate (%/s) | |
|---|---|---|---|---|
| | | | 25 mm from weld start | Weld mid-length |
| No crack | 0 | 0 | –0.18 | –0.20 |
| Crack | 0 | 0.033 | –0.09 | –0.06 |
| No crack | 17 | 0.033 | –0.05 | +0.00 |
| Crack | 17 | 0.083 | +0.26 | +0.55 |

### *Negative Strain Rate*

As can be seen in any of the strain rate-dilution maps, cracks that occur at low 4043 dilutions involves a negative critical strain rate (i.e. inward movement of material to feed shrinkage). At first hand this appears counter intuitive, but may actually reflect upon the material's very poor weldability. Even with the inward movement of base material, the solidification shrinkage is apparently not sufficiently compensated to avoid cracking. Data from previous work has shown similar behavior, where cracking may occur even with inward movement of base material [5].

### *Crack Boundary*

The nature of the crack-no crack boundary for the 6060/4043 alloy system is such that variations in critical dilution of 0–20% corresponds to a maximum difference in critical strain rate, expressed in terms of cross-head speed, of roughly 0.1 mm/s. In terms of test sensitivity, it would be preferable that this number be much larger. As it is, this means that test increments approaching 0.01 mm/s are needed to accurately identify the location of a boundary (i.e. one tenth of the full range). In this work, increments of 0.017 mm/s were used, which only roughly determined the boundary for each experimental condition. Thus, the use of finer increments should be considered for future tests on this alloy. Other alloy systems will likely exhibit their own unique sensitivity to strain rate.

### *Interacting Factors*

As is usual for welding experiments, it is difficult to vary one parameter systematically and not have numerous other factors change indirectly. For CTW testing, the use of 4043 filler changes not only the silicon content of the weld metal, but increases the size and shape of the weld bead. Thus, it becomes difficult to separate the metallurgical effects from the mechanical. Likewise, variations in travel speed or changes in shielding gas will alter the heat input and, although corrected with adjustments to current, there may be subtle changes to pool shape that can affect grain structure. Thus, it becomes difficult to know if the effect of travel speed on weldability is due to a change in local strain rate or grain structure. The application of modelling may help to answer these questions in the future.

## Conclusion

The newly developed CTW test, evaluated in terms of critical strain rate – dilution maps, has been successfully used to quantify the weldability of the aluminium 6060/4043 alloy system. In particular, the effect of minor alloy additions and travel speed on solidification cracking susceptibility has been studied by observing how these changes shift the crack-no crack boundary in a strain rate-dilution map. Small additions of oxygen or iron were found to have little or no effect on weldability. Improvements in weldability were experienced with the addition of hydrogen or a grain refiner, or with the use of faster travel speeds.

## Acknowledgement

The authors are grateful to BAM for internal funding of this project, and specifically wish to thank Th. Böllinghaus and Th. Kannengiesser for their insightful support of this research effort. In addition, technical support at BAM from A. Hannemann, P. Friedersdorf, K. Scheideck, R. Breu, M. Lammers, and M.Richter was greatly appreciated. The authors are also grateful to Metallurg Aluminum for donation of master alloys.

## References

1. Dudas JH, Collins FR (1966) Preventing Weld Cracks in High-Strength Aluminum Alloys. Welding Journal 45:241s–249s
2. Metzger GE (1967) Some Mechanical Properties of Welds in 6061 Aluminum Alloy Sheet. Welding Journal 46(10):457s–469s

3. Mousavi MG, Cross CE, Grong Ø, Hval M (1997) Controlling Weld Metal Dilution for Optimized Weld Performance in Aluminum. Sci. Tech. Weld. Joining 2:275–278
4. Liptak JA, Baysinger FR (1968) Welding Dissimilar Aluminum Alloys. Welding Journal 47:173s–180s
5. Coniglio N, Cross CE, Michael Th, Lammers M (2008) Defining a Critical Weld Dilution to Avoid Solidification Cracking in Aluminum. Welding Journal (in review)
6. Pellini WS (1952) Strain Theory for Hot Tearing. Foundry 80:125–199
7. Prokhorov NN (1956) The Problem of the Strength of Metals While Solidifying During Welding. Svarochnoe Proizvodstvo 6:5–11
8. Prokhorov NN, Bochai MP (1958) Mechanical Properties of Aluminum Alloys in the Crystallization Temperature Range during Welding. Svarochnoe Proizvodstvo 2:1–6
9. Prokhorov NN, Gavrilyuk MN (1971) Strain Behaviour of Metals during Solidification after Welding. Welding Production 13(6):8–13
10. Zacharia T (1994) Dynamic Stresses in Weld Metal Hot Cracking. Welding Journal 73(7): 164s–172s
11. Feng Z (1994) A Computational Analysis of Thermal and Mechanical Conditions for Weld Metal Solidification Cracking. Welding in the World 33:340–347
12. Kannengiesser T, McInerney T, Florian W, Böllinghaus T, Cross CE (2002) The influence of local weld deformation on hot cracking susceptibility. In: Mathematical modelling of weld phenomena 6, Maney, pp 803–817
13. Feurer U (1977) Influence of Alloy Composition and Solidification Conditions on Dendrite Arm Spacing, Feeding, and Hot Tear Properties of Aluminum Alloys. In: Proc. international symposium engineering alloys, Delft, 131–145
14. Senda T, Matsuda F, Takano G, Watanabe K, Kobayashi T, Matsuzaka T (1971) Experimental Investigations on Solidification Crack Susceptibility for Weld Metals with Trans–Varestraint Test. Trans. of JWRI 2(2):141–162
15. Arata Y, Matsuda F, Nakata K, Shinozaki K (1977) Solidification Crack Susceptibility of Aluminum Alloy Weld Metals (Report II) – Effect of Straining Rate on Cracking Threshold in Weld Metal during Solidification. Trans. of JWRI 6(1):91–104
16. Matsuda F, Nakagawa H, Nakata K, Okada H (1979) The VDR Cracking Test for Solidification Crack Susceptibility on Weld Metals and its Application to Aluminum Alloys. Trans. of JWRI 8:85–95
17. Tamura H, Kato N, Ochiai S, Katagiri Y (1977) Cracking Study of Aluminum Alloys by the Variable Tensile Strain Hot Cracking Test. Trans. JWS 8:16–22
18. Rappaz M, Drezet JM, Gremaud M (1999) A New Hot–Tearing Criterion. Met. Mat. Trans. 30A:449–455
19. Rindler W, Kozeschnik E, Enzinger N, Buchmayr B (2002) A Modified Hot Tearing Criterion for Steels. In: H Cerjak (ed) Mathematical Modelling of Weld Phenomena 6, Woodhead Publishing Limited, pp 819–835

20. Jennings PH, Singer ARE, Pumphrey WI (1948) Hot-Shortness of some High-Purity Alloys in the Systems Al-Cu-Si and Al-Mg-Si. J. Inst. Metals 74:227–248
21. Mondolfo LF (1976) Aluminum Alloys–Structure & Properties, Butterworths, London
22. Coniglio N, Cross CE (2006) Characterization of Solidification Path for Aluminum 6060 Weld Metal with Variable 4043 Filler Dilution. IIW-1755–06, Welding in the World 50(11/12):14–23
23. The Aluminum Association (2006) International Alloy Designations and Chemical Composition Limits for Wrought Aluminum and Wrought Aluminum Alloys
24. Van Horn KR (1967) Aluminum Properties, Physical Metallurgy, and Phase Diagrams, vol.1. American Society of Metals, Ohio
25. Lu L, Dahle AK (2005) Iron-rich Intermetallic Phases and their Role in Casting Defect Formation in Hypoeutectic Al-Si Alloys. Met. Mat. Trans. 36A:819–835
26. Wang L, Makhlouf M, Apelian D (1995) Aluminum Die Casting Alloys : Alloy Composition, Microstructure, and Properties-Performance Relationships. International Materials Reviews 40(6):221–238
27. Mbuya TO, Odera BO, Ng'ang'a SP (2003) Influence of Iron on Castability and Properties of Aluminum Silicon Alloys: Literature Review. International Journal of Cast Metals Research 16(5):451–465
28. Singer ARE, Jennings PH (1947) Hot-Shortness of some Aluminum-Iron-Silicon Alloys of High Purity. J. Inst. of Metals 73:273–284
29. Mousavi MG, Cross CE, Grong Ø (1999) Effect of Scandium and Titanium-Boron on Grain Refinement and Hot Cracking of Aluminum Alloy 7108. Science and Technology of Welding and Joining 4(6):381–388
30. Chihoski RA (1972) The Character of Stress Field Around a Weld Arc Moving on Aluminum Sheet. Welding Journal 51(1):9s–18s
31. Chihoski RA (1979) Expansion and Stress Around Aluminum Weld Puddles. Welding Journal 58(9):263s–276s
32. Hunziker O, Dye D, Roberts SM, Reed RC (2005) A Coupled Approach for the Prediction of Solidification Cracking during the Welding of Superalloys. In: Böllinghaus Th, Herold H (eds) Hot Cracking Phenomena in Welds, Springer-Verlag Berlin Heidelberg, pp 299–319
33. Shibahara M, Serizawa H, Murakawa H (2005) Finite Element Method for Hot Cracking Analysis using Temperature Dependent Interface Element. In: Böllinghaus Th, Herold H (eds) Hot Cracking Phenomena in Welds, Springer-Verlag Berlin Heidelberg, pp 253–267
34. Cicala E, Duffet G, Andrzejewski H, Grevey D, Ignat S (2005) Hot Cracking in Al-Mg-Si alloy laser welding – operating parameters and their effects. Materials Science and Engineering 395A:1–9
35. Savage WF, Aronson AH (1966) Preferred orientation in the weld fusion zone. Welding Journal 45:85s–89s
36. Campbell J (1991) Castings, Butterworth-Heinemann, Oxford, Boston
37. Westengen H (1982) Formation of Intermetallic Compounds During DC Casting of a Commercial Purity Al-Fe-Si Alloy. Z. Metallkde. 73:360–368

38. Cross CE, Olson DL, Edwards GR (1993) The Role of Porosity in Initiating Weld Metal Hot Cracks. In: Zacharia Th (ed) Int. Conf. Proc. On Modeling and Control of Joining Processes, pp 549–557
39. Talbot DEJ (2004) The Effects of Hydrogen in Aluminum and Its Alloys, Maney Publishing, London
40. M′Hamdi M, Mo A (2005) On Modelling the Interplay between Microporosity Formation and Hot Tearing in Aluminium Direct-Chill Casting. Materials Science and Engineering A:105–108
41. Arata Y, Matsuda F, Nakata K, Shinozaki K (1977) Solidification Crack Susceptibility of Aluminum Alloy Weld Metals (Report III) - Effect of Straining Rate on Crack Length in Weld Metal. Trans. of JWRI 6(2):47–52
42. Borland JC (1960) Generalized Theory of Super Solidus Cracking in Welds (and Castings). British Welding Journal: 508–512
43. Pumphrey WI, Jennings PH (1948) A Consideration of the Nature of Brittleness and Temperature Above the Solidus in Castings and Welds in Aluminum Alloys. J. Inst. Metals 75:235–256


## Appendix

The following tables reveal the different measurements realized for all the studied experimental conditions. They explicitly relates the calculated dilution to the 4043 filler feeding speed, the weld pool shape measurements according to Fig. 8, and dimensional cross sectional characteristics of the weld pools. Each table correspond to one experimental condition at low and high 4043 filler dilutions.

**Table A.1.** 4043 filler dilution and dimensional characteristics for CTW welds of alloy 6060-T4 [5]. Distances (A), (B), (C) are according to Fig. 8

| Base metal | Filler Wire Speed (mm/s) | 4043 Dilution (%) | Top Width (A) (mm) | Distance behind electrode (B) (mm) | Distance in front of electrode (C) (mm) |
|---|---|---|---|---|---|
| 6060-T4 | 0 | 0 | 7.5 | 6.1 | 4.7 |
| 6060-T4 | 41.7 | 16 | 8.1 | 5.9 | 4.8 |

| Cross Section Area ($mm^2$) | Bead Thickness (mm) | Root Width (mm) | Over-Bead Width (mm) | Over-Bead Curvature ($mm^{-1}$) |
|---|---|---|---|---|
| 25.1 | 4.2 | 5.5 | 7.5 | –0.068 |
| 31.7 | 5.0 | 6.5 | 8.1 | +0.011 |

**Table A.2.** 4043 filler dilution and dimensional characteristics for CTW welds of alloy 6060-T6. Distances (A), (B), (C) are according to Fig. 8

| Base metal | Filler Wire Speed (mm/s) | 4043 Dilution (%) | Top Width (A) (mm) | Distance behind electrode (B) (mm) | Distance in front of electrode (C) (mm) |
|---|---|---|---|---|---|
| 6060-T6 | 0 | 0 | 7.5 | 6.3 | 5.6 |
| 6060-T6 | 41.7 | 17 | 7.8 | 5.9 | 5.3 |
| **Cross Section Area ($mm^2$)** | **Bead Thickness (mm)** | **Root Width (mm)** | **Over-Bead Width (mm)** | **Over-Bead Curvature ($mm^{-1}$)** | |
| 26.1 | 4.8 | 5.8 | 7.8 | –0.066 | |
| 31.5 | 5.0 | 6.9 | 8.1 | +0.009 | |

**Table A.3.** 4043 filler dilution and dimensional characteristics for CTW welds of alloy 6060-T6 with 6060 insert. Distances (A), (B), (C) are according to Fig. 8

| Base metal | Filler Wire Speed (mm/s) | 4043 Dilution (%) | Top Width (A) (mm) | Distance behind electrode (B) (mm) | Distance in front of electrode (C) (mm) |
|---|---|---|---|---|---|
| 6060-T6 | 0 | 0 | 8.0 | 5.6 | 6.1 |
| 6060-T6 | 41.7 | 17 | 7.9 | 6.0 | 5.0 |
| **Cross Section Area ($mm^2$)** | **Bead Thickness (mm)** | **Root Width (mm)** | **Over-Bead Width (mm)** | **Over-Bead Curvature ($mm^{-1}$)** | |
| 24.2 | 4.3 | 5.7 | 8.1 | –0.056 | |
| 28.7 | 5.1 | 6.0 | 8.0 | +0.025 | |

**Table A.4.** 4043 filler dilution and dimensional characteristics for CTW welds of alloy 6060-T6 with 6060+1.7%Fe insert. Distances (A), (B), (C) are according to Fig. 8

| Base metal | Filler Wire Speed (mm/s) | 4043 Dilution (%) | Top Width (A) (mm) | Distance behind electrode (B) (mm) | Distance in front of electrode (C) (mm) |
|---|---|---|---|---|---|
| 6060-T6 | 0 | 0 | 8.3 | 6.0 | 6.4 |
| 6060-T6 | 41.7 | 20 | 7.5 | 5.1 | 5.5 |
| **Cross Section Area ($mm^2$)** | **Bead Thickness (mm)** | **Root Width (mm)** | **Over-Bead Width (mm)** | **Over-Bead Curvature ($mm^{-1}$)** | |
| 26.7 | 4.3 | 5.7 | 8.5 | –0.053 | |
| 29.6 | 5.3 | 6.5 | 7.6 | +0.025 | |

**Table A.5.** 4043 filler dilution and dimensional characteristics for CTW welds of alloy 6060-T6 with 6060+0.8%Ti insert. Distances (A), (B), (C) are according to Fig. 8

| Base metal | Filler Wire Speed (mm/s) | 4043 Dilution (%) | Top Width (A) (mm) | Distance behind electrode (B) (mm) | Distance in front of electrode (C) (mm) |
|---|---|---|---|---|---|
| 6060-T6 | 0 | 0 | 8.9 | 6.2 | 6.3 |
| 6060-T6 | 41.7 | 15 | 9.0 | 6.0 | 5.6 |
| **Cross Section Area ($mm^2$)** | **Bead Thickness (mm)** | **Root Width (mm)** | **Over-Bead Width (mm)** | **Over-Bead Curvature ($mm^{-1}$)** | |
| 25.2 | 4.3 | 4.5 | 8.4 | –0,025 | |
| 28.7 | 5.1 | 5.1 | 7.9 | +0.015 | |

**Table A.6.** 4043 filler dilution and dimensional characteristics for CTW welds of alloy 6060-T6 welded at a torch travel speed of 2 mm/s. Distances (A), (B), (C) are according to Fig. 8

| Base metal | Filler Wire Speed (mm/s) | 4043 Dilution (%) | Top Width (A) (mm) | Distance behind electrode (B) (mm) | Distance in front of electrode (C) (mm) |
|---|---|---|---|---|---|
| 6060-T6 | 0 | 0 | 8.2 | 6.6 | 7.2 |
| 6060-T6 | 20.8 | 18 | 8.3 | 6.3 | 6.7 |
| **Cross Section Area ($mm^2$)** | **Bead Thickness (mm)** | **Root Width (mm)** | **Over-Bead Width (mm)** | **Over-Bead Curvature ($mm^{-1}$)** | |
| 25.7 | 4.0 | 6.3 | 8.3 | –0.074 | |
| 28.5 | 5.2 | 6.1 | 8.1 | +0.000 | |

**Table A.7.** 4043 filler dilution and dimensional characteristics for CTW welds of alloy 6060-T6 welded at a torch travel speed of 6 mm/s. Distances (A), (B), (C) are according to Fig. 8

| Base metal | Filler Wire Speed (mm/s) | 4043 Dilution (%) | Top Width (A) (mm) | Distance behind electrode (B) (mm) | Distance in front of electrode (C) (mm) |
|---|---|---|---|---|---|
| 6060-T6 | 0 | 0 | 7.3 | 5.4 | 5.6 |
| 6060-T6 | 62.5 | 16 | 7.2 | 5.4 | 4.3 |
| **Cross Section Area ($mm^2$)** | **Bead Thickness (mm)** | **Root Width (mm)** | **Over-Bead Width (mm)** | **Over-Bead Curvature ($mm^{-1}$)** | |
| 25.1 | 4.2 | 6.5 | 7.6 | –0.107 | |
| 30.1 | 5.1 | 6.7 | 7.9 | –0.015 | |

**Table A.8.** 4043 filler dilution and dimensional characteristics for CTW welds of alloy 6060-T6 welded with He+(Ar+1%$O_2$) shielding gas. Distances (A), (B), (C) are according to Fig. 8

| Base metal | Filler Wire Speed (mm/s) | 4043 Dilution (%) | Top Width (A) (mm) | Distance behind electrode (B) (mm) | Distance in front of electrode (C) (mm) |
|---|---|---|---|---|---|
| 6060-T6 | 0 | 0 | 8.2 | 5.8 | 6.4 |
| 6060-T6 | 41.7 | 18 | 7.6 | 6.0 | 6.4 |
| **Cross Section Area ($mm^2$)** | **Bead Thickness (mm)** | **Root Width (mm)** | **Over-Bead Width (mm)** | **Over-Bead Curvature ($mm^{-1}$)** | |
| 27.6 | 4.2 | 7.1 | 8.4 | –0.108 | |
| 29.1 | 5.1 | 6.4 | 7.6 | +0.000 | |

**Table A.9.** 4043 filler dilution and dimensional characteristics for CTW welds of alloy 6060-T6 welded with He+(Ar+2%$H_2$) shielding gas. Distances (A), (B), (C) are according to Fig. 8

| Base metal | Filler Wire Speed (mm/s) | 4043 Dilution (%) | Top Width (A) (mm) | Distance behind electrode (B) (mm) | Distance in front of electrode (C) (mm) |
|---|---|---|---|---|---|
| 6060-T6 | 0 | 0 | 8.7 | 5.9 | 6.2 |
| 6060-T6 | 41.7 | 18 | 8.4 | 5.8 | 6.0 |
| **Cross Section Area ($mm^2$)** | **Bead Thickness (mm)** | **Root Width (mm)** | **Over-Bead Width (mm)** | **Over-Bead Curvature ($mm^{-1}$)** | |
| 26.4 | 4.3 | 6.7 | 8.8 | –0.084 | |
| 31.4 | 5.1 | 6.4 | 8.5 | +0.015 | |

# Using Simulation for Investigations of Hot Cracking Phenomena in Resistance Spot Welding of 6xxx Aluminum Alloys (AA6016 and AA6181)

A. Eder, S. Jaber, N. Jank

Fronius International, Wels-Thalheim, Austria

## Abstract

This study presents how Numerical Simulation, by using the Finite Element Method (FEM), can help to avoid hot cracking phenomena for resistance spot welding of 6xxx AlMgSi alloys. It is well known that aluminum spot welding is a difficult task especially for 6xxx alloys. Small silicon contents in these alloys increase the cracking sensitivity dramatically in comparison with 5xxx alloys. In order to reduce tedious experimental work and to understand such problems SORPAS®, a commercial available Finite Element Method software for resistance welding, was used. Virtual simulation experiments for the innovative resistance spot welding process DELTA SPOT™ led to optimized welding parameters in order to get the desired nugget size without surface cracks.

## Introduction

Using the innovative resistance spot welding process DELTA SPOT™ [1] one can weld a wide range of metal combinations like three-sheet aluminum, three-sheet steel and dissimilar metals like aluminum-steel. All this is made possible by using a coated process tape in combination with a resistance spot welding gun – Figs. 1 and 2.

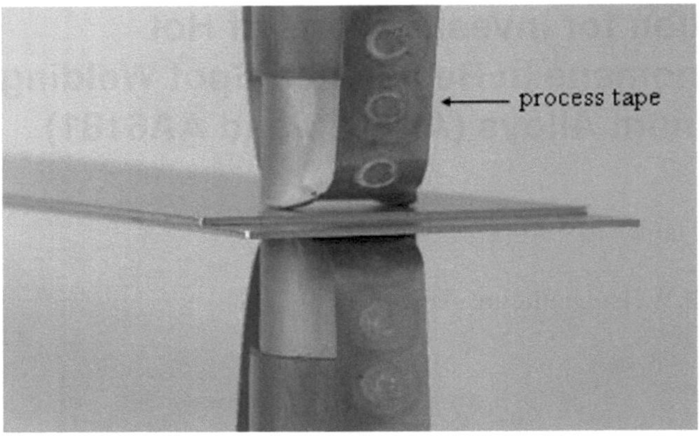


**Fig. 1.** DELTASPOT™ electrodes with a process tape (courtesy of FRONIUS Inc. Welding technology. Austria)

The tape is posed between the electrodes and the sheets to be joined. The resistance caused by the tape generates an additional heat input into the work piece. It not only allows dissimilar material welding but also eliminates work piece-coating and contaminant effects. It leads to ideal contact, splash-free welds and high electrode lifetime. After every weld the process tape moves to its next position. The contact surface therefore always remains clean. Additionally the fine coating of the tape optimizes the electrical contact to the work piece, especially for welding aluminum where it prevents spatter formation. Each welding spot is reproducible – there is always the same quality as at the start.

DELTA SPOT™ was designed with the automotive and supplier industries very much in mind. One goal is to solve difficult welding tasks arising from the use of new materials in the automotive industry. Therefore welding aluminum, especially high strength aluminum alloys, is an important topic. For resistance spot welding of aluminum the process tape has another important effect. It can be used to control the heat input into the workpiece. The additional heat due to the resistance of the process tape compensates the negative effect of the high conductivity of aluminum. The tape leads to high process efficiency and reliability.

This study concentrates on the problems arising by resistance spot welding of aluminum and its alloys – especially the problem of hot cracking. The difference between welding with and without process tape is demonstrated for AlMg3 (AA 5754). As an example for difficult to weld 6xxx aluminum alloys, AlMg0.4Si alloys are investigated. For these alloys it is demonstrated how simulation can be used to find an optimal process curve for resistance spot welding in order to avoid hot cracking.

**Fig. 2.** DELTASPOT™ spot welding gun (courtesy of FRONIUS Inc. Welding technology. Austria)

## Welding High Strength Aluminum Alloys

Aluminum is widely used for light weight constructions in automotive and aerospace industries due to its perfect properties: low density, third most abundant element, highly formable (face centered cubic crystal structure). The low strength and stiffness is compensated by alloys with other elements. A brief overview of aluminum alloys used for automotive applications is given in [2]. See [3] for aerospace applications.

But, most high strength aluminum alloys are considered to be unweldable. One of the major metallurgical problems of high strength aluminum alloys is hot cracking. Cracks arise when the thermal stresses generated during cooling exceed the strength of the almost solidified metal. The tendency of cracking grows with the solidification range of the aluminum alloys – Fig. 3.

The cracking sensitivity is depending on the alloying element contents. Si and Mg have a strong effect on the weld cracking sensitivity [4]. Simple phase diagrams for crack prediction are not possible because real commercial aluminum alloys contain many alloying elements.

Strategies for avoiding hot cracking are e.g. preheating, a controlled shrinkage of the parts or low heat input. Spot welding with a process tape is another possibility which is demonstrated in the following section.

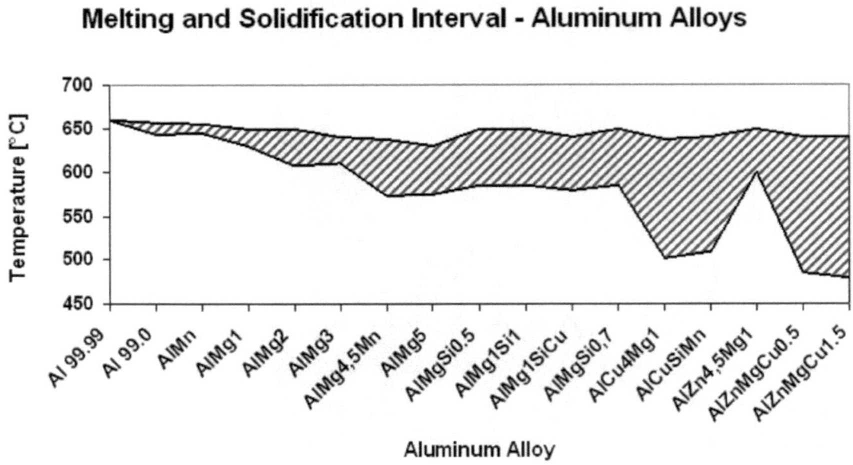


**Fig. 3.** Melting and solidification interval of aluminum alloys

## Spot Welding of Aluminum With and Without a Tape

In order to demonstrate the difference between welding aluminum with and without a process tape we made resistance spot welding experiments for AlMg3 (AA5754). The process parameters for the 1 mm sheets were 3 kN electrode force and approximately 16 kA welding current in one pulse mode. In the case without a process tape a higher welding current than in the process tape case was necessary. During the welding series the current was held constant independent of the electrode contamination. The welding time was set to 400 ms (incl. upslope and down-slope time). R40 electrodes of 12 mm in diameter were used.

Due to the rapid cooling of the molten weld pool hot cracking and sometimes solidified eutectics appear when joining with the conventional spot welding process. Figure 4 shows the macro section for spot welded AlMg3 sheets without a process tape. The weld nugget has an elliptic form and is of approximately 4 mm in diameter. The cracks start at the weld nugget and are directed outward in a typical angle [5]. Within a few spot welds there is a contamination of the electrodes with aluminum. The contamination dramatically affects the appearance of the welded spot. This means there has to be done a cleaning of the electrodes after three to five spot welds.

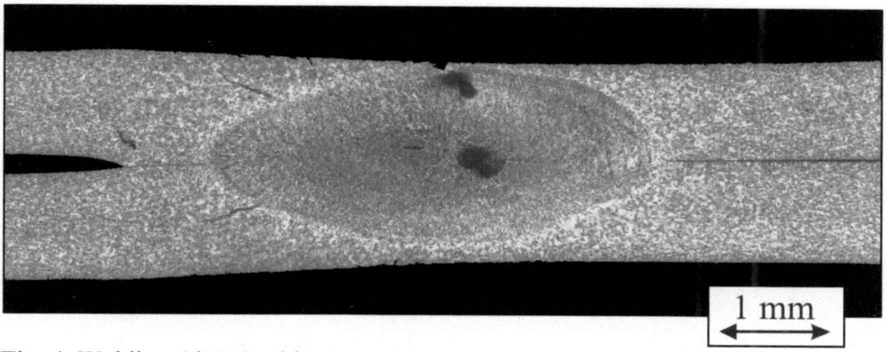


**Fig. 4.** Welding AlMg3 without a process tape

These phenomena do not appear for welding AlMg3 with a process tape (in this case PT1407). The process parameters where chosen as in the conventional spot welding example, except the current which could even be reduced. In the macro-section there are just some small eutectics visible. Compare Fig. 5. The nugget size is larger and it has a cylindrical form which leads to a higher tensile strength – Fig. 6 [6]. The cylindrical form of the weld nugget is caused by the additional heat from the process tapes. This changes the cooling behavior and the stress distribution in the sheets which helps reduce cracking sensitivity. It also reduces the gap adjacent to the weld nugget.

As mentioned above for spot welding with a process tape less current is necessary due to the additional heat from the tape. It makes a weld more economically. Additionally the electrodes stay clean due to the protection of the tape. Therefore a high quality of the sheet surface appearance can be reached.

**Fig. 5.** Welding AlMg3 with a process tape

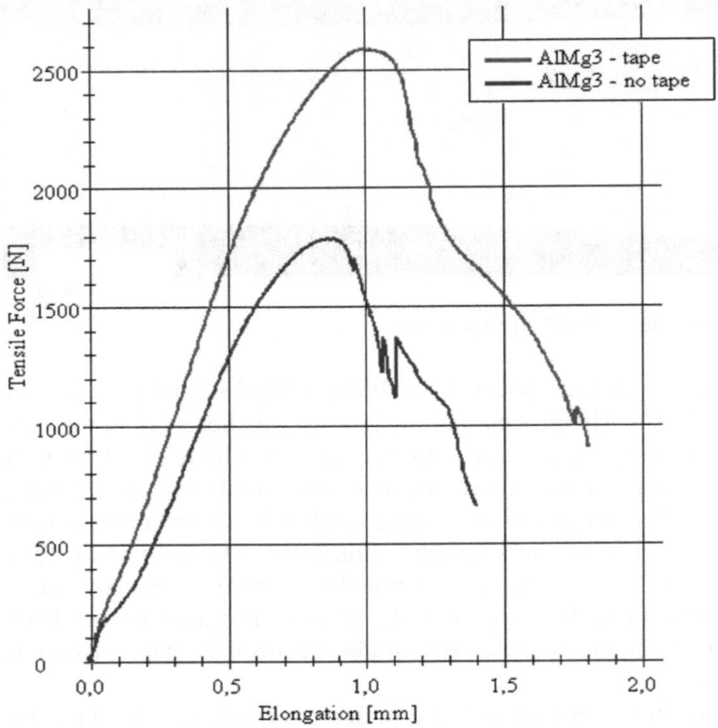


**Fig. 6.** Tensile test results for AlMg3

## Hot Cracks in Spot Welding of 6xxx Alloys

Spot welding of the high strength aluminum alloys is much more sophisticated to solve. Test welds of 6xxx (AA6016) alloys with the process tape PT 1200 were made. The welding parameters were 3 kN with one pulse of approx. 17 kA for two 1 mm sheets. The standard electrode R70 was used. Even with the aid of a process tape cracks appeared at the surface and in the nugget – see Fig. 7. Due to the large solidification interval of AA6016 and the critical concentration of the alloying elements a high risk of cracking is present.

In order to understand and to solve this problem the temperature profile during cooling of the spot weld was needed. Therefore a series of Finite Element simulations with the commercial available resistance welding program SORPAS® were made.

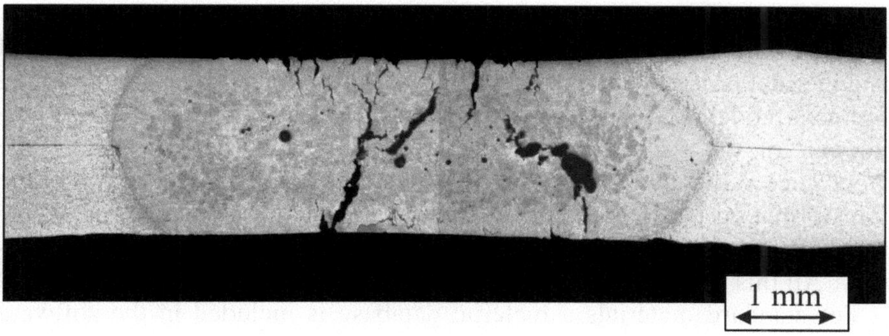


**Fig. 7.** Hot cracks in spot welded AA 6016 sheets

## Resistance Spot Welding Simulation with a Tape

The simulation of the resistance spot welding process involves strong interactions between the work pieces, electrodes, interfaces, process tape, coatings and machine settings. Figure 8 [7, 8] shows the main components which have to be considered for resistance spot welding with a process tape.

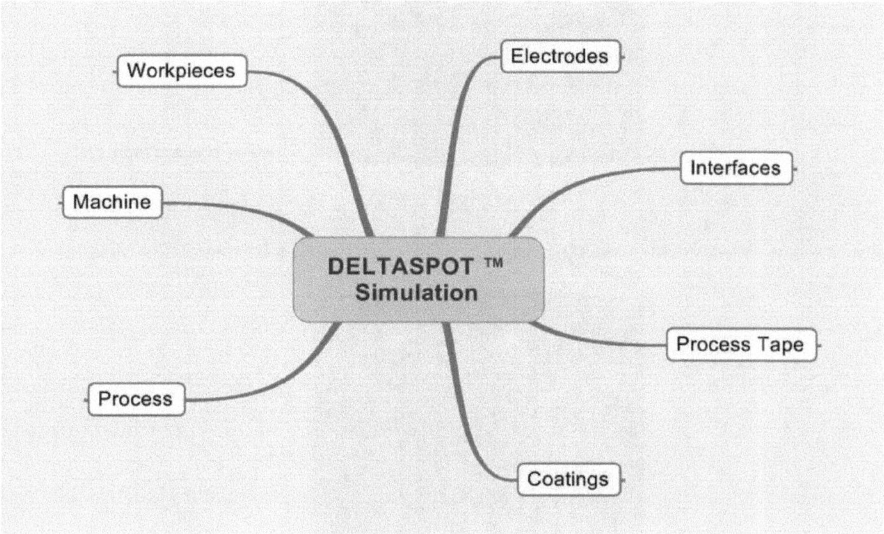


**Fig. 8.** Simulation components for the DELTASPOT™ process

SORPAS® takes into account the metallurgical, electrical, mechanical and thermal process that will occur during the welding process using Finite Element modeling. The software simulates the heat generated by the welding current, the resistance of the electrodes, the tape, the coatings and the work piece, the heat transfer across materials, the metallurgical phase transformation (solid to liquid) caused by temperature change, the deformation and strain distribution across the contact areas and the process window. All this information is then presented in graphic form to the user.

For standard materials a material database is included in the software package, but also new materials can be added if necessary. Special care has to be taken for the process tape and the coatings since they are relatively thin and have a different physical behavior compared to the work-piece and the electrodes. The coatings have an influence on the contact resistance and are usually evaporated during the process. Additionally some machine setting parameters like electrodes, and current and force curves are needed to complete the process model.

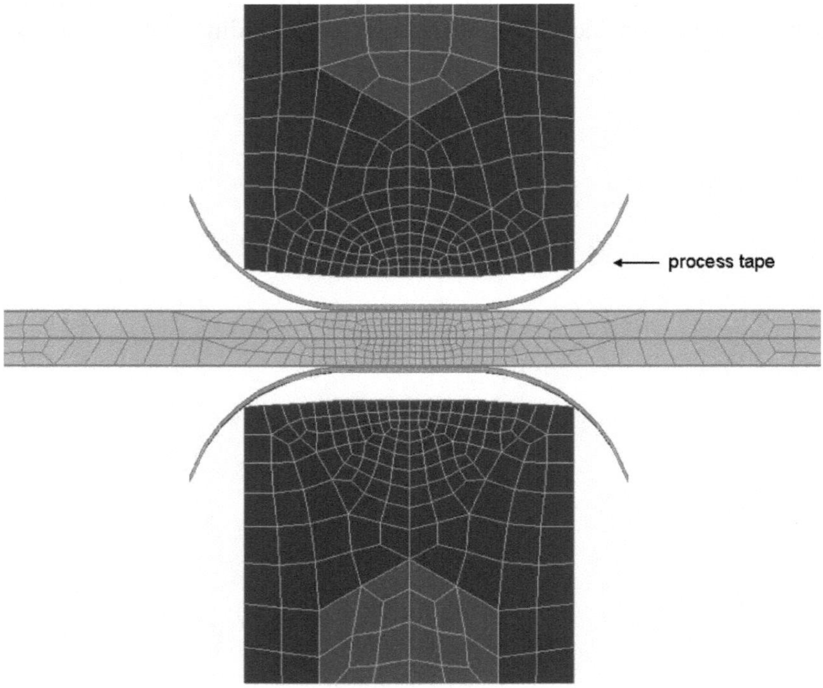


**Fig. 9.** FEM mesh for resistance spot welding with a process tape

A typical FEM mesh for solving a resistance spot welding process with an additional tape is shown in Fig. 9. An adaptation of the standard program settings was necessary to be able to include the thin tape and coatings.

## Mathematical Models for Spot Welding Simulation

In order to calculate the time dependent distributions of voltage, current and temperature in materials and electrodes we need the following governing axi-symmetric differential equations [7] using cylindrical coordinates $(r,\phi)$.

The differential equation for the electrical potential $\phi$ is

$$\frac{1}{r}\frac{\partial}{\partial r}\left(r\sigma_r\frac{\partial\phi}{\partial r}\right)+\frac{\partial}{\partial z}\left(\sigma_z\frac{\partial\phi}{\partial z}\right)=0, \tag{1}$$

where $\sigma$ is the electrical conductivity of the materials.

The temperature distribution $T$ is a result of the transient heat transfer differential equation with internal heat source $\dot{Q}$

$$\frac{1}{r}\frac{\partial}{\partial r}\left(r\lambda_r\frac{\partial T}{\partial r}\right)+\frac{\partial}{\partial z}\left(\lambda_z\frac{\partial T}{\partial z}\right)+\dot{Q}=\rho\,C\frac{\partial T}{\partial t}, \tag{2}$$

where $\lambda$ is the thermal conductivity, $\rho$ is the mass density and $C$ is the heat capacity. All material data is given in a temperature dependent way, which allows phase transformation calculations.

The deformation, the stress, strain distribution and the contact areas at interfaces are calculated by the variation approach for the functional of the potential energy

$$\pi=\int_V\bar{\sigma}\dot{\bar{\varepsilon}}dV-\int_S FvdS,\qquad \frac{\partial\pi}{\partial v}=0 \tag{3}$$

The first term on the right hand side is the potential energy of the bulk deformation. The second term respects the boundary conditions due to external load or velocity and friction etc. Wanheim and Bay's friction theory for real contact areas [9] is used for the contact resistivity:

$$\rho_{contact}=3\left(\frac{\sigma_{s\_soft}}{\sigma_n}\right)\left(\frac{\rho_1+\rho_2}{2}+\rho_{calibration}\right), \tag{4}$$

where $\sigma_{s\_soft}$ is the flow stress of the softer metal, $\sigma_n$ is the contact normal pressure at the interface. The subscripts 1 and 2 indicate the two base metals in contact. To take account for surface contaminants such as oil, water, dirt and oxides etc. a term called $\rho_{calibration}$ is introduced in the model. The quantitative influence of the surface contaminants on the contact resistance is usually not known. Therefore the parameter $\rho_{calibration}$ in Eq. (4) can be used for calibration. In order to complete the mathematical model, appropriate boundary and initial conditions are used for the electrical potential, temperature field and mechanical model.

### Solver Strategy

Equations (1–4) represent strongly coupled models, an electrical, a thermal, a metallurgical and a mechanical one. These models are all influenced by the dynamic behavior of the materials, interfaces, the machines and the processes.

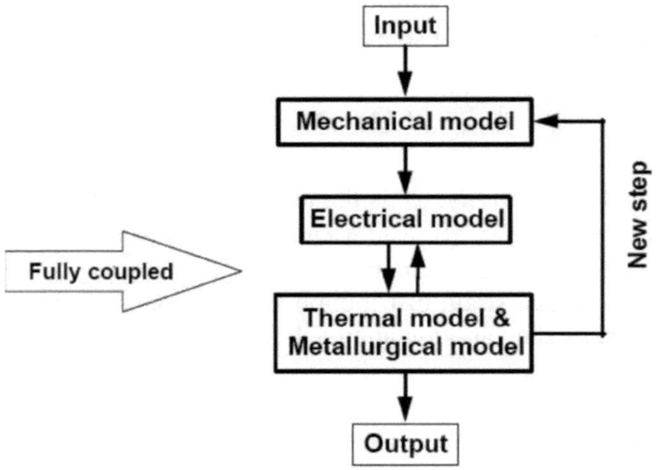


**Fig. 10.** SORPAS® procedure to solve the governing equations

FEM techniques are applied for Eqs. (1–4) to obtain a numerical discretization. Figure 10 shows the solver strategy, designed to obtain efficient spot welding simulations [7].

### Parameter Optimization Process

During resistance spot welding the main process parameters, welding current and electrode force, can be varied in order to get high quality welds. At the beginning of a spot weld the force increases in a linear manner which leads to a good electrical contact between the work pieces. After this preload the current pulse starts. Depending on the material there can be used one or several current pulses. For example it can be useful to set a preheating pulse to avoid critical cooling rates [10].

The process engineer has to choose the right process curve for the materials to be welded. Due to the variety of process curves this can be a difficult and time consuming task. SORPAS® allows any force and current form as well as material combinations which makes it suitable for process optimization. Several simulations can be run successively in a batch process for parametric studies. The optimization loop applied for solving difficult spot welding problems is shown in Fig. 11.

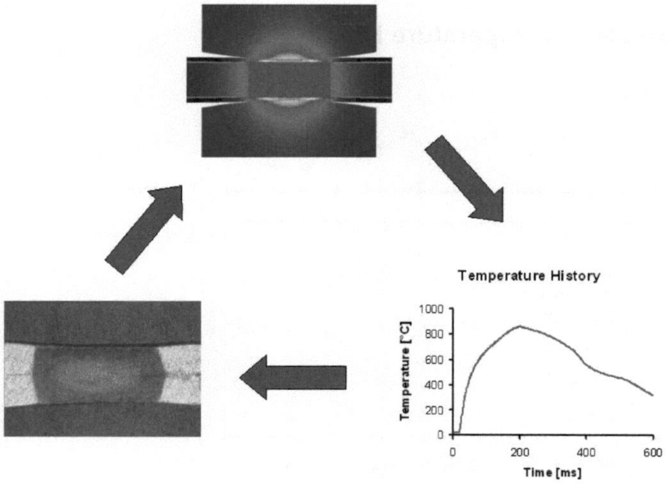


**Fig. 11.** Virtual resistance spot welding optimization loop

Usually the simulation starts on the basis of a first practical trial. According to a spot weld macro-section resulting from a first guess of a current-force pulse the simulation model is calibrated. The calibration usually is done via the contact resistance model – compare Eq. (4). Then virtual parameter studies by variation of the shape of the current-force pulse have to be done. This results in a series of work piece temperature profiles. These temperature histories in combination with metallurgical knowledge (e.g. CCT-diagrams, optimal cooling rates) can be used to get suggestions for a new practical trial. If the following welding experiment is of sufficient quality the loop is finished, otherwise it starts again (new adapted model, other strategies, and so on). In order to get reliable results a strong collaboration between Numerical Simulation and experimentation engineers has to be guaranteed.

## Parameter Studies for High Strength Aluminum Alloys

The goal for resistance spot welding the aluminum alloys AA 6016 and AA 6181 was to minimize hot cracking as well as to get the desired nugget size. A virtual parameter study for different types of welding process curves (current and force) was done in SORPAS®. Exemplary are some resulting temperature histories close to the work piece surface on the symmetry axis for AA 6016 as shown in Fig. 12.

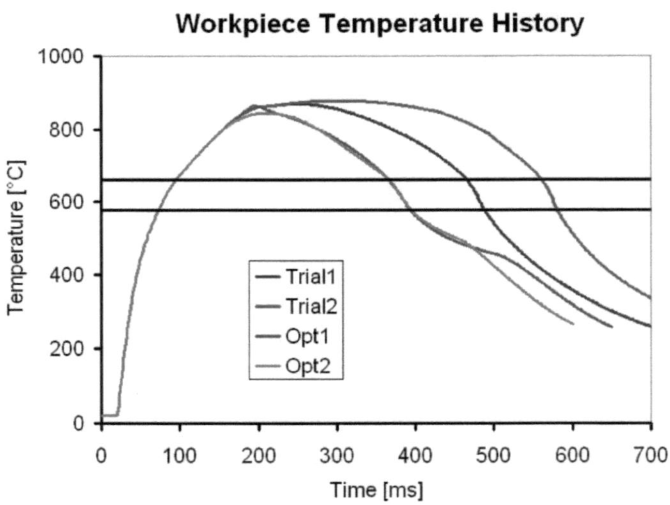


**Fig. 12.** Simulation of temperature history close to the work piece surface

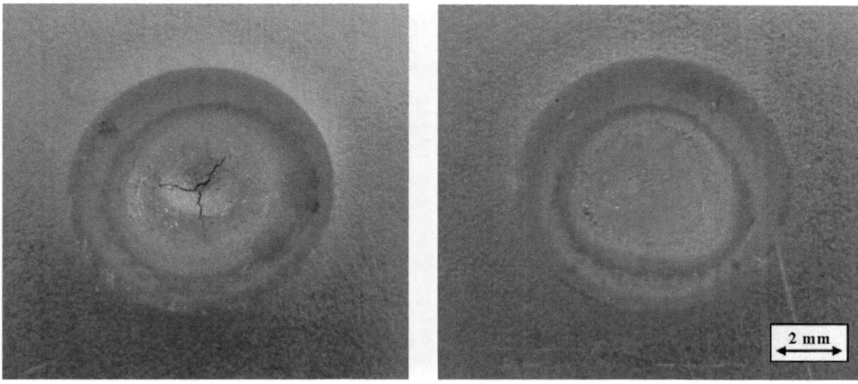


**Fig. 13.** Cracks at the surface for an optimized AA 6016 weld

The simulated temperature curves are results of different current-force welding process curves. In Fig. 12 the solidification interval is marked by bold black horizontal lines. Trials 1 and 2 show the temperature history of the first simulated trial process curves which were not sufficient for crack prevention. At the end of the parameter study the lower cooling rates of the optimized curves Opt1 and Opt2 are expected to prevent from crack formation. The optimized cooling slope allows the healing of the surface cracks due to the force applied during solidification. Figure 13 shows the surface cracks of an AA 6016 spot weld at the beginning of the parameter study and the crack-free weld with an optimized process curve. The real and the virtual macro-sections of the spot welding are in good agreement as can be seen in Fig. 14.

**Fig. 14.** Validation of the simulation results

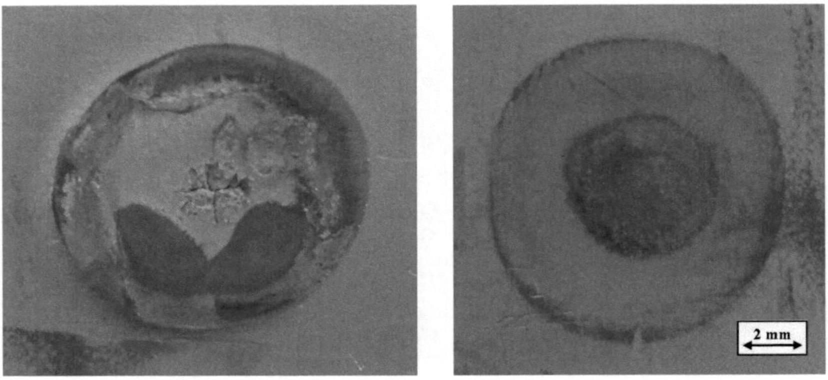


**Fig. 15.** Surface of an AA 6181 weld before and after optimization

For resistance spot welding of 2 mm AA 6181 sheets with the process parameters 24 kA, 220 ms, 5 kN, electrode R150 and process tape PT1407 the same procedure was performed. The successful optimization of the welding task can be seen in Fig. 15. Even the extreme cracks at the surface disappear.

## Conclusion

Resistance spot welding of aluminum alloys is a difficult task. Hot cracking is one of the main problems to be solved. The goal in this study was to minimize cracking phenomena especially for the 6xxx alloys AA6016 and AA 6181. The solution strategy was based on numerical simulation together with an innovative resistance spot welding process including a process tape.

First the difference between spot welding with and without a process tape was demonstrated by the use of DELTA SPOT™ for resistance spot welding of AlMg3. Additionally SORPAS®, a commercial resistance welding simulation software that uses finite element techniques, was introduced as a tool for spot welding simulation including a process tape.

For the difficult to weld AA 6016 and AA 6181 aluminum alloys virtual parameter studies for resistance spot welding with a process tape were performed. Validation and analysis of the simulations led to optimized temperature histories and finally to crack free welding process curves.

Solving difficult spot welding problems is not anymore just a question of trial and error experiments. Simulation can help to investigate difficult spot welding problems by supplying fast and reliable answers.

## References


1. Stieglbauer W (2006) DeltaSpot – Spot welding with process tape. IIW Doc.III-1381-06
2. Irving B (2000) The auto industry gears up for aluminum. Welding Journal, 11:63–68
3. Robson J (2002) Designing high strength aluminum alloys for aerospace applications. Manchester Materials Science Centre.
4. Krüger U (1994) Weldability. TALAT Lecture 4202, Schweißtechnische Lehr- und Versuchsanstalt Berlin. EAA – European Aluminium Association.
5. Senkara J, Zhang H (2000) Cracking in spot welding aluminium alloy AA5754. Welding Journal 7:194–201
6. Sun X, Stephens EV, Davies RW, Khaleel MA, Spinella DJ (2004) Effects of fusion zone size on failure modes and static strength of aluminum resistance spot welds. Welding Journal 11:308–318
7. Zhang W (2003) Design and implementation of software for resistance welding process simulations. SAE 2003 World Congress, Detroit, USA
8. Zhang W (2005) Recent advances and improvements in the simulation of resistance welding processes. IIW Doc.III-1345-05
9. Wanheim T, Bay N (1987) A model for friction in metal forming processes. Annals of the CIRP 27:189–194
10. Matthes KJ, Richter E (2006) Schweißtechnik, Schweißen von metallischen Konstruktionswerkstoffen. Carl Hanser Verlag München Wien. ISBN-10: 3-446-40568-2

# Part IV
# Liquation Cracking

# Evaluating Hot Cracking Susceptibility of Ni-Base SAW Consumables for Welding of 9% Ni Steel

L. Karlsson, E-L Bergquist, S. Rigdal, N. Thalberg

ESAB, Göteborg, Sweden

## Abstract

A study of Ni-base weld metal hot cracking resistance was made comparing results from Modified Varestraint-Transvarestraint Testing (MVT), metallographic evaluation and mechanical testing. Submerged arc weld metals were produced with Alloy 625, Alloy 59 and Alloy C-276 filler wires at two levels of heat input. Welds were subjected to MVT testing, three-point and wrap-around bend testing as well as metallographic inspection of weld cross-sections. All test methods generally ranked the lower heat input welds as more crack resistant than those produced with a higher heat input. However, MVT testing ranked alloy types in a different order than mechanical testing and metallographic evaluation. Alloy 59 type weld metals were judged most crack resistant with MVT testing and C-276 welds were ranked as slightly better than those produced with Alloy 625 wires. Mechanical testing and metallographic evaluation, on the other hand, ranked the Alloy 625 type weld metals as most crack resistant and Alloy C-276 weld metals as least resistant. Discrepancies between results of the different test methods and practical experience can most likely be attributed to a number of factors including metallurgical aspects and test procedures. It is concluded that test methods simulating actual application welding conditions as closely as possible are most likely to accurately predict hot cracking susceptibility of real weldments. An assessment of the hot cracking susceptibility of welding consumables should therefore preferably be based on results from a combination of test methods.

## Introduction

Large vessels and tanks for transportation and storage of liquefied natural gas (LNG) at cryogenic temperatures are since many years commonly constructed from 9% Ni steel. One of the most important steps in the fabrication of storage tanks is welding. Obviously both productivity and consistent high weld quality are key issues. Submerged arc welding (SAW) offers the highest productivity and is therefore usually the preferred method when geometry and welding position permits [1, 2].

The weld metal should provide low temperature properties matching those of the 9% Ni base material. Since many years, it is therefore standard practice to use Ni-base consumables. These safely comply with requirements on mechanical properties, including strength and ductility, and have a thermal expansion coefficient closely matching that of 9% Ni steels. Several types of Ni-based consumables have been developed and the principal compositions are now standardised (e.g. solid wires for SAW according to AWS A5.14 and EN 18274). However, in practice only a few are able to meet the very high requirements for 9% Ni applications. Basic submerged arc welding fluxes are considered mandatory to produce a clean deposit with a low amount of microslag inclusions improving toughness and minimising the risk of hot cracking. Many of these Ni-base alloys are also used for welding of highly austenitic stainless where overalloying is needed to match parent material corrosion resistance [3].

Several hundreds of large storage tanks and vessels fabricated from 9% Ni steel and welded with Ni-base consumables are in operation. Many years of failure-free experience from these show that fit-for-purpose welds can be produced but different opinions exist as to which filler materials are most suitable.

Hot cracking tests are rarely parts of welding consumable qualification and resistance to hot cracking is generally not specifically discussed. It is nevertheless an important aspect in the selection and qualification of consumables and is sometimes identified as a problem through evaluation of metallographic specimens. The main influence is however more indirect sometimes causing unacceptable cracking of transverse and longitudinal bend test specimens or low elongation and premature fracture of tensile test specimens. A complication is that both the filler materials inherent hot cracking susceptibility and the welding procedure influence these results. Since a badly designed welding procedure can cause hot cracking in any type of weld metal it can be difficult to agree on whether the consumable or the procedure is to be blamed. Reliable methods for evaluation of hot cracking susceptibility producing results comparable to those from fabrication experience are therefore needed.

Many different hot cracking test methods have been developed over the years [4]. In principle all of these can be classified into two major groups:

- Self Restrained Tests where the specimen loading is produced by stresses developed during welding of a restrained weldment and
- Externally Loaded Tests that require the use of equipment to impose a strain on the specimen during welding or some other thermal cycle.

Some tests are fully standardised whereas others are only described in EN/ISO technical reports [4, 5]. Externally loaded tests such as the Varestraint and the Transvarestraint tests have, although not being fully standardised, the advantage of rapid testing, low scatter of results and good reproducibility (with a single machine). Useful data can also be generated from a small number of tests.

The chapter presents results from a study comparing results from Modified Varestraint-Transvarestraint Testing (MVT), metallographic evaluation and mechanical testing. SAW weld metals produced with three types of Ni-base filler wires (Alloy 625, Alloy 59 and Alloy C-276) at two levels of heat input were tested. Test results are presented and the influence of test method on hot cracking susceptibility ranking is discussed.

## Experimental

Three types of Ni-base SAW weld metals were subjected to MVT testing, metallographic evaluation and bend testing.

### Weld Metal

SAW weld metals were produced with three types of 1.6 mm diameter Ni-base filler wires (Alloy 625, Alloy 59 and Alloy C-276). Welding was done with a basic flux at two levels of heat input for each filler wire type. Typical all-weld metal composition and properties for the three SAW wire/flux combinations are presented in Tables 1 and 2.

**Table 1.** SAW consumables for 9% Ni-steel. Typical all-weld metal composition (wt-%)

| Filler wire alloy type | C | Si | Mn | Cr | Ni | Mo | Nb | W | Fe |
|---|---|---|---|---|---|---|---|---|---|
| 59 | <0.01 | 0.2 | 3 | 22 | Bal. | 14 | – | – | 3 |
| 625 | 0.01 | 0.2 | 1.5 | 21 | Bal. | 8.5 | 3 | – | 2 |
| C-276 | 0.01 | 0.2 | 1.9 | 15 | Bal. | 14 | – | 3.5 | 7 |

**Table 2.** Consumable classification and typical all-weld metal properties

| Filler wire alloy type | Consumable classification AWS A5.14 / EN 18276 | Tensile properties $R_{p0.2}$ [MPa] | $R_m$ [MPa] | $A_5$ [%] | Charpy-V -196°C [J] |
|---|---|---|---|---|---|
| 59 | ERNiCrMo-13 / S Ni 6059 | 470 | 675 | 45 | 70 |
| 625 | ERNiCrMo-3 / S Ni 6625 | 440 | 720 | 40 | 90 |
| C-276 | ERNiCrMo-4 / S Ni 6276 | 480 | 700 | 35 | 75 |

A standard AWS joint configuration with 60° included angle, 13 mm root opening and a backing plate was used. Joint faces were buttered prior to deposition of the welds to minimise dilution with parent material. Welds were produced with a heat input of 0.9 kJ/mm and 1.5 kJ/mm. The weld metals were designated starting with the type of filler wire alloy followed by L or H for the lower and the higher heat input level, respectively. For example 625-H stands for a weld metal produced with an Alloy 625 wire with a heat input of 1.5 kJ/mm.

One set of welds was subjected to MVT testing followed by additional metallographic evaluation whereas the other was studied by three-point and wrap-around bend testing as well as inspection of weld cross-section.

## Metallographic Evaluation

Weld cross-sections were examined for the presence of cracks. Three specimens were extracted from each of the six welds, near the start and stop ends and from the centre. A limited study was also done on some specimens subjected to MVT testing.

The cross-sections were prepared using standard metallographic techniques, electrolytically etched in 70% $H_3PO_4$ and examined using a stereo microscope and at higher magnification in a light optical microscope. Cracks were counted and classified based on their location relative to bead fusion boundaries. Those connected to a fusion boundary of a bead were classified as liquation (and /or ductility dip) cracks and cracks in the interior of beads as solidification cracks (Fig. 1).

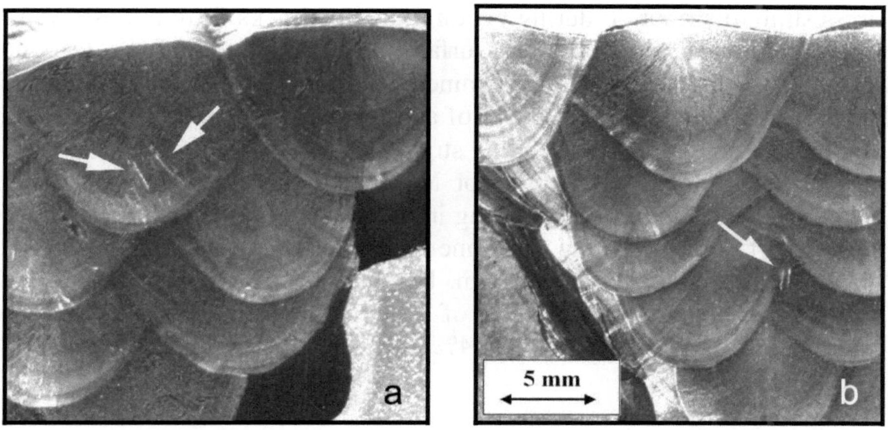


**Fig. 1.** Weld cross sections illustrating classification of cracks based on location relative to bead fusion boundaries
(**a**) Solidification cracks in the interior of beads in weld 276-L
(**b**) Liquation cracks at bead fusion boundary in weld 59-L

## Bend Testing

The presence of cracks was also evaluated using three-point bend and wrap around bend testing. Three-point bend testing was done with 10 mm thick, transverse side bend specimens and longitudinal face bend test specimens. The diameter of the former was in both cases 40 mm and the bending angle was 180°.

Wrap around testing was done using 5 mm thick, transverse side bend test specimens. The diameter of the former was 24 mm and the bending angle was 180°. All testing was done in accordance with SS-EN 910.

## MVT Testing

In MVT testing a GTA torch is used to remelt the machined surface of the weld metal to be tested. As the arc passes a predefined point the specimen is bent at a constant rate around a die. The specimen surface is thereby subjected to a strain of well-defined level and rate. As a result of the stress/strain distribution during bending the hot cracks form almost exclusively at or near the specimen surface. On completion of the test the specimen surface is visually examined for cracks at a magnification of X25. The number and total length of all visible cracks are determined and plotted as a function of the bending strain [5].

Specimens (10×40×100 mm) for MVT testing were machined 2 mm below the weld top surface resulting in a test surface located largely in the top layer of beads. Welding was done under Ar and with a nominal energy input of approximately 0.75 kJ/mm. Bending was done transverse to the welding direction at a bending rate of 1.8 mm/s with bending radii selected to produce strains of 1%, 2% and 4%. One specimen was tested per weld and strain level.

## Results

Mechanical testing and metallographic evaluation of hot cracking tendency produced the same ranking of filler wires irrespective of heat input. However, MVT testing resulted in an entirely different ranking.

### Metallographic Evaluation

The metallographic evaluation of cross-section produced the same very clear ranking of the three types of filler wires regardless of whether solidification, liquation (and ductility dip) or the total number of cracks was considered (Table 3). 625 welds contained only a few small cracks (1-2 per cross-section), the 59-type welds an average of 4-5 cracks and the C-276 welds typically 10 cracks.

A representative cross section is shown for each type of filler wire in Fig. 2. As seen in the figure, the C-276 type welds not only contained a larger number of cracks but the average size also tended to be larger.

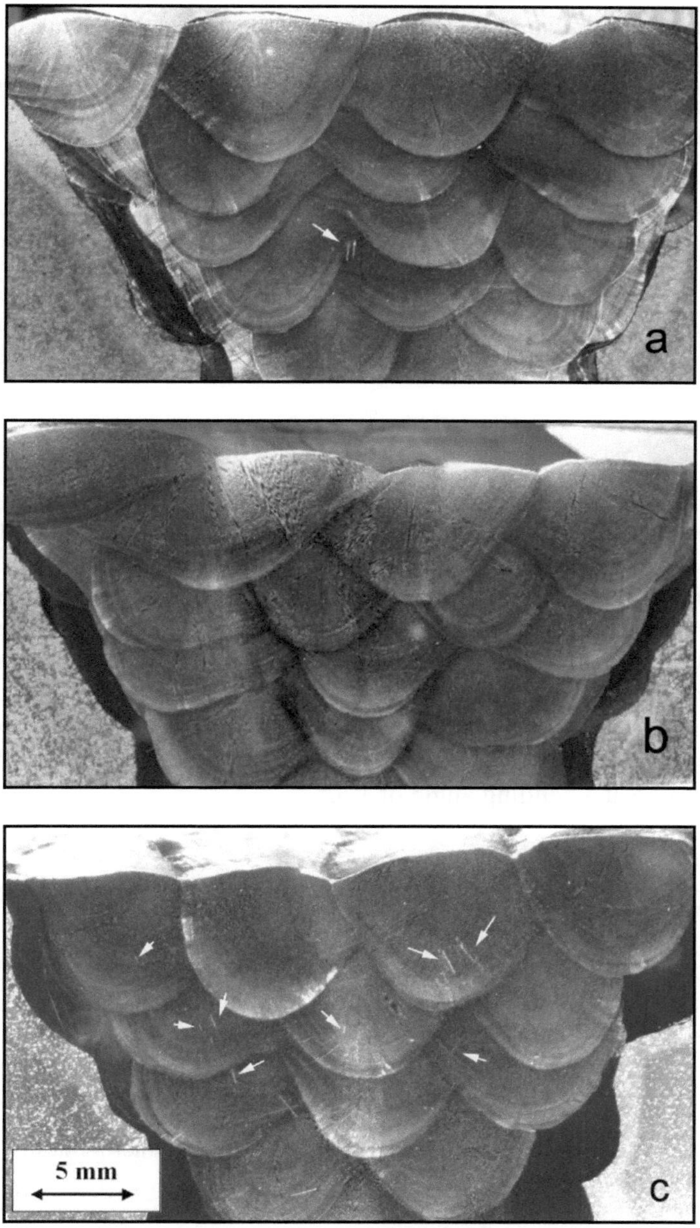


**Fig. 2.** Examples of cross-sections used for metallographic evaluation of hot cracking tendency. (**a**) 59L, (**b**) 625-L, (**c**) 276-L

**Table 3.** Number of hot cracks as determined by metallographic evaluation on three weld cross-sections from each weld

| Weld | Number of cracks (on 3 cross-sections) | | |
|---|---|---|---|
| | Solidification cracks | Liquation/ductility dip cracks | Total |
| 59-L | 4 | 8 | 12 |
| 59-H | 0 | 15 | 15 |
| 625-L | 1 | 2 | 3 |
| 625-H | 1 | 4 | 5 |
| 276-L | 22 | 9 | 31 |
| 276-H | 10 | 21 | 31 |

## Bend Tests

Results from bend testing are summarised in Table 4. The two types of side bend tests and the longitudinal bend test were in complete agreement. Welds 625-L and 625-H passed without remarks except in side bend wrap around testing where one small crack was detected at the edge of the specimen. Only a limited number of small cracks shorter than 0.5 mm (here called micro-cracks) were seen in 59-L and 59-H specimens (Fig. 3). All 276-L and 276-H specimens contained numerous micro-cracks and in most cases also larger cracks (Fig. 4). One 276-H 3-point side bend specimen even broke at a bending angle of 120°.

**Table 4.** Bend test results

| Weld | Side bend, 3-point | Longitudinal face bend | Side bend, wrap around |
|---|---|---|---|
| 59-L | micro-cracks | micro-cracks | micro-cracks |
| 59-H | micro-cracks | micro-cracks | micro-cracks |
| 625-L | no remark | no remark | no remark |
| 625-H | no remark | no remark | 1 micro-crack at edge |
| 276-L | large cracks and micro-cracks | large cracks | several micro-cracks |
| 276-H | large crack and micro-cracks, one specimen broken at 120° | micro-cracks | several micro-cracks |

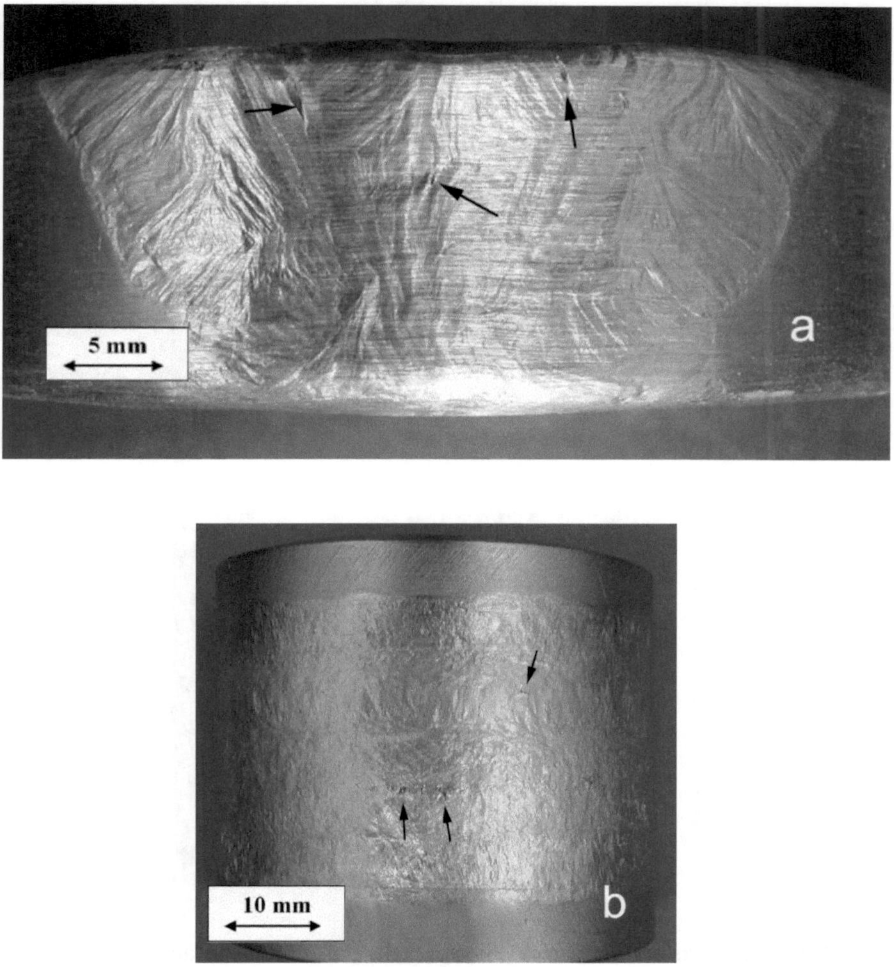


**Fig. 3.** Micro-cracks on strained surface of 59-H bend test specimens
(**a**) 3-point side-bend specimen
(**b**) Longitudinal face-bend specimen

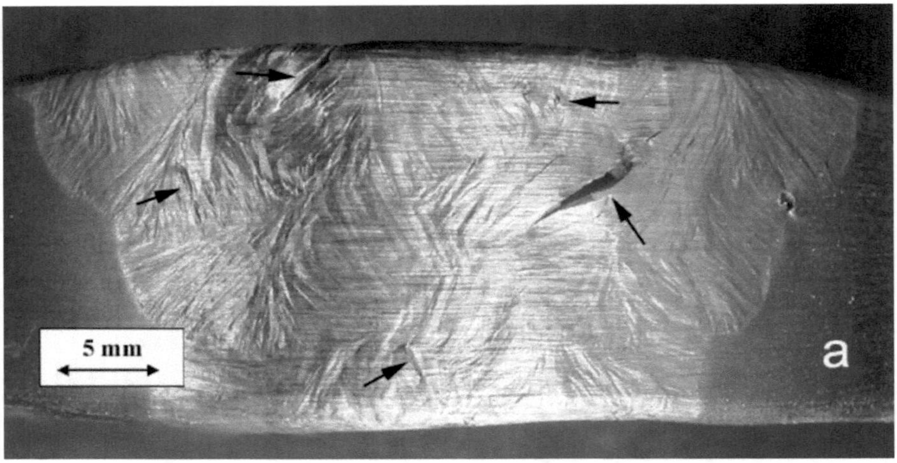


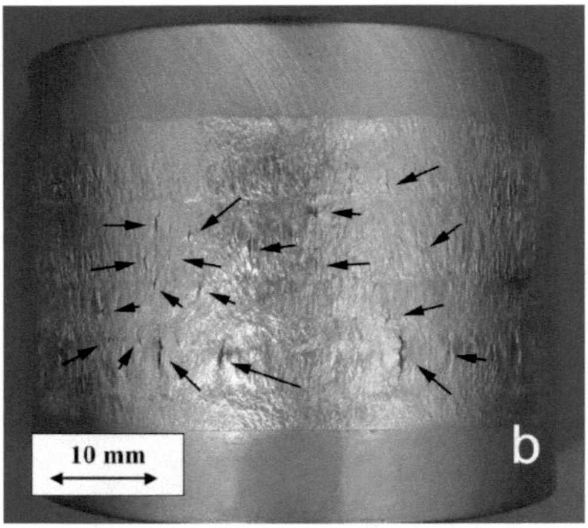


**Fig. 4.** Large cracks and numerous micro-cracks on strained surface of 276-L bend test specimens
(**a**) 3 point side-bend specimen
(**b**) Longitudinal face-bend specimen

## MVT Tests

Results from MVT testing is presented as tables giving total cracks lengths (Table 5) and total number of cracks (Table 6) for all strain rates, as a standard crack length versus strain diagram (Fig. 5) and as a diagram presenting number of cracks against strain (Fig. 6).

**Table 5.** Results of MVT tests presented as combined crack lengths for all strains

| Weld | Combined crack length for 1%, 2% and 4% strain (mm) | | | | |
|---|---|---|---|---|---|
| | GTA WM | Weld metal HAZ | GTA WM + Weld metal HAZ | "Distant cracks" | Total |
| 59-L | 10.2 | 3.8 | 14.0 | 5.3 | 19.3 |
| 59-H | 6.8 | 4.1 | 0.9 | 2.1 | 13.0 |
| 625-L | 16.8 | 3.1 | 19.9 | 1.6 | 21.5 |
| 625-H | 17.6 | 2.6 | 20.2 | 2.2 | 22.4 |
| 276-L | 12.1 | 5.6 | 17.7 | 4.0 | 21.7 |
| 276-H* | 15.3 | 6.8 | 22.1 | 10.3 | 32.4 |

* only tested at 2% and 4% strain

**Table 6.** Results of MVT test presented as the total number of cracks for all strains

| Weld | Number of cracks | | | | |
|---|---|---|---|---|---|
| | GTA WM | Weld metal HAZ | GTA WM + Weld metal HAZ | "Distant cracks" | Total |
| 59-L | 21 | 10 | 31 | 18 | 59 |
| 59-H | 16 | 12 | 28 | 7 | 35 |
| 625-L | 29 | 14 | 43 | 10 | 53 |
| 625-H | 38 | 19 | 57 | 4 | 61 |
| 276-L | 22 | 14 | 36 | 17 | 53 |
| 276-H* | 20 | 18 | 38 | 37 | 75 |

*only tested at 2% and 4% strain

Newly formed cracks, at a distance from the GTA bead fusion boundary (ductility dip cracks), could not easily be distinguished from hot cracks already present in the weld metals before MVT testing. The "distant" cracks are therefore included in the tables but excluded from the data used to construct diagrams.

Using combined crack lengths or total number of cracks can produce various rankings of hot cracking sensitivity depending on whether cracks in weld metal, HAZ or distant cracks are considered. Looking only at weld metal cracks the 59 type weld metals were least crack sensitive and the 625 welds most prone to cracking. Ranking in HAZ depends on whether number of cracks or crack length is used as criteria. However, including or excluding distant cracks has the most dramatic effect on ranking. The C-276 type weld metals were clearly most crack prone at a distance from the GTA bead whereas the 625 were least crack sensitive in this region.

Including all cracks thereby shifted the ranking suggesting the 625 or 59 type weld metals to be most resistant and the C-276 type more likely to crack.

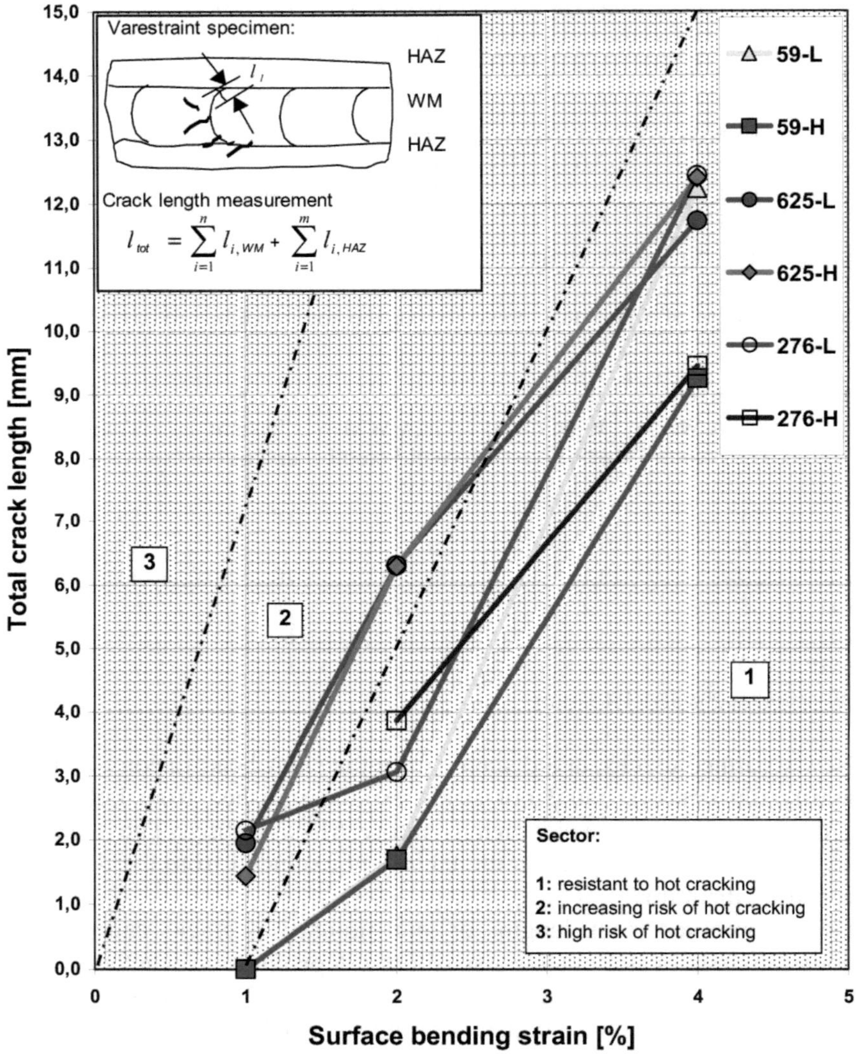


**Fig. 5.** Results of MVT testing. Cracks formed at a distance from the GTA remelted bead were not included.

Results are presented in Fig. 5 in the standard format presenting total crack length versus strain. The ranking is varying with strain but the tendency is that the 59 type weld metals are most crack resistant and the

625 weld metals most prone to hot cracking. According to the classification normally used when interpreting MVT diagrams the 625 type weld metals would be considered restricted weldable and the 59 and C-276 types weldable or restricted weldable depending on strain rate and heat input [5].

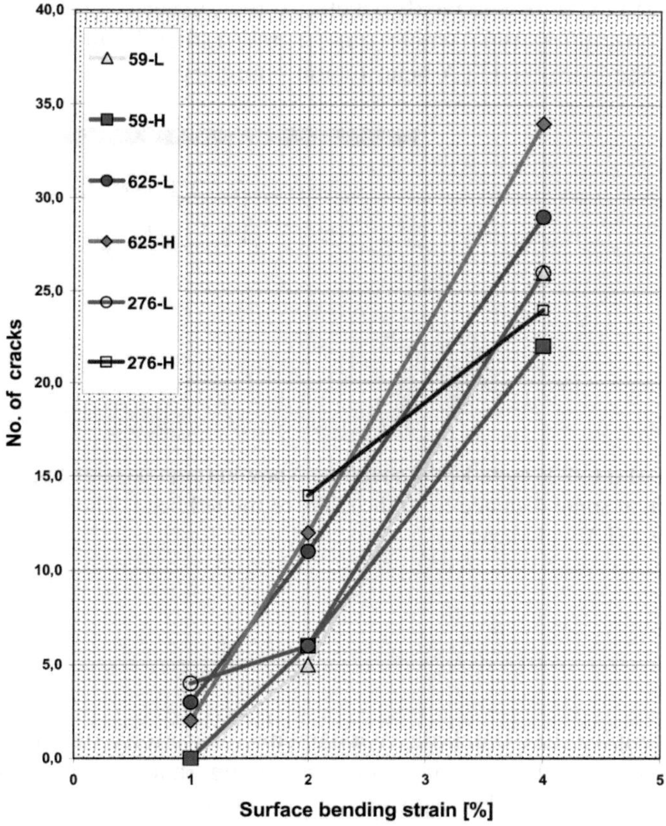


**Fig. 6.** Results of MVT testing presented as number of cracks versus surface strain. Only cracks in GTA re-melted bead and adjacent heat affected material were considered

A presentation of the results as the number of cracks against bending strain (Fig. 6) resulted in virtually the same ranking as in the previous figure. The general tendency was again that the 625 weld metals were most prone to hot cracking and the 59 type least. The main difference was at the intermediate strain were the largest number of cracks were found in the 276-H weld metal.

A limited metallographic study of cross-sections of MVT specimens revealed that most but far from all cracks were surface breaking. Although statistics is limited there appeared to be a tendency that relatively fewer of the hot cracks formed in C-276 type welds penetrated the top surface (Fig. 7). This was also the only alloy type where quite a few rather large cracks were seen at a distance from the GTA bead fusion boundary.

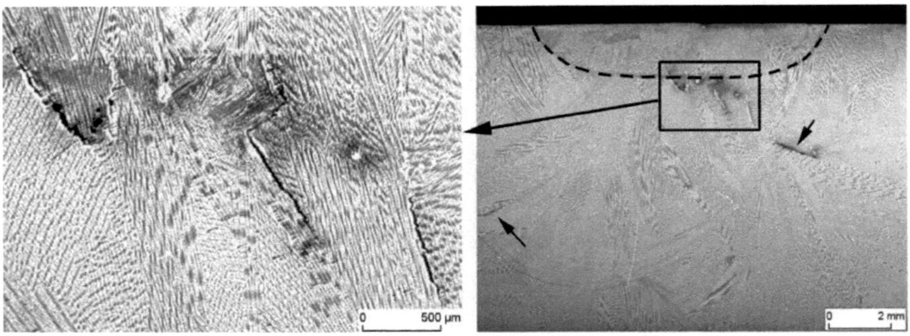


**Fig. 7.** Cross section of a C-276 type MVT specimen showing non-surface breaking cracks and a crack at some distance from the GTA re-melted bead. The broken line outlines the MVT GTA weld bead

## Discussion

Considering the complex hot cracking mechanisms, variations of stress and strain fields within welds etc. it is not surprising that various test methods or tests on different welds with nominally identical compositions can give different results. However, it is necessary to understand when well-established test methods can be used as a tool for consumable selection and when not. The following paragraphs will therefore discuss possible reasons for discrepancies between the hot cracking test results presented above.

### Ranking of Hot Cracking Susceptibility

Weld metals were ranked with respect to hot cracking susceptibility for each test (Table 7). As noted earlier, evaluation based on metallography and bend testing was in very good agreement ranking the 625 weld metals as best and C-276 weld metals as worst. All methods also ranked the lower heat input welds as more crack resistant. As noted earlier these results were in poor agreement with MVT test results.

**Table 7.** Comparison of hot cracking tendency ranking based on results from metallographic evaluation, bend testing and MVT testing. One is most cracking resistant and six most prone to cracking

| Test method | Criteria | Weld | | | | | |
|---|---|---|---|---|---|---|---|
| | | 59 L | 59 H | 625 L | 625 H | 276 L | 276 H |
| Metallography | Solidification cracks | 4 | 1 | 2 | 2 | 6 | 5 |
| | Liquation & ductility dip cracks | 3 | 5 | 1 | 2 | 4 | 6 |
| | Total # of cracks | 3 | 4 | 1 | 2 | 5 | 5 |
| Side bend, 3-point | # and size (semi-quantitative) of cracks | 3 | 3 | 1 | 1 | 5 | 6 |
| Side bend, wrap around | | 3 | 4 | 1 | 2 | 5 | 5 |
| Face bend | | 3 | 4 | 1 | 1 | 6 | 5 |
| Av. ranking of metallographic and bend tests | | 3.2 | 3.5 | 1.2 | 1.7 | 5.2 | 5.3 |
| MVT: all strain rates | # of cracks in GTA WM | 3 | 1 | 5 | 6 | 4 | 2 |
| | # of cracks in GTA HAZ | 1 | 2 | 3 | 6 | 3 | 5 |
| | # of cracks in GTA WM & HAZ | 2 | 1 | 5 | 6 | 3 | 4 |
| | # of "distant" cracks | 5 | 2 | 3 | 1 | 4 | 6 |
| | Total # of cracks | 4 | 1 | 2 | 5 | 2 | 6 |
| | Crack length in GTA WM | 3 | 1 | 5 | 6 | 4 | 2 |
| | Crack length in GTA HAZ | 1 | 2 | 3 | 6 | 3 | 5 |
| | Crack length in GTA WM & HAZ | 2 | 1 | 5 | 6 | 3 | 4 |
| | Crack length "distant" cracks | 5 | 2 | 3 | 1 | 4 | 6 |
| | Total crack length | 4 | 1 | 2 | 5 | 2 | 6 |
| MVT: Cracks in GTA WM and HAZ | # of cracks at 1% strain | 1 | 1 | 4 | 3 | 5 | n.d |
| | # of cracks at 2% strain | 1 | 2 | 4 | 5 | 3 | 6 |
| | # of cracks at 4% strain | 3 | 1 | 5 | 6 | 4 | 2 |
| | Crack length at 1% strain | 1 | 1 | 4 | 3 | 5 | n.d |
| | Crack length at 2% strain | 2 | 1 | 6 | 5 | 3 | 4 |
| | Crack length at 4% strain | 4 | 1 | 3 | 5 | 6 | 2 |
| Av. ranking MVT tests | | 2.6 | 1.3 | 3.9 | 4.7 | 3.6 | 4.2 |

n.d. = no data

Evaluation of MVT test results did show more scatter in the ranking compared to metallography of cross-sections or bend testing. However, there was fairly good agreement between different criteria in that the 59 type weld metals were most crack resistant. On an average the C-276 welds were ranked as slightly better than the 625 type weld metals although the difference was small and ranking varied significantly with criteria. With the exception of the 59 welds, also MVT testing ranked lower heat input specimens as less crack susceptible than those produced with higher heat input.

It is notable than none of the criteria used to evaluate the MVT specimens produced a similar ranking as metallography of cross-sections or bend testing.

## Hot Cracking Test Methods and Cracking Mechanisms

The MVT ranking of the alloy types is in good agreement with previous results from MVT testing predicting Alloy 59 base material to be most crack resistant and 625 as most likely to crack [6]. Weld metals of 625 and C-276 type were ranked in the same order as parent material and higher heat input during testing produced generally more and longer cracks but did not change the ranking order. However, a more recent comparison reported Alloy 59 parent material to be significantly more susceptible to cracking than Alloy 625 at higher strains although the difference was very small at 1% or 2% strain [7]. Obviously care should be taken when comparing tests performed on different heats of the same material.

Alloy 59 and Alloy 625 type weld metals have good reputation for resistance to hot cracking in practical applications although problems are sometimes encountered [3]. Opinions about Alloy C-276 type filler metals vary but also these are successfully used in numerous applications. Comparing practical experience to test results it appears that there is agreement on 59 type weld metals being relatively crack resistant. On the other hand, metallography, bend testing and practical experience suggests a reasonably good hot cracking resistance whereas MVT ranks 625 type weld metals as the least resistant of the three alloys tested.

A number of factors contribute most likely to discrepancies between results of the different test methods and practical experience. These include metallurgical aspects such as solidification, segregation and precipitation behaviour of the alloys, tendencies to form non-surface breaking cracks and differences in heat input between actual welding procedures and MVT testing.

Both Alloy 625 and C-276 are known to have rather complex solidification behaviour [8, 9, 10]. Significant segregation occurs and solidification is completed by precipitation of topologically close-packed P-phase in Alloy C-276 whereas a eutectic-type reaction between the austenitic matrix and various Nb-rich phases such as NbC and Laves phase occurs in Alloy 625 [3, 11]. However, Alloy 59 is expected to solidify largely as single-phase austenite with less segregation of alloying elements and thereby less extension of the solidification range. Alloys 625 and C-276 are also significantly more prone to precipitation during heat treatment as often seen in reheated regions of weldments [12].

A larger solidification temperature range of is contributing to a greater risk of solidification cracking as is often observed when comparing 625 and 59 type weldments [3]. Back-filling can on the other hand contribute to healing of solidification cracks thereby making the effect on cracking tendency of real weldments smaller. This could partly explain the larger cracking tendency in MVT testing of 625 welds compared to metallography and bend testing. Strains are larger and solidification more rapid during MVT testing, than in actual SAW applications, due to the lower heat input. The strain rate will also be different during MVT testing as compared to in actual welding. Another consequence of differences in heat input is that slower cooling and larger weld pools lead to more pronounced segregation. The segregated regions will be those where liquation cracking occurs most easily due to a lower melting temperature. The importance of heat input was confirmed by test results for all weld metals, with one exception, ranking lower heat input welds as more crack resistant (Table 7).

Apart from effects of cracking mechanisms the relatively greater number of non-surface penetrating cracks in Alloy C-276 welds will clearly confuse results in MVT testing. The reason is presently not understood since the fraction of liquation cracks, identified from metallographic cross-sections (Table 3 and Fig. 2), was equal or higher in C-276 weld metals than for the other two types.

In conclusion there are indications that the MVT test weld pool size and cooling rate provokes relatively more solidification cracking in Alloy 625 compared to in submerged arc welding and thereby predicts a higher cracking sensitivity than observed in practice. Quite contrary significant subsurface cracking in C-276 weld metals renders MVT test results misleading in the opposite direction. Although being a technique capable of producing useful data from a small number of tests it is clear that the MVT method should be complemented with other tests simulating actual application welding conditions to permit a fair assessment of welding consumables.

## Conclusions

- The hot cracking resistance of submerged arc weld metals, produced with Alloy 625, Alloy 59 and Alloy C-276 Ni-base filler wires at two levels of heat input, was evaluated with MVT testing, metallographic evaluation and bend testing.
- Metallography and bend testing ranked the 625 weld metals as most crack resistant and C-276 weld metals as least resistant. All methods generally ranked the lower heat input welds as more crack resistant. However, MVT testing judged 59 type weld metals as being most crack resistant and ranked C-276 welds as slightly better than those produced with Alloy 625 wires.
- Several factors are believed to contribute to discrepancies between results of the different test methods. These include metallurgical aspects such solidification, segregation and precipitation behaviour of the alloys, tendencies to form non-surface breaking cracks and differences in strain, strain rate and heat input between actual welding procedures and MVT testing.
- Test methods simulating actual application welding conditions as closely as possible are most likely to accurately predict hot cracking susceptibility of real weldments. MVT testing should therefore preferably be complemented with other methods to permit a fair assessment of welding consumables.

## Acknowledgement

BAM (Bundesanstalt für Materialforschung und -prüfung, Berlin) is gratefully acknowledged for performing MVT tests and for commenting on results.

## References

1. L.-E. Stridh, L. Karlsson, S. Rigdal and N. Thalberg, “New methods in Welding of 9% Nickel Steel for LNG Applications”, Proc. IIW 58th Annual Assembly, International Conference “Benefits of new methods and trends in welding to economy, productivity and quality”, 14–15 July 2005, Prague, pp. 39–53.

2. L. Karlsson, S. Rigdal L.-E. Stridh and N. Thalberg, "Efficient Welding of 9% Nickel Steel for LNG Applications", Proc. Stainless Steel World 2005 Conference & Expo, 8–10 Nov 2005, Maastricht, pp. 138–142.
3. L. Karlsson, S. Rigdal and S.L. Andersson, "Welding of highly alloyed austenitic and duplex stainless steels", Welding in the World, 1997, 39 (2) pp. 99–110.
4. Hot cracking phenomena in welds, Editors Th. Böllinghaus and H. Herold, Springer-Verlag Berlin Heidelberg, 2005.
5. CEN ISO 17641 Part 1–3, Destructive tests on welds in metallic materials – Hot cracking tests for weldments – Arc welding processes.
6. U. Brill, T. Hoffman and K. Wilken, "Solidification cracking: Super stainless steels and nickel base alloys", Proc. Materials weldability symposium, Materials week 1990, Detroit, ASM International, 1990, pp. 99–105.
7. T. Kannengiesser, M. Wolf and H. Schobbert, "Recent developments in nickel base material welding considering the influence of shielding gas on the hot cracking resistance", IIW Doc. II-1554–05.
8. M.J. Cieslak, T.J. Headley and A.D. Jr. Romig, "The welding metallurgy of HASTELLOY Alloys C-4, C-22 and C-276", Met. Trans. A, 1986, 17, pp. 2035–2047.
9. M.J. Cieslak, "The welding and solidification metallurgy of Alloy 625", Welding Journal, 1991, 70, pp. 49s–56s.
10. J.N. DuPont, "Solidification of an Alloy 625 weld overlay", Metallurgical and Materials Transactions A, 1996, 27A, pp. 3612–3620.
11. M. Raghavan, B.J. Berkowitz and J.C. Scanlon, "Electron microscopic analysis of heterogeneous precipitates in Hastelloy C-276", Met. Trans. A, 1982, 13A, pp. 979–984.
12. U. Heubner and M. Köhler, "Das Zeit-Temperatur-Ausscheidungs- und das Zeit-Temperatur-Sensibilisierungs-Verhalten von hochkorrosionsbeständigen Nickel.Chrom-Molybdän-Legierungen", Werkstoffe und Korrosion, 1992, 43 (5) pp. 181–190.

# Assessment of HAZ Hot Cracking in a High Nitrogen Stainless Steel

K. Stelling, M. Lammers, D. Meinel

Federal Institute for Materials Research and Testing, Berlin, Germany

## Abstract

Fully austenitic materials, such as nickel base alloys and several stainless steels, can be prone to hot cracking in the heat affected zone during welding. Such cracks, due to their small size commonly denoted as micro-cracks, were found in laser hybrid welds of a high nitrogen stainless steel (UNS S 34565 / German material no. 1.4565).

Mechanical testing, such as tensile and Charpy V-notch tests as well as cyclic loading, has been carried out on hot crack afflicted laser plasma hybrid welds. The test results show that the ductility of the material is decreased, whereas the strength is not influenced at all. SEM micrographs of the fracture surfaces revealed that micro-cracks in the HAZ were associated with failure.

For a thorough evaluation of the cracking mechanism, the orientation of the micro-cracks alongside the fusion line of the hybrid welds is of great importance. Thus, computer tomography using micro focus X-ray has been carried out. Using this new approach, the orientation of the cracks towards the fusion boundary was ascertained. It turned out that the critical strain emerged at specific sites of the hybrid weld. The critical strain was oriented tangentially to the fusion boundary and perpendicularly to the welding direction. X-ray cone beam computer tomography turned out as a very useful tool for the investigation of the crack distribution and hence, for getting qualitative information about the critical stress/strain distribution in real welds.

## Introduction

Fully austenitic stainless steels offer high performance for many applications, such as offshore technology, chemical industries and wherever high corrosion resistance and/or special mechanical, e.g. high ductility at cryogenic temperatures, or special physical properties, e.g. low magnetic permeance, are required. In these steels, nitrogen as an alloying element can effectively reduce the material costs if it is used as a substitute for nickel. Thus, the nitrogen alloyed stainless steels represent a very cost-saving option compared to nickel base alloys. Furthermore, nitrogen additions strongly increase the yield strength of these materials [1, 2].

However, fully austenitic materials can suffer from high sensitivity to hot cracking, i.e. solidification cracking, liquation cracking and ductility-dip cracking (DDC) [3-6].

Hot cracking occurs during welding if a critical strain is imposed on an interface within a so called brittle temperature range (BTR) or ductility dip temperature range (DTR) in case the of DDC [5, 7].

In addition to solidification cracking, cracking within the BTR implies also material separation in the partially melted zone. This type of cracking is denoted as liquation cracking because it arises from low melting phases on the grain boundaries. It is the temperature as a function of time, i.e. heat input, which controls diffusion and thus dissolution and/or precipitation of low melting phases. If the peak temperature enters the BTR, then the material cannot sustain critical localised strain during the cooling cycle which results in hot tearing [8, 9].

Different mechanisms for liquation cracking are described in literature. [11, 12, 13]

In contrast to that, DDC occurs below the solidus temperature and thus, is rather a solid-state cracking mechanism. DDC sensibility is increased if long, straight grain boundaries prevail which have low resistance to grain boundary sliding [7].

It is clear that both, mechanical and metallurgical factors must be present to generate hot cracking. The welding process, the welding parameters and the thermal properties of the material govern the thermal cycle, the weld pool shape, and hence, the thermo-mechanical loading of the material adjacent to the weld pool. Another important factor regarding mechanical loading would be the internal or external restraint.

However, the entire mechanism controlling crack formation in the heat affected zone (HAZ) is still not completely understood. Besides this, it is not definitely known whether these micro-cracks can have an unfavorable effect on the in-service behavior of the weld. Tensile tests carried out on a micro-fissured austenitic weld metal showed that micro-cracks influence especially the ductility of the weld metal [14].

Furthermore, it is not clear how liquation cracking can successfully be avoided by adapting the welding parameters. It is commonly recommended to lower the heat input. But in this context, the weld pool shape is rarely taken into consideration.

However, it is clear that the weld pool shape has a strong influence on the stress/strain pattern [8, 12].

For example, a necked weld pool shape can have a detrimental effect on the hot cracking behaviour in the HAZ of fully austenitic materials. This weld pool shape is typically found in laser and laser arc hybrid welds. Some authors already started to use numerical simulation of temperature and strain distribution in the partially melted zone in order to quantify critical conditions for liquation cracking in mere laser welds [15].

The term “hybrid welding” describes the combination of two welding processes in a common process zone. A laser welding process is coupled with arc welding processes in order to profit by the special attributes of the single processes. This means that higher welding speed can be achieved compared to mere arc welding due to the deep penetration of the laser. Compared to mere laser welding, gap bridging is improved because of the broader weld pool and the addition of filler metal. During laser arc hybrid welding of fully austenitic materials it was often observed that micro-cracks occurred along the fusion line, obviously depending on the characteristic weld geometry.

However, in order to get a deeper insight into the mechanism of hot cracking during welding it would be important to know where critical conditions prevail in real welds. Additionally, it has to be established how the cracks are oriented relative to the weld pool because that would indicate in which direction critical strain/stress act during welding.

In this study, computer tomography with micro focus X-ray has been used for obtaining more information about the distribution and the orientation of the liquation cracks in a laser plasma hybrid weld of a high nitrogen stainless steel.

Additionally, tensile testing, Charpy V-notch testing and cyclic tensile loading has been used for assessing the mechanical behaviour of liquation crack afflicted welds. Subsequently, scanning electron microscopy (SEM) was used for examination of the fracture surface after testing in order to make clear whether these cracks were involved in specimen failure.

## Welding Experiments

Laser plasma hybrid welds of high nitrogen stainless steel plates of 6.6 mm plate thickness were produced for the subsequent examinations. The composition of the base metal UNS S 34565 (German material no. 1.4565) and the filler powder are summarised in Table 1.

Hybrid welding has been carried out using a diode pumped Nd:YAG laser. A scheme of the arrangement of laser beam and plasma torch can be seen in Fig. 1. The parameters for the welds are summarised in Table 2.

**Table 1.** Chemical composition of the UNS S 34565/1.4565 stainless steel plate and the filler powder, % by weight

| Material | C | Cr | Ni | Mn | Mo | Si |
|---|---|---|---|---|---|---|
| UNS S 34565 6.6 mm plate | 0.026 | 23.7 | 18.66 | 6.084 | 5.565 | 0.162 |
| filler powder | 0.015 | 27.25 | 22.4 | 7.75 | 4.45 | 0.53 |
| | S | P | N | Nb | others | |
| UNS S 34565 6.6 mm plate | <.001 | 0.016 | 0.466 | 0.010 | Al: 0.024 | |
| filler powder | 0.004 | 0.003 | 0.340 | – | W: 0.34<br>Cu: 1.01 | |

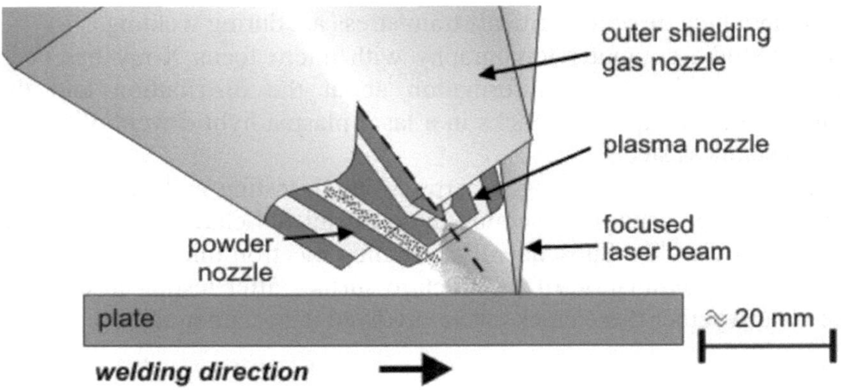


**Fig. 1.** Arrangement of laser beam and plasma torch in the hybrid welding process

**Table 2.** Welding parameters

| | Tensile test specimen | Charpy test specimen |
|---|---|---|
| Current | 270 A | 230 A |
| Voltage | 23 V | 20 V |
| Laser power | 4.3 kW | 4.2 kW |
| Welding speed | 1.4 m/min | 1.5 m/min |
| Nominal heat input | 4.5 kJ/cm | 3.5 kJ/cm |
| Focal point position | 2.5 mm under the plate surface | 3.0 mm under the plate surface |
| Plasma gas | Argon | Argon |

## Microscopic Examination

The cross sections of the welds are shown in Figs. 2a and 3a. The shape of the hybrid weld is typically divided into two parts. The broader upper part results from the plasma arc, whereas the lower part is tapered by the deep penetration of the laser. The particular shape of the weld pool is influenced by the applied welding parameters, such as arc current, focal point position and plasma gas flow. In hybrid welding, these parameters can be varied within a comparatively wide range, thus, a variety of weld pool geometries can be produced.

a) b)

**Fig. 2.** (**a**) Cross section of the laser plasma hybrid weld used for the tensile test and cyclic tensile loading, (**b**) hot cracks in the HAZ of the base metal

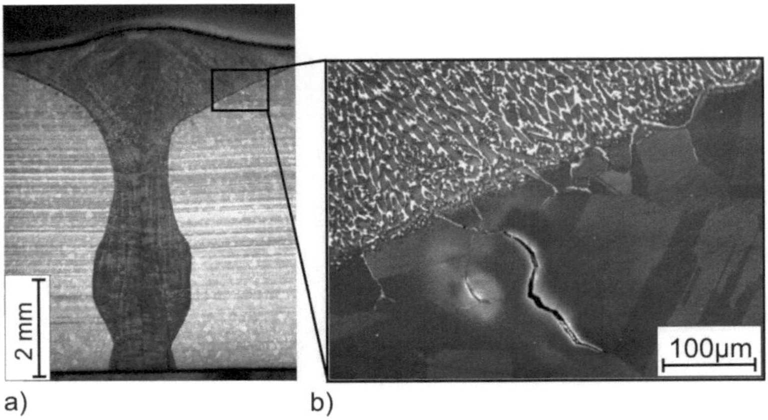


**Fig. 3.** (**a**) Cross section of the laser plasma hybrid weld used for the Charpy test, (**b**) hot cracks in the HAZ of the base metal

However, the microscopic examination revealed that hot cracks developed in the base metal adjacent to the fusion line, see Fig. 2b and Fig. 3b. The length of the cracks is about 200 µm, and thus in the dimension of the grain size. These intergranular cracks typically emerged in the plasma part of the weld, close to the transition area towards the laser part. The crack distribution was found to be asymmetric in welds with an asymmetric shape. This presumably results from different stress/strain patterns which depend on the orientation and/or the curvature of the fusion boundary. It could also be seen from the microscopic examination that some of the hot cracks have been in contact with the molten pool, Fig. 2b, others seem to emerge in a distance from the fusion line, Fig. 3b. The cracks appear to be liquation cracks and/or ductility-dip cracks resulting from critical thermo-mechanical strain within a BTR and DTR, respectively.

In the next step, mechanical testing was carried out to determine whether these micro-cracks have a detrimental effect on the mechanical properties of the weld.

## Mechanical Testing

### Tensile Test

Tensile test specimens were prepared according to Fig. 4. The specimen surface has been ground until a uniform specimen thickness was reached in order to avoid notch effects.

For comparison, base metal specimens and weld metal specimens were fabricated. The loading direction was perpendicular to the rolling direction of the plate and also to the welding direction regarding the weld metal specimens (transverse tensile test). The tensile test was performed at room temperature according to DIN EN 10 002. [16]

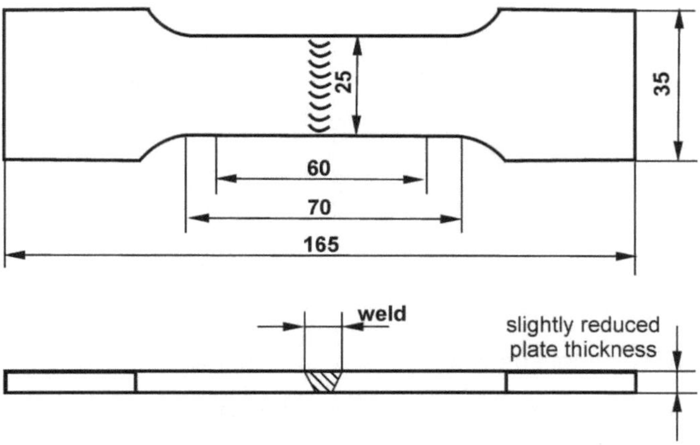


**Fig. 4.** Specimen geometry for the tensile test, loading direction was perpendicular to the welding direction

Figure 5 shows the average values for the base metal specimens and the weld metal specimens. The measured 0.2%-proof stress ranged from 400 to 450 MPa and the tensile strength ranged from 830 to 860 MPa, respectively. There was no significant difference in the mechanical strength between base metal and welded specimen.

In contrast, the ductility, represented by the reduction of area, appreciably decreased from 65% in the base metal specimens to 40% in both welded specimens. It has to be pointed out that the absolute value for the ductility is still high. However, the welded specimens failed along the fusion line and that would actually be a disqualifying criterion for quality assurance.

Figure 6 shows the fractured weld metal tensile test specimen. From Fig. 6b which shows the base metal side of the fracture, it is obvious that the fracture surface has two different appearances. The upper part of the fracture surface shows shiny sites which may indicate brittle fracture, whereas the lower part shows ductile fracture. From the SEM micrographs in Fig. 7, the different surfaces, i.e. the flat appearance of the shiny fracture areas which might indicate portions of brittle fracture (a) and the dimple structure of the ductile fracture (b), become apparent.

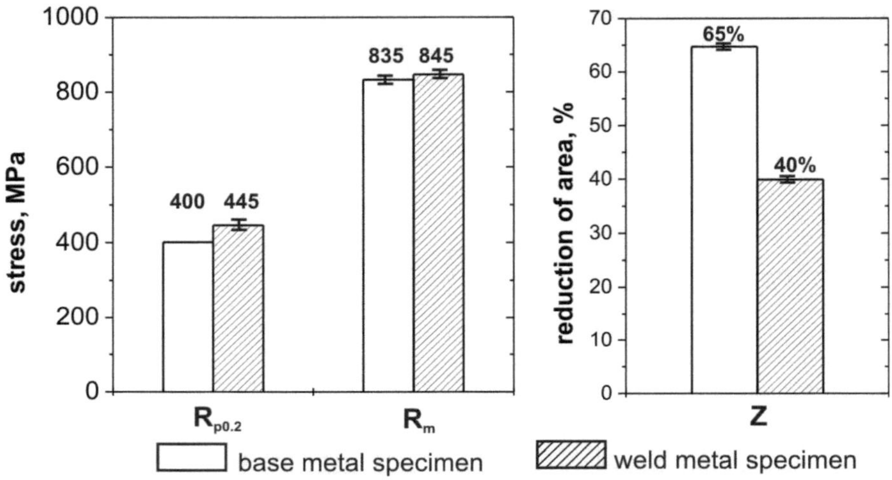


**Fig. 5.** Results of the tensile test, weld metal specimens fractured along the fusion line

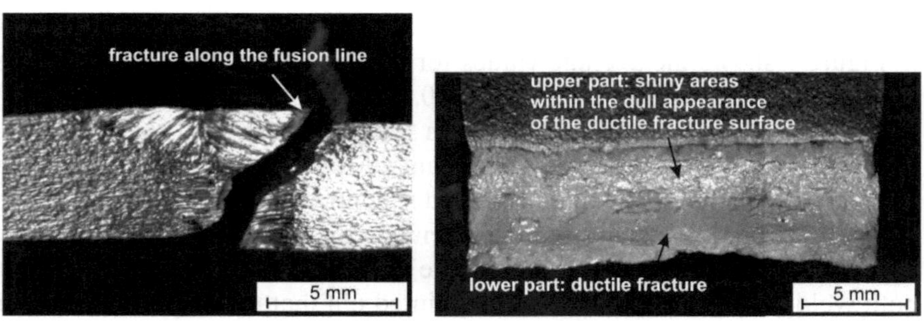


a) b)

**Fig. 6.** (**a**) Fractured tensile specimen, (**b**) fracture surface of (a), base metal side

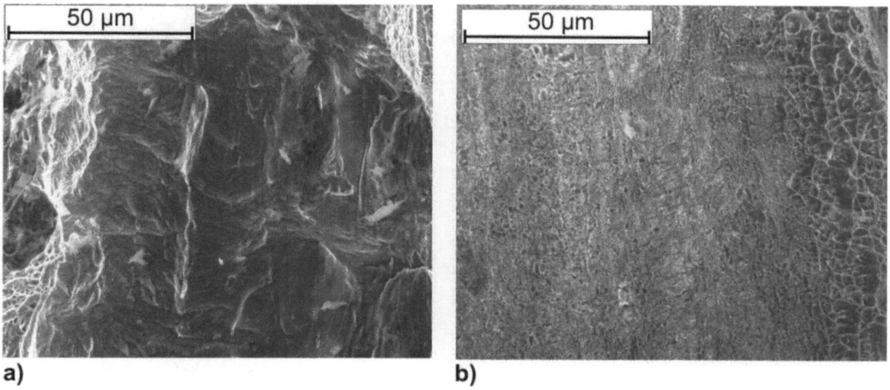


**Fig. 7.** SEM micrographs of the surface in Fig. 9b, (**a**) fracture surface within the upper part, (**b**) ductile (dimple) fracture in the lower part

## Charpy Test

The ductility of materials can be assessed by the Charpy V-notch test. Normally, face centred cubic (fcc) metals such as austenitic steels provide high ductility, especially at cryogenic temperatures. However, weld imperfections can reduce low temperature toughness. Thus, the notch impact resistance, i.e. the absorbed energy measured in the Charpy test, was used to evaluate the influence of micro-cracks on the fracture toughness. Preparation of the V-notch specimens has been carried out according to DIN EN 10045 [17].

In Fig. 8a, the specimen geometry and the position of the notch in relation to the weld metal zone are shown. At least three specimens were tested at three different temperatures, i.e. RT, –60°C and –120°C.

From the diagram in Fig. 9 it becomes obvious that the toughness, i.e. the absorbed energy in the Charpy test, decreased towards low test temperatures. The examination of a specimen tested at –120°C revealed that the crack proceeded along the fusion line in the upper part of the weld, i.e. the plasma part.

The path along which the specimens fractured is marked in the cross section of the weld in Fig. 8b.

In order to ascertain the reason for the toughness drop at low temperature, fractographic examination of both fracture surfaces of a V-notch specimen has been carried out.

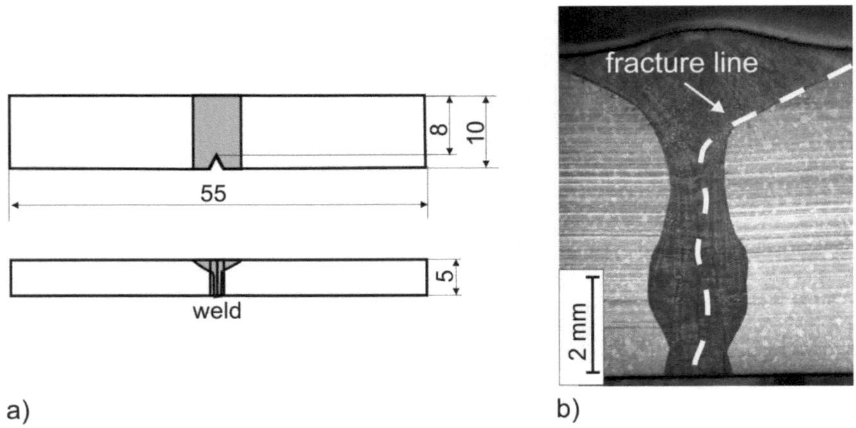


**Fig. 8.** (**a**) V-notch specimen according to DIN EN 10045 and (**b**) estimated fracture line during testing at –120°C

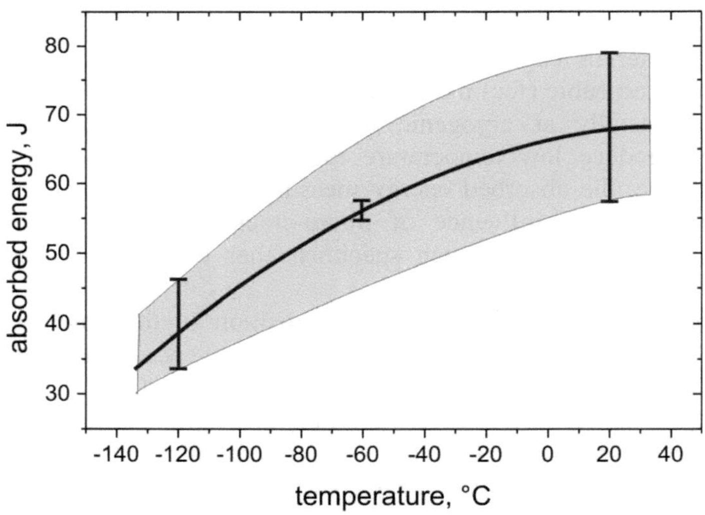


**Fig. 9.** Absorbed energy of the test specimen at different temperatures

**Fig. 10.** Survey of the fracture surface of the V-notch specimen which cracked partially along the fusion line, on the *left*: weld metal side, on the *right*: base metal side

Figure 10 shows the corresponding fracture surfaces of a specimen tested at -120°C. Since the specimen fractured along the fusion line in the upper part (plasma part of the weld), the two halves of the specimen represent the weld metal side (on the left) and the base metal side (on the right) of the fusion line region.

SEM examination revealed that the major portion of the surface has a dimple structure which is typical of ductile fracture. However, at the top side of the specimen, i.e. in the plasma part of the weld, sites could be found which indicate pre-existing intercrystalline cracks in the specimen. These cracks are supposed to be hot cracks that had already been observed in microscopic examination. The appearance of these cracks can be seen in detail in Figs. 11 and 12.

Due to the fact that hot cracking within BTR involves surface liquation, it can normally be distinguished from DDC by the freely solidified surface structure, whereas the structure of DDC usually shows sharper ridges and also reveals a slight surface roughness at higher magnification. [18]

Considering this, just some of these cracks can clearly be identified as liquation cracks. A freely solidified surface inside the crack can be seen in the detail of section no. 2. This kind of surface structure was observed especially on the weld metal side of the specimen.

Other cracks show a more even structure which seems neither completely freely solidified nor completely cracked in the solid-state, see details of sections no. 1 and 3. This structure is not typical of liquation cracks. It more resembles the surface of a ductility-dip crack.

Concluding from the different surface structures of the hot-cracks, i.e. freely solidified surface and also smooth surface with just slight indications of melted material, cracking presumably occurred at different stages/temperatures within the BTR and/or DTR. The degree of liquation and thus, the surface appearance of the cracks obviously depends on the distance from the fusion zone.

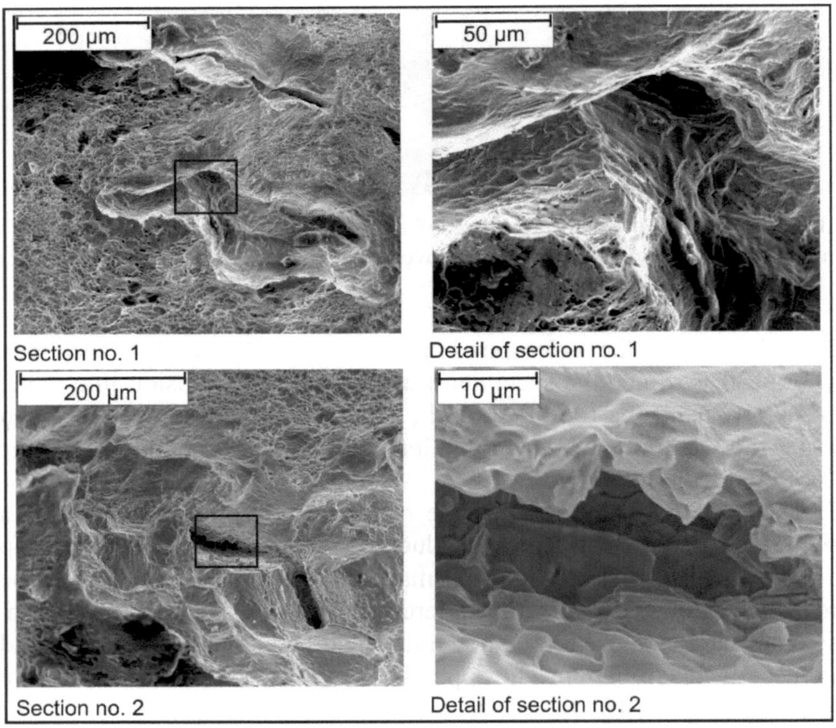


**Fig. 11.** SEM pictures, section sites are marked in Fig. 10, weld metal side

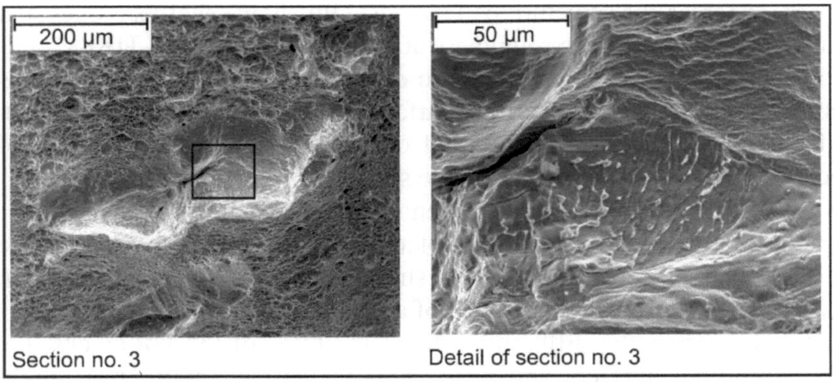


**Fig. 12.** SEM pictures, section site is marked in Fig. 10, base metal side

The above described hot cracks are surrounded by an area which has a flat, transcrystalline appearance without pronounced dimple structure. This surface structure supposedly results from a complex stress state which was induced by the micro-cracks during the impact test.

Therefore, at least four different surface structures can be distinguished:

1. transcrystalline fracture with dimple structure (major part)
2. transcrystalline fracture without dimples, probably due to a complex stress state around the hot cracks,
3. intercrystalline hot crack with a smooth surface structure
4. intercrystalline hot crack with a freely solidified surface

The first three of the different surface types are indicated in section no. 4, Fig. 13. The fourth one was shown in detail of section no. 2, Fig. 11.

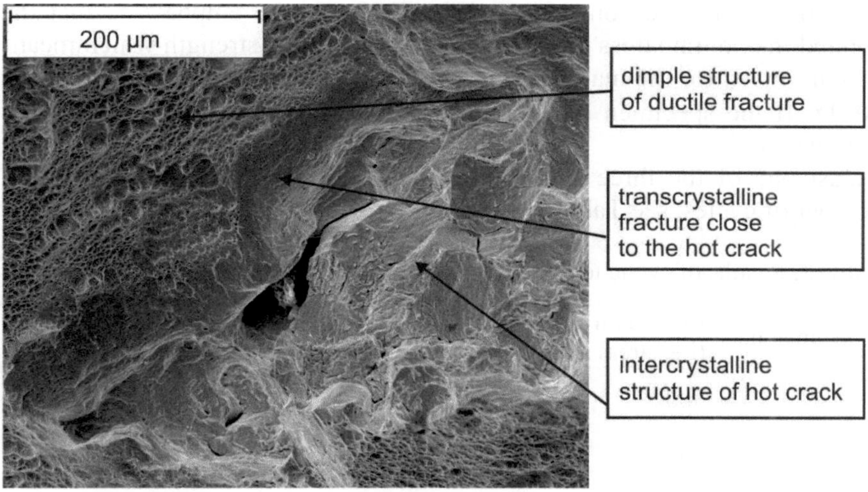


**Fig. 13.** SEM micrograph of section no. 4, demonstration of the different types of fracture surfaces, section site is marked in Fig. 10, base metal side

The SEM examination revealed that the reduced ductility of the weld at lower temperatures can be attributed to the micro-cracks along the fusion line. Although ductile dimple fracture represents the major part of the fracture surface, the micro-cracks were obviously involved in failure and governed the fracture path during testing, especially at low temperatures.

### Cyclic Tensile Loading

Fatigue failure can be induced by surface imperfections such as notches or cracks. Fatigue represents the most dangerous failure mechanisms due to the often unnoticed progress.

In a very first attempt, the hot crack afflicted hybrid welds were subjected to cyclic loading using oscillating tensile stress in order to see whether it is possible to provoke fatigue failure starting from a micro-crack.

The specimen geometry according to Fig. 4 was the same geometry as for the tensile test. Here too, the specimen surface was ground until a plane specimen surface was reached without any surface notches visible with the naked eye.

Three specimens were tested under the following conditions: Oscillating tensile stress with a peak stress of 300 MPa ($\sigma_{min}$ = 0 MPa or stress ratio R = 0) was imposed on the specimen. It has to be emphasised that the applied maximum stress was about 75% of the yield strength which means a high impact for the material. The oscillation frequency was chosen to be 30 Hz. If the specimen did not fail after $2 \times 10^6$ cycles, then testing was terminated.

Just one of the three specimens failed before reaching the maximum number of cycles, see Table 3.

**Table 3.** Results of cyclic tensile loading

| Specimen no. | Maximum stress (i.e. stress amplitude), MPa | Number of load cycles | After test |
|---|---|---|---|
| 1 | 300 | $2.00 \times 10^6$ | no failure |
| 2 | 300 | $2.00 \times 10^6$ | no failure |
| 3 | 300 | $0.73 \times 10^6$ | cracking, initiation at hot crack |

The major part of the fracture surface was at a right angle to the direction of the tensile stress which indicates fatigue failure. In Fig. 14b and Fig. 15, the portion of fatigue fracture and of final catastrophic failure characterised by strong plastic deformation can clearly be distinguished. An examination with the naked eye revealed that cracking initiated on the top side of the specimen surface close to the fusion line. The macroscopic fracture propagation direction is indicated by directional markings on the fracture surface.

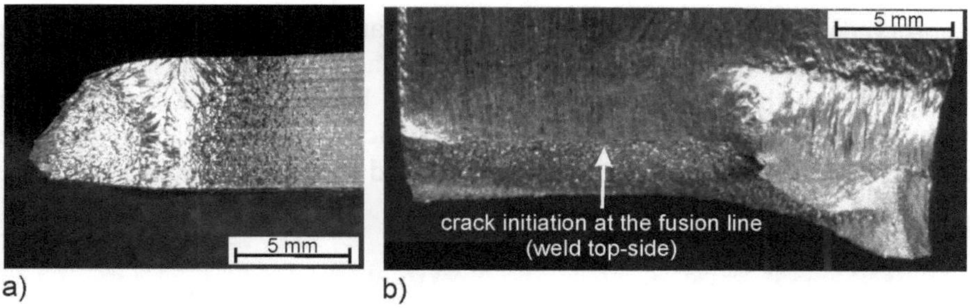


**Fig. 14.** Macrographs of the failed test specimen, (**a**) side view, position of the weld is distinguished by means of a characteristic surface structure of the specimen after extensive plastic deformation, (**b**) top view

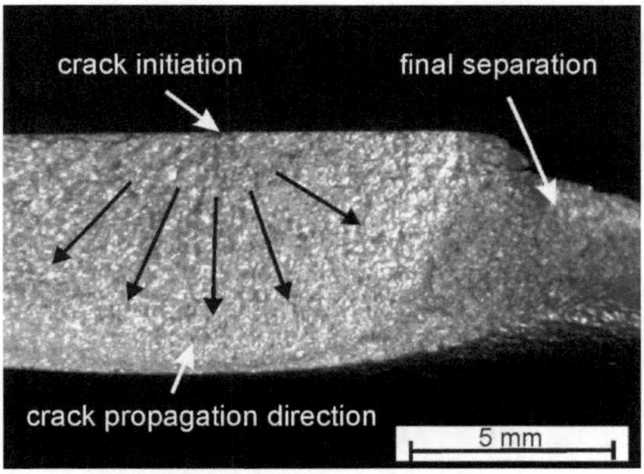


**Fig. 15.** Directional markings on the fracture surface of the failed specimen indicate the crack initiation and the propagation direction of the fracture.

At higher magnification, the initiation site is identified as a micro-crack which was opened towards the ground specimen surface. The micro-crack obviously spans two to three grains. A single exposed grain becomes clearly visible in the SEM micrograph in Fig. 16b. The surface of this micro-crack reveals a smooth and even structure. Slip line formation occurred on the free crack surface. The slip lines are oriented in a distinct manner depending on the crystallographic orientation of the grain, see Fig. 16c. Ductility dip cracks can have such surface appearance. So this could be an indication for a DDC mechanism. However, it is also possible that the slip lines originate from cyclic loading during testing.

In Fig. 17a, the transition area of the micro-crack towards the fatigue fracture surface is shown. The transcrystalline fatigue fracture in the base metal can be seen in Fig. 17b. Since the fracture has propagated on distinct multiple fracture paths, fatigue striations with different orientations can be observed on the fracture surface, i.e. they are oriented according to the crystallographic planes of the underlying metal.

Resuming the results of the cyclic tensile loading tests it can be stated that under extreme loading conditions, which have been applied to the welded specimens, micro-hot cracks can lead to fatigue failure if they are opened towards the surface.

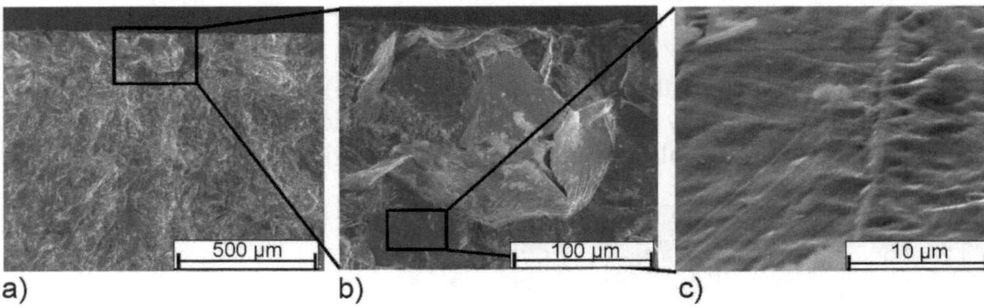


**Fig. 16.** SEM micrographs showing the initiation site of the fracture at different magnification, (**a**) survey, (**b**) surface of the intergranular micro-crack, (**c**) smooth surface structure revealing slight slip line formation

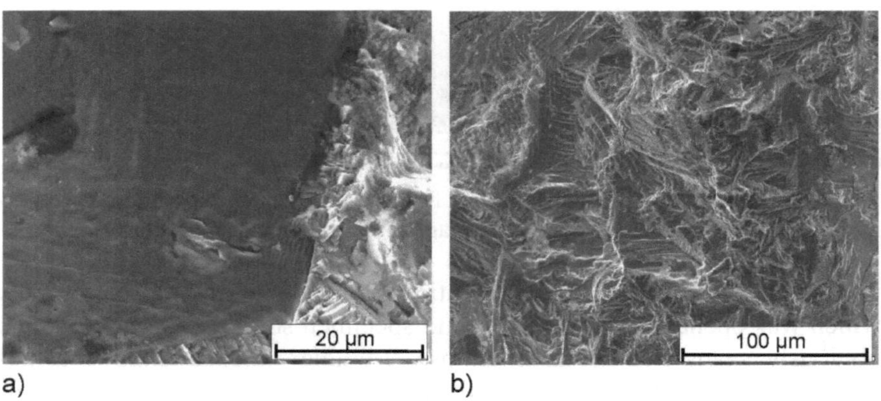


**Fig. 17.** (**a**) transition area of the micro-crack and the transcrystalline fatigue fracture surface, (**b**) transcrystalline fatigue fracture surface

## X-ray Cone Beam Tomography

From the examination of the fractured specimens it was possible to get information about the surface structures of the micro-cracks, and also about the orientation towards the fracture surface. However, the orientation of the cracks relative to the fusion line which would indicate the direction of the critical strain during the welding process did not become clear either from the SEM examination or from the two dimensional micrographs.

For that reason, a weld section was examined using three-dimensional computer tomography. This technique is known to be a very useful tool for revealing the distribution of imperfections inside a component.

The principle of three-dimensional or cone beam tomography (CT) is shown in Fig. 18. The tomograph was developed at BAM in the department for non-destructive testing.

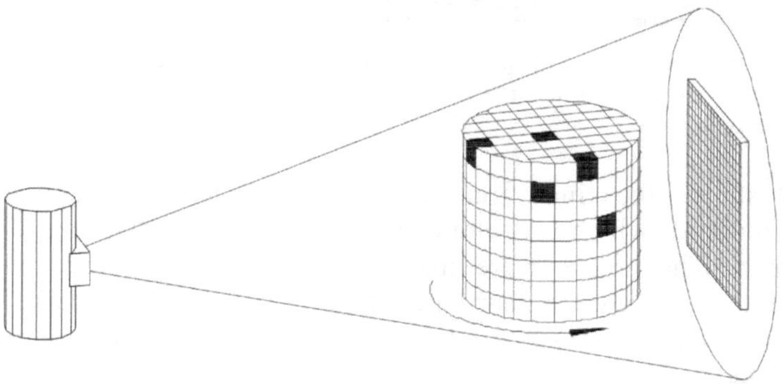


**Fig. 18.** Principle of three-dimensional or cone beam tomography

During testing, the sample is turned in the cone beam of a micro focus X-ray tube and manifold slices are measured using an area detector. With the usage of magnification techniques, a spatial resolution of less than 10 μm could be reached depending on the object size.

In order to get highest resolution and quality of the CT images, the specimen should have an almost cylindrical geometry and a low geometric measure in the radiation direction.

For this reason, a small piece has been cut off from the weld. Then the specimen was oriented such that the lowest specimen thickness was in radiation direction, i.e. the turning axis of the specimen was perpendicular to the welding direction as well as to the plate normal. For the analysis of crack distribution, the cracks which can be identified in the tomography due to their lower attenuation were manually marked in every single plane of the CT in order to create a three-dimensional picture of the crack-distribution.

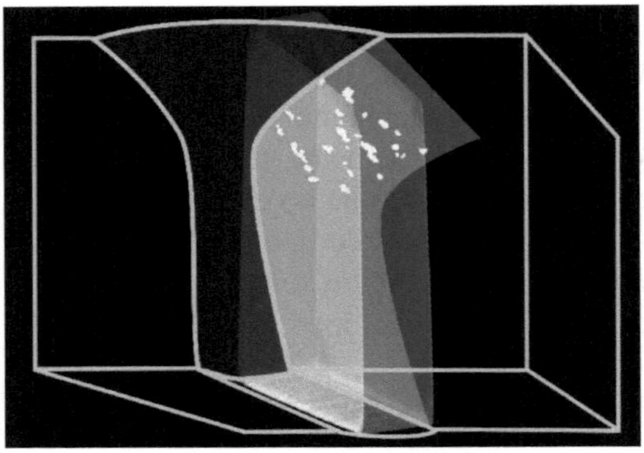

**Fig. 19.** Three-dimensional drawing of the weld with embedded transparent CT segment showing the distribution of the liquation cracks along the fusion boundary of the weld

Figure 19 shows the final, transparent three-dimensional computer tomography (cuboid) which was embedded in a schematic drawing of the weld specimen. The cracks were marked with white color so that it becomes visible how they are distributed across the fusion boundary. The distribution as well as the size and orientation of the cracks are more clearly displayed in Fig. 20.

The distribution of the cracks and their orientation in relation to the fusion boundary becomes clear from the viewing direction perpendicularly to the fusion boundary which is shown in Fig. 20b. This sight reveals that the crack planes are oriented almost perpendicularly to the fusion line and parallel to the welding direction. In other words, the critical strain acted tangentially to the fusion boundary and perpendicularly to the welding direction. Schematically, this is demonstrated in Fig. 20c. No cracks could be detected in the lower, i.e. the laser part, of the weld.

The critical conditions obviously prevail in the broader, upper region of the weld. Suggested reasons for that are:

1. This part of the weld undergoes heat accumulation because it is the region of the highest heat input and because of the broad, slightly and inwardly curved shape of the weld bead. Thus, the partially melted zone or the zone of grain boundary brittleness is broader than in the lower part of the weld.
2. The volume of the liquid and, thus, the solidification shrinkage as well as further cooling shrinkage of the weld bead is increased towards the upper part of the weld.

3. The ratio of the fusion boundary area and the underlying base metal volume is highest in the inwardly curved regions of the weld. If shrinkage starts during cooling tangential strains are imposed on these regions.

However, inwardly curved regions are found in the lower part of the weld, i.e. the laser part, too. Nevertheless, hot cracks in these regions are rarely found. Therefore, it must be a combination of increased local shrinkage strains and adverse temperature fields, i.e. adverse metallurgical conditions, causing the typical crack distribution.

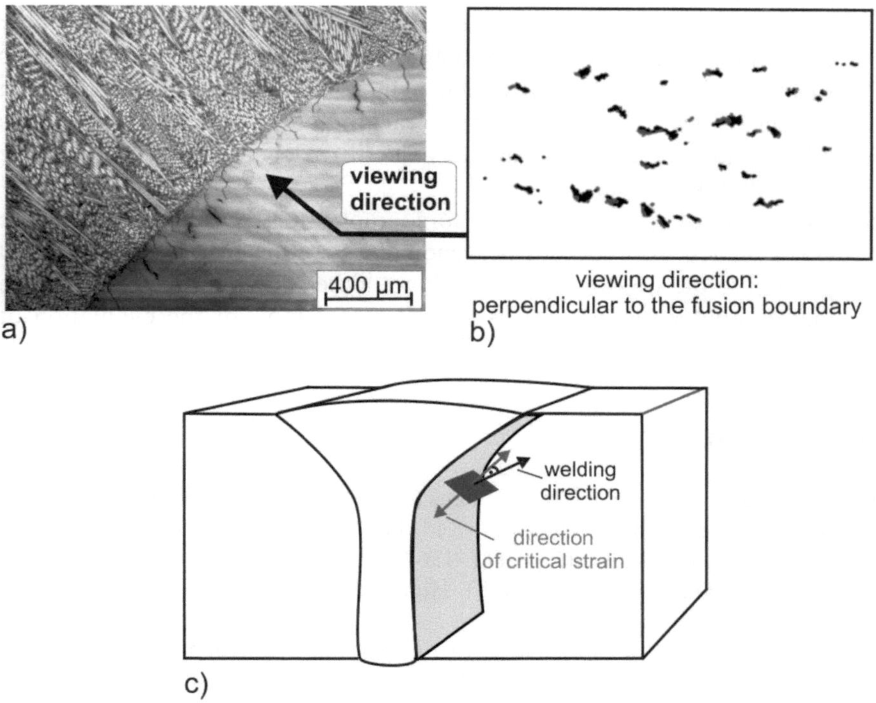


**Fig. 20.** (**a**) Crack distribution along the weld flank in the upper weld part and (**b**) cracks visualized by computer CT with the viewing direction perpendicular to the fusion boundary; (**c**) schematic illustration of the direction of critical strain towards the fusion boundary

## Conclusions

1. The high nitrogen stainless steel UNS S 34565 is susceptible to hot cracking in the heat affected zone of the base metal during laser plasma hybrid welding.
2. Cracking typically occurred in the upper part of the weld produced by the trailing plasma process during hybrid welding.
3. Transverse tensile testing and Charpy V-notch testing of welded specimens indicate a decrease in ductility. But the strength of the material is obviously not affected.
4. SEM surface examination of fractured specimens revealed that hot cracks were involved in failure. The intergranular hot cracks show different surface structures. Some grains exhibited an almost dendritic surface, whereas others revealed a more even but still very smooth surface. Thus, it is concluded that cracking occurred at different stages of liquation and/or grain boundary ductility depending on the distance from the fusion line.
5. By means of X-ray computer tomography it was found out that the crack planes were located perpendicularly to the fusion boundary and parallel to the welding direction. It could be shown in this contribution that computer tomography is a very useful tool for the evaluation of critical regions for hot cracking inside the material.

## References

1. Brooks JA, Thompson AW, Williams JC (1980) Weld Cracking of Austenitic Stainless Steels. In: R. Kossowsky (ed) Physical Metallurgy of Metal Joining, Warrendale, Pennsylvania, pp 117–136
2. Berns H, Gavriljuk VG (1999) High Nitrogen Steels. Springer, Berlin Heidelberg New York
3. Lin W, Lippold JC, Baeslack WA III (1993) An Evaluation of Heat-Affected Zone Liquation Cracking Susceptibility, Part I: Development of a Method for Quantification. Welding Research Supplement, April:135s–153s
4. Savage WF, Lundin CD (1965) The Varestraint Test. Welding Journal 34(10):433s–442s
5. Nissley NE, Collings MG, Guaytima G, Lippold JC (2002) Development of the Strain-to-Fracture Test for Evaluating Ductility-Dip Cracking in Austenitic Stainless Steels and Ni-Base Alloys. Welding in the World:32–40
6. Lundin CD, Lee CH, Menon R, Osorio V (1988) Weldability Evaluation of Modified 316 and 347 Austenitic Stainless Steels: Part I – Preliminary Results. Welding Research Supplement February:35s–46s

7. Collins MG, Ramirez AJ, Lippold JC (2004) An Investigation of Ductility-Dip Cracking in Nickel-Based Weld Metals – Part III. Welding Journal February:39s–49s
8. Borland JC, Younger RN (1960) Some Aspects of Cracking in Welded Cr-Ni Austenitic Steels. British Welding Journal January:22–59
9. Kou S (2003) Solidification and Liquation Cracking Issues in Welding. JOM June:37–42
10. Kou S (1987) Welding Metallurgy. John Wiley and Sons, New York, pp 239–262
11. Radhakrishnan B, Thompson RG (1992) A Model for the Formation and Solidification of Grain Boundary Liquid in the Heat-Affected Zone (HAZ) of Welds. Metallurgical Transactions A 23A:1783–1799
12. Pepe JJ, Savage WF (1967) Effects of Constitutional Liquation in 18-Ni Maraging Steel Weldments. Welding Journal 46(9): 411s–422s
13. Radhakrishnan B, Thompson RG (1991) Modeling of Microstructural Evolution in the Weld Heat Affected Zone (HAZ). In: Cieslak MJ, Perepezko JH, Kang S, Glicksman ME (eds) The Metal Science of Joining, TMS, Ohio, pp 31–40
14. Cui Y, Lundin CD, Hariharan V (2006) Mechanical behaviour of austenitic stainless steel weld metals with microfissures. Journal of Materials Processing Technology 171:150–155
15. Nishimoto K, Woo I, Shirai M(2002) Analyses of temperature and strain distributions in laser welds – Study of the weldability of Inconel 718 cast alloy (Report 6). Welding International 16(4):284–292
16. DIN EN 10 002 (2001): Tensile Test – Part 1: Test Method (at Room Temperature)
17. DIN EN 10 045 (1991): Charpy impact test – Part 1: Test Method
18. Matsuda F, Nakagawa H (1977) Some Fractographic Features of Various Weld Cracking and Fracture Surfaces with Scanning Electron Microscope. Transactions of JWRI 6(1):81–90

# Crack Appearance in Hot Rolled Billets

S.T. Mandziej[1], G. Krallics[2]

[1] Advanced Materials Analysis, Enschede, The Netherlands
[2] Budapest University of Technology and Economics, Budapest, Hungary

## Abstract

Hot cracks may appear in metal alloys on heating or on cooling when the tensile strains and related stresses, caused by thermal expansion or shrinkage and usually enhanced by various restraints, cannot be compensated by a local plastic deformation of an alloy. In general, these metal alloys, which have large thermal expansion coefficients, are susceptible to hot cracking during casting, welding or e.g. re-heating for hot rolling. The metallurgical quality of an alloy affects its susceptibility to hot cracking, in particular the chemical and microstructural inhomogeneities influence this susceptibility. Additional plastic strains, applied in hot forming, usually extend the cracks leading to damage. In this work an attempt is made to describe the micro- and macro-mechanisms of the cracking as well as microstructural and mechanical factors assisting the damage, based on observations of continuously cast ingots in which internal hot cracks and cavities were formed during initial steps of hot rolling. Characteristic of these ingots was chemical segregation resulting in differences of hot ductility in different parts of the ingots. Thus during plastic deformation, due to interaction between the microstructurally and mechanically different regions, cracks and cavities did nucleate and grow. Discussed here are physical data necessary to adequately describe the behaviour of material during deformation and cracking, as well as physical simulation methods to gain these data. The gained data and identification of the damage processes are then used for computer modelling with an aim to determine critical conditions of controlling the application process in order to avoid the cracking and to assure the manufacture of sound products. The micro-scale conditions characteristic of e.g. welds and the macro-scale situations typical for rolling of steel billets are discussed.

## Introduction

During hot rolling after continuous casting of steel various cracks may occur resulting from the size and shape of the product as well as from temperature, strain and strain rate of the applied processing. Thus during rolling of 410 mm diameter continuously cast ingots into square 200 mm cross section billets, periodic large internal cavities (bursts) were observed. They often appeared in a medium-carbon Cr-Mo construction steel, however never in any plain carbon steel of similar carbon content, although the casting and rolling conditions were identical.

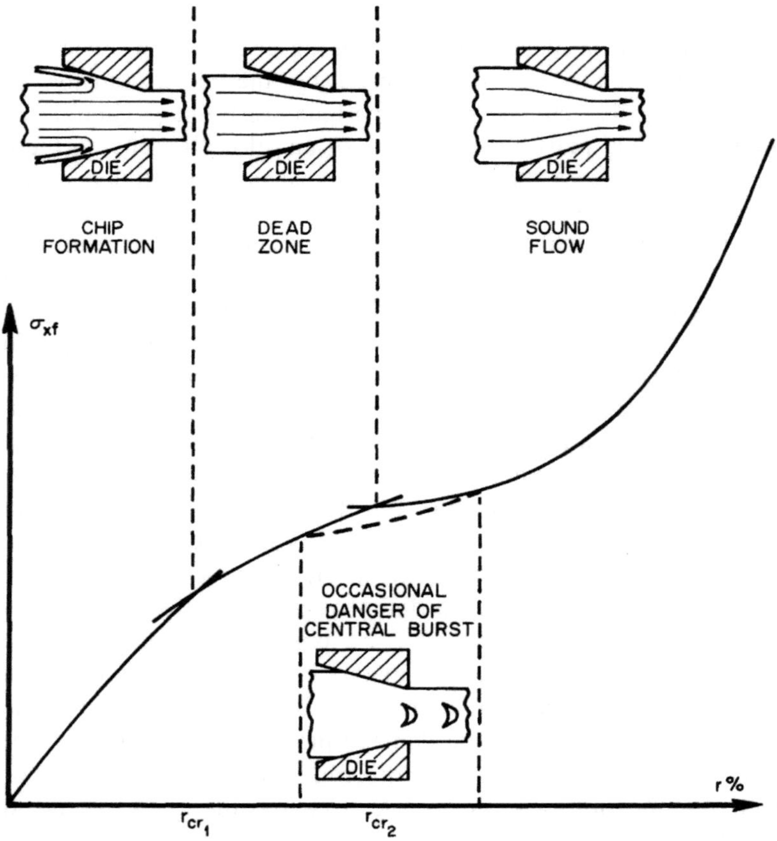


**Fig. 1.** Drawing stress ($\sigma_{xf}$) as a function of reduction (r%), with indicated burst-sensitive zone [1]

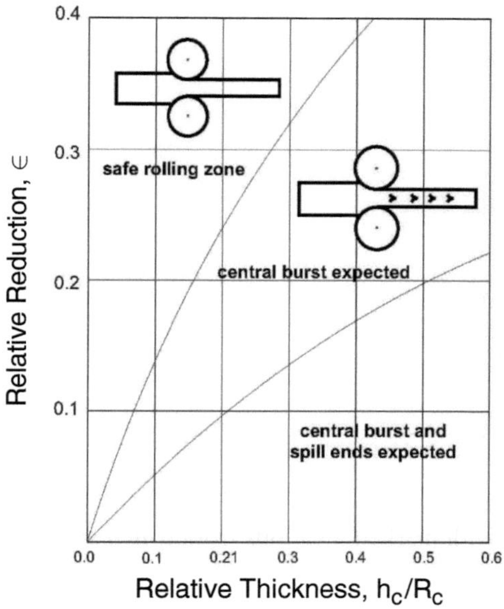


**Fig. 2.** Relation between cross-reduction and size of rolls as a factor affecting the central bursting in bar rolling [2]

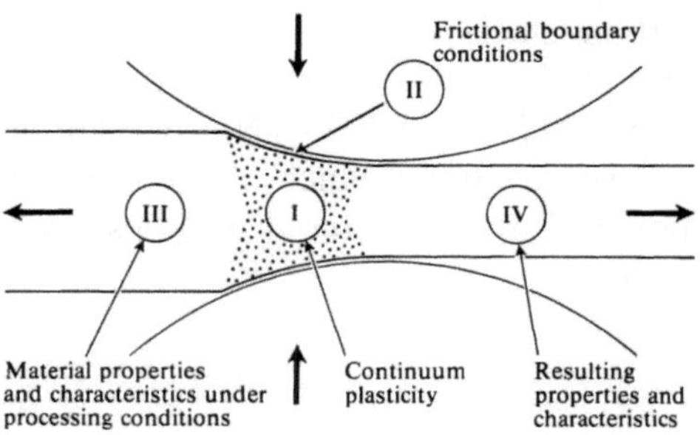


**Fig. 3.** Schematic of rolling with marked area where the pulling force is created due to friction [3]

The central bursts have been characteristic of drawing processes run at transient cross-reduction conditions, as schematically presented in Fig. 1 [1]. They were also reported in the case of hot rolling, at conditions like these presented in Fig. 2 [2]. During rolling, the axial force pulling in the direction of material's flow is mainly due to the friction between rolls and billet, Fig. 3 [3]. Accordingly, in the situation shown in Fig. 3, the burst in rolling is expected to form rather on entry to the rolls than on exit. The magnitude of this type of central internal crack is shown in Fig. 4; the picture taken from the cross-section of 200 mm hot-rolled billet.

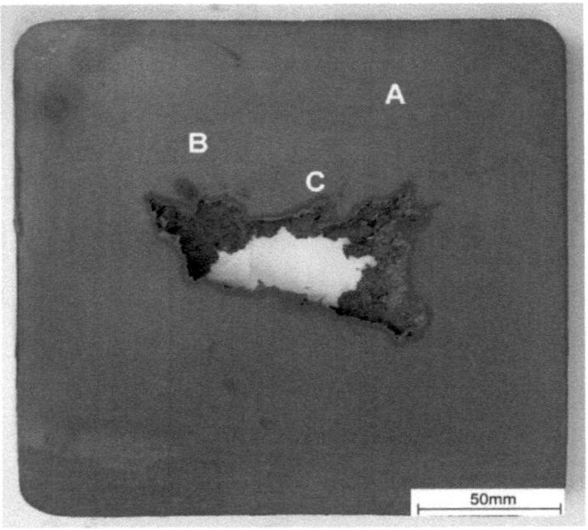


**Fig. 4.** Cross-section of 200 mm billet with marked locations from which samples were taken for hot tensile testing

What is shown in Fig. 4 appeared in the hot rolling process designed according to the rules described in the professional literature, which has been always valid for the plain carbon steels but not always for the considered here low-alloy 34CrMo4 grade steel, containing 0.35%C-0.81%Mn-0.24%Si-1.01%Cr-0.21%Mo-0.029%Al-0.0073%N. Certainly additional factors operated, like e.g. segregation, which could not be predicted from only chemical composition of the material and designed technological parameters of its processing.

## The Hot Ductile Fracture

Fracture in bulk deformation processing usually appears as ductile and rarely as brittle fracture. Depending on temperature and strain rate, the details of the ductile fracture mechanism vary. Figure 5 illustrates in a schematic way the "classical" modes of ductile fracture obtained in tensile tests over a wide range of strain rates and temperatures. At temperatures below approximately one-half the melting point of given material, a typical dimpled type of ductile fracture prevails. At very high temperatures, a rupture type of fracture occurs in which the material recrystallises rapidly and pulls down to point, with nearly 100% reduction in area. A transgranular creep type of failure occurs at temperatures less than those causing rupture. Intragranular voids form, grow, and coalesce into internal cavities, which result in a fracture with a finite reduction in area.

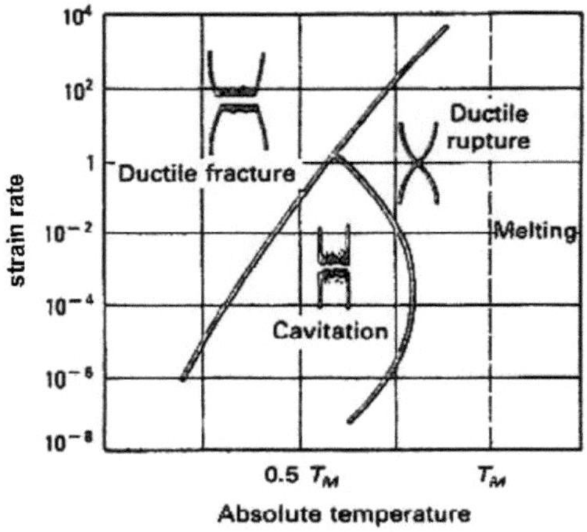


**Fig. 5.** "Classic" modes of ductile fracture

A commonly used representation of possible fracture mechanisms is the (Ashby) map, Fig. 6. Such a map shows the area of dominance in terms of normalized stress versus normalized temperature for the dominant fracture mechanism. The maps are constructed chiefly by using the best mechanistic models of each fracture process.

Failure in deformation processing around one-half of the melting temperature 0.5 $T_M$, occurs by ductile mode I. The three stages of such ductile fracture are shown schematically in Fig. 7. The first stage is void initiation, which usually occurs at second-phase particles or inclusions. Voids are initiated because particles do not deform, and this causes the ductile matrix around the particles to deform more than average. This in turn produces more strain hardening, thus creating higher stress in the matrix near the particles. When the stress becomes sufficiently large, the interface may separate or the particle may crack. As result, ductility is strongly dependent on the size and density of the second-phase particles, as shown in Fig. 8 [4]. The second stage of ductile fracture is void growth, which is a strain-controlled process. Voids elongate as they grow, and the ligaments of matrix between the voids became thinner. The final stage of ductile fracture is the cavity formation by the separation of the ligaments that link the growing voids.

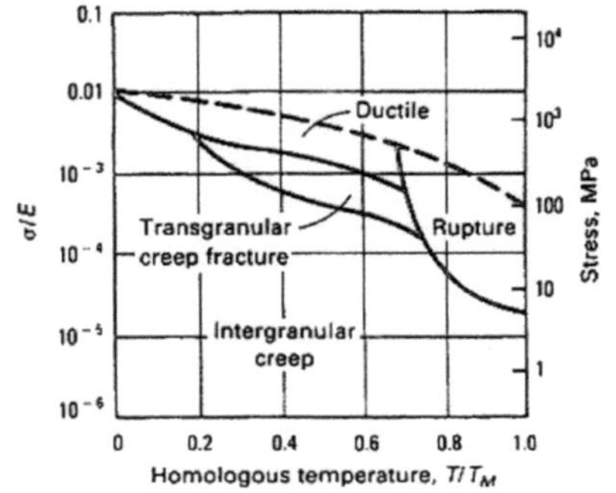


**Fig. 6.** Fracture mechanism map

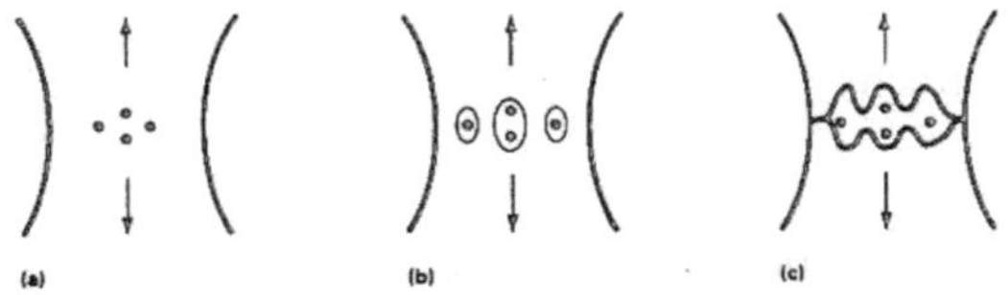


**Fig. 7.** Three stages of ductile fracture: (**a**) Void initiation at hard particles (**b**) Void growth (**c**) Void linking

The described fracturing by fibrous tearing (mode I) occurs by void growth in the crack plane that is essentially normal to the tensile axis. In mode II of the ductile failure the voids grow in arrays at an oblique angle to the crack plane under the influence of shear strain. This type of shear-band tearing can be observed on the surface of the cone in ductile cup-and-cone tensile fracture. It commonly occurs in deformation processing in which friction and/or geometric conditions procure inhomogeneous deformation, leading to local shear bands. Localization of deformation in these shear bands leads to an adiabatic temperature increase that causes local softening.

Compressive stresses superimposed on tensile or shear stresses by the deformation process can have a significant influence on closing small cavities and crack or limiting their growth and thus enhancing workability. Because of this important role of the stress state, it is not possible to express workability in absolute terms. Workability depends not only on material characteristics but also on process variables, such as strain, strain rate, temperature, and stress state.

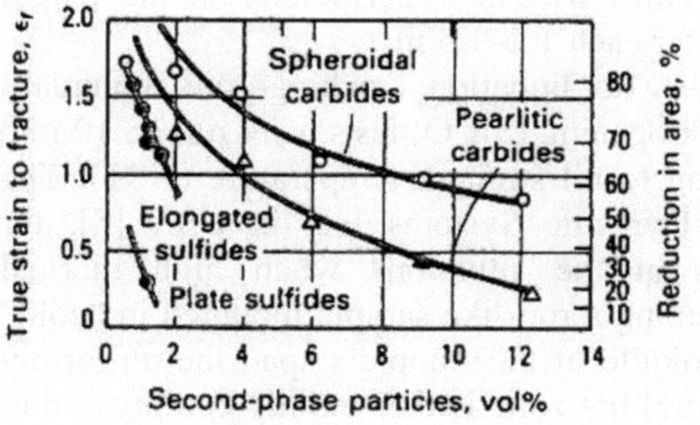


**Fig. 8.** Effect of volume fraction of second–phase particles on the ductility of plain low-carbon steels [4]

Many deformation processes for metals and alloys are performed at temperatures above 0.5 $T_M$ i.e. in the hot working range. Three principal benefits accrue from hot working. First is the lower flow stress at higher temperatures, which offers an economy method for size reduction of large work pieces. Second is the fact that metals at high temperatures are generally capable of achieving larger deformation strains without fracture than at lower temperatures. However, for many alloy systems, the temperature and strain rate must be appropriately chosen. Lastly, high-temperature deformation assists homogenization of the ingot structure (chemical segregation) and the closing of internal voids. These benefits come at the cost of surface oxidation, poorer dimensional accuracy, and the need to appropriately heat and cool the work piece. In general, the mechanisms of hot working are rather complex and may vary considerably from alloy to alloy.

## Characteristic of Central Burst-Susceptible Steel

The material under consideration here in its initial form is a round ingot of 410 mm diameter made of 34CrMo4 grade steel, which has been continuously cast in vertical direction with travel speed of 0.5 m/min at a casting temperature 1390°C. Cutting off a 6 m long piece from the con-cast ingot appeared after cooling to about 1050°C and transferring to re-heating/soaking oven at inlet temperature of about 900°C. The exit temperature from this furnace to the rolling track was 1280–1290°C, and rolling with the rate of about 5/s at the temperature of 1240°C into 200 mm square billet. After this procedure, internal burst cavities were found by NDT, and typical for these cavities was their periodicity in the length direction; they were repeating after each 1.2–1.5 m.

To check for the susceptibility to liquation cracking/grain boundary sensitization, strain-induced crack opening (SICO) tests were run on 10 mm diameter samples after re-heating to nil strength temperature (NST). The SICO test was developed by Dynamic Systems Inc. in USA [5], for determining formability limits at the situations when high thermal gradients occur. It comprises heating a rod-like sample mounted in "cold" Gleeble's jaws, so that in the middle of the sample's span the uniformly heated zone is formed. This central hot zone is then axially compressed to form a bulge. The test has been dedicated to Gleeble physical simulators, in which the use of high-rate electric resistance heating may cause a non controlled overheat of anvil-work piece interface in a conventional uniaxial compression test. The SICO procedure was in the first instance applied here for determining the melting and solidification parameters. For this 25 mm diameter bars were used and from them the melting (solidus) temperature was found to be ~1390°C, the nil strength temperature ~1350°C and the nil ductility temperature ~1300°C. Then samples for the 10 mm SICO tests were taken from different locations on the cross section of the 200 mm square billet:

Sample A - taken ~30 mm from the central burst cavity,
Sample B - taken ~15 mm from the central burst cavity,
Sample C - taken exactly at the central burst cavity.

These locations have been marked on the cross-section of the billet shown in Fig. 4. The results obtained from SICO tests are given in Table 1.

**Table 1.** SICO tests for hot/reheat crack susceptibility

| Material | Peak temp (NST) | Deformation temperature | Strain | Strain rate | Results: number of cracks |
|---|---|---|---|---|---|
| Sample A | 1345 | 1290 | 1.25 | 6/s | 6s-cracks |
| Sample B | 1345 | 1290 | 1.25 | 6/s | No cracks |
| Sample C1 | 1345 | 1290 | 1.25 | 6/s | 2s-cracks |
| Sample C2 | 1345 | 1290 | 1.25 | 10/s | 4m-cracks + 17s-cracks |

The highest susceptibility to the liquation cracking appeared on the samples taken from the close vicinity of the burst cavity, especially for the strain rate exceeding the parameters of the real process. Then additionally cracks were found on sample taken at about 30 mm from the burst cavity, i.e. from the transition region of globular to columnar primary crystals. No crack sensitivity was found in the region of about 15 mm from the burst cavity. Representative SICO samples from locations A and C are presented in Figs. 9 and 10.

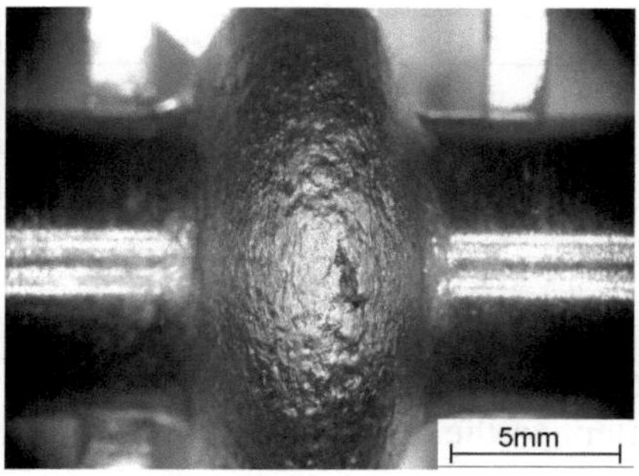


**Fig. 9.** SICO sample A showing a small crack near to perimeter of the bulge

**Fig. 10.** SICO sample C2 showing a chain of small cracks along perimeter

Following the above SICO results, additional samples for hot tensile testing were taken from the as continuously cast 410 mm diameter ingot, from the locations adequate to these which showed different hot crack sensitivity in the SICO tests. These samples were marked as given in Table 2.

In the hot tensile tests the nil strength temperature (NST) of 1345°C was taken as the maximum of the heating cycle and from it the samples were cooled down to the pulling temperatures. Two samples were taken from the center of the ingot, location (1) where a small shrinkage cavity as well as visible porosity appeared with dendritic crystals inside. The mechanical results from these samples are of course irrelevant nevertheless the samples were useful for observation of fracture mode.

**Table 2.** Results of hot tensile tests

| Sample number | Distance from centre | Temperature [°C] | | YS [MPa] | Strain $\varepsilon_l$ | RA [%] | remarks |
|---|---|---|---|---|---|---|---|
| | | reheat | test | | | | |
| 1 | 0 mm | 1345 | 1240 | 43 | 0.37 | 16 | s. cavity |
| 1A | 0 mm | 1345 | 1200 | 81 | 0.56 | 62 | porosity |
| 2 | 30 mm | 1330 | 1200 | 95 | 0.60 | 86 | |
| 2A1 | 30 mm | 1345 | 1240 | 85 | 0.61 | 88 | |
| 2A2 | 30 mm | 1345 | 1200 | 96 | 0.65 | 94 | |
| 2A3 | 30 mm | 1345 | 1240 | 79 | 0.54 | 83 | |
| 3 | 60 mm | 1330 | 1200 | 95 | 0.64 | 96 | |
| 3A1 | 60 mm | 1345 | 1240 | 76 | 0.55 | 75 | |
| 3A2 | 60 mm | 1345 | 1200 | 11 | 0.07 | -8 | brittle |
| 3A3 | 60 mm | 1345 | 1240 | 83 | 0.66 | 78 | |

The results of the hot tensile testing on the sound samples showed that at the temperature of 1240°C, i.e. equivalent to the application conditions, the material from location (3) deformed slightly slower in the crosswise direction than this from location (2), while their yield strengths were comparable. As the procedure of checking for susceptibility to liquation cracking comprises heating-up a sample with fixed Gleeble's jaws that due to thermal expansion causes slight bulging and increase of sample's diameter in the central uniformly heated zone, it appeared on one sample from the location (3), that it begun already cracking on heating in a brittle manner, failing during subsequent pulling at very small extension ($\varepsilon_l$) and with "negative" reduction in area. This odd behaviour was caused by a liquid phase, which already on heating under the compression due to thermal expansion spread along the grain boundaries as a semi-continuous film. Except for this, all other samples tested at 1200°C showed high ductility, i.e. the RA of more than 85%. The case of brittle failure and then the scatter of results indicate the influence of segregation and/or porosity on the results of the hot tensile testing.

## Metallographic Examination

In the first instance the central burst cavities were examined. As the internal defects were found in the ingots, one might suspect that the formation of the burst cavity might be related to the porosity in the middle of the ingot, however it is not. The walls of the burst cavity appeared to be entirely of shear fracture type, Fig. 11, with numerous fibre-like "pillars" breaking by tensile necking, Fig. 12.

**Fig. 11.** Shear-type separation surfaces dominating inside the burst cavity

**Fig. 12.** Fibre-like pillar with defined necking portion in the burst cavity

When a sample containing shrinkage porosity is taken from the middle of the ingot and then hot tensile tested, in the cavity extending during the test numerous dendrites remain intact, while the walls of the cavity show the tensile shear mode, Fig. 13. The same type of shear can be observed on the outer surface of hot tensile samples, Fig. 14, and almost exactly the same shear-slip features can be found on all inner walls of the burst cavity. In the last, due to tri-axiality of the generated strains, a number of fibre-like pillars occur standing up from the main inner wall planes, Figs. 15 and 16.

**Fig. 13.** Dendritic crystals in the cavity of hot tensile sample 1

**Fig. 14.** Outer surface of the tensile sample 1 showing shear-slip lines

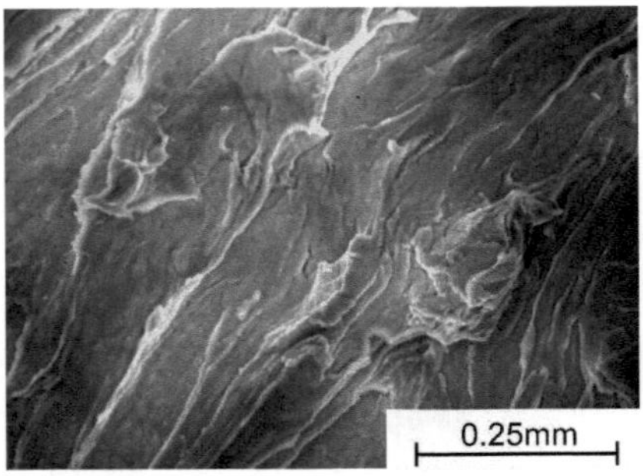


**Fig. 15.** Shear-slip lines on the surface of the burst cavity in the 200 mm square billet

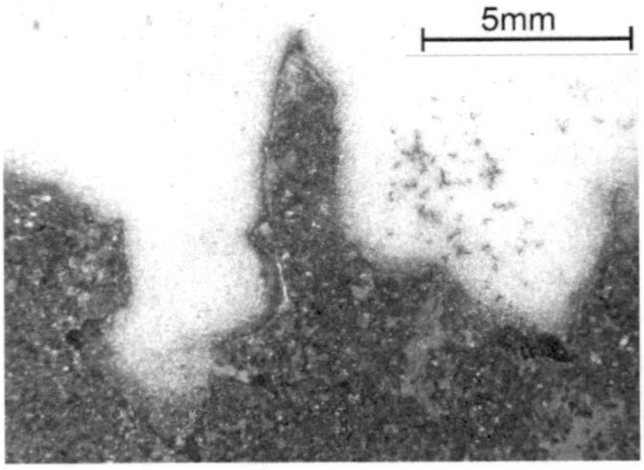


**Fig. 16.** A cross section through a fibre-pillar protruding from wall of the burst cavity; nital etched - DF

Characteristic of these pillars in the investigated medium-carbon Cr-Mo steel is their austenite grain size; it is much coarser than a few millimetres below their roots, which indicates that they were formed already on the entry to the rolls and this way could not accumulate the major amount of strain exerted during the rolling. The following two pictures taken in a dark-field mode (DF) show the austenite grain size in the pillar, Fig. 17, and at the distance of ~10 mm below the pillar's root, Fig. 18.

**Fig. 17.** Coarse prior austenite grains in the fibre-like pillar; nital etched – dark field image

**Fig. 18.** Finer prior austenite grains at about 10 mm below the pillar's root; nital etched - DF

On the shear-slip type walls of the burst cavity evidences can be found that during their formation hard particles existed which assisted the deformation and separation process of the fibres and the walls, as well as that some fibres had different ductility than the others. In the subsequent Fig. 19 this situation is shown. After a slight electro-polishing and etching of such shear-slip surface, the particles, which left the sharp shear traces on the surfaces, can be revealed, in the size and amount adequate to explain the appearance of "scratches" on the burst cavity walls, Fig. 20.

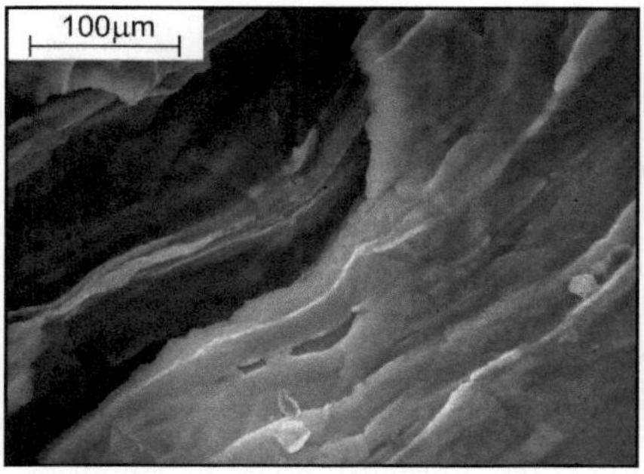


Fig. 19. Shear-slip type of the burst cavity wall with sharp "scratches" caused by solid hard inclusion particles

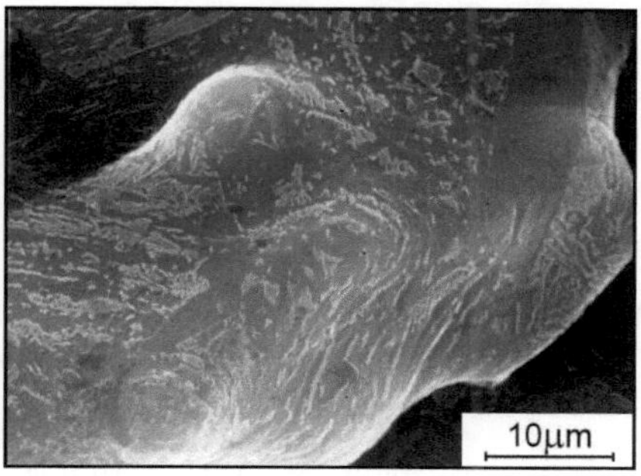


Fig. 20. Fragment of polished and etched surface of the burst cavity showing microstructure with inclusion

In general, as seen on the metallographic samples taken from the vicinity of cavity of the 200 mm square billet, the material after deformation still consists of coarser grain patches embedded in finer-grained surrounding, Fig. 21, and the cracks seem to easier propagate through the coarser-grained material while retard when enter the finer-grained regions. Near to such cracks in the pearlitic-ferritic microstructure of the billet, finer than ferrite grains bright particles could be found, Fig. 22.

Fig. 21. Region of coarser prior austenite grains near to burst cavity in 200 mm square billet; nital etched, dark-field image

Fig. 22. Pearlitic-ferritic microstructure with bright AlN particles in ferrite grains and black spheroidal inclusions; nital etched

During examination of the grip portion of sample 3A2, which sample was taken from the crack-sensitive zone of the ingot, and which sample failed during the hot tensile test in a brittle manner i.e. at little elongation and without necking, the AlN particles tracing boundaries of the primary columnar and globular crystals were also found, Fig. 23, as well as some porosity, Fig. 24.

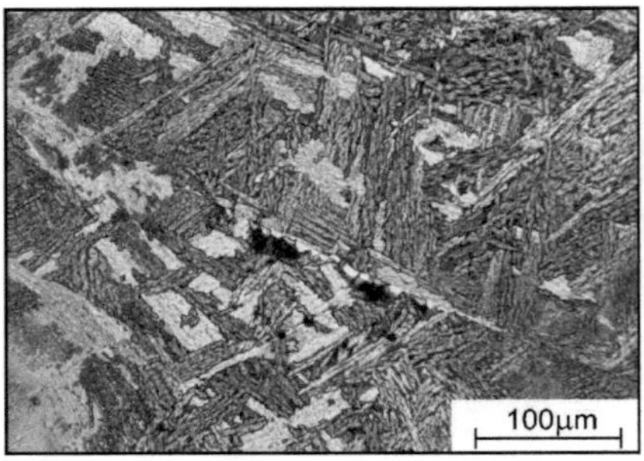


**Fig. 23.** A chain of AlN particles tracing primary dendrite boundary in the crack-sensitive zone (3) of the ingot; nital etched

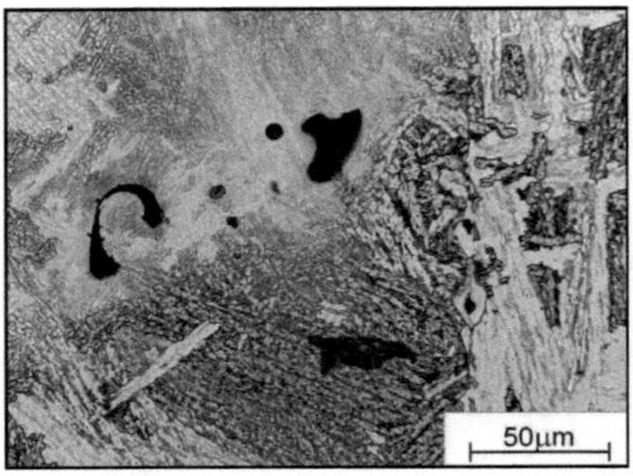


**Fig. 24.** Porosity near to dendrite boundary in as-cast microstructure of the ingot in zone (3); nital etched

## Discussion

The cavitation described here, in particular the formation of the central burst cavity is evidently controlled by the large strains occurring in the metal forming process. It is also evidently assisted by inhomogenity of the primary microstructure i.e. segregation, as well as by the solidification or reheat defects of porosity type. The columnar crystals in the as-solidified structure of the continuously cast ingots are large – they reach the lengths of up to 50 mm, and during their crystallization various contaminants are moved with the crystallization front towards the center of the material. Among the contaminants also the grain-refining elements and particles are moved. Accordingly, the region of dendrites appears "cleaner" and coarse-grained. After homogenizing annealing before hot rolling the intensity of segregation becomes certainly reduced on micro-scale, nevertheless it does not seem so as regards the zone segregation of the AlN and other inclusions. In result, during rolling the cavitation occurs, which initiates in the regions where higher density of these inclusions can be found.

In general, the material under consideration with its 0.013% P and 0.007% S is clean, most probably much cleaner than the materials, which were available at the time when the first books and papers on hot ductility and central bursting in drawing, extrusion and rolling were produced (compare with standards in e.g. Metals Handbook 1961, permitting 0.040% P and 0.040% S for a similar grade of steel [6]). For sure the investigated here material contained less sulfide inclusions than the ones manufactured in 1960s. So the historical models of the ductile cracking, by the formation of voids around inclusions, in particular the sulfides, and then coalescence of these voids, not always can be applied without reserve. The same remark may refer to the deformability maps.

Nevertheless, the cavitation during metal forming is still an existing important phenomenon, which often causes premature failures of the material during processing. It appears in a large number of steels, which generate microscopic voids (or cavities) when subjected to large strains under tensile loading. The formation of microscopic cavities often occurs along the grain boundaries of primary i.e. as-cast structure during the initial operations of high-temperature deformation. It is believed that the cavitation may lead to premature failure at levels of deformation far less than those at which the flow localization controlled failure would occur [7]. For a given material, the extent of cavitation depends on the specific deformation conditions like the amount of crosswise strain, strain rate and temperature, as well as on the surface condition, which may affect the friction between tool (rolls) and workpiece (billet) and this way the level of tensile strains and dynamic stresses. An important requirement for cavitation during flow under hot working is the presence of tensile stress. On the other hand, under conditions of homogeneous compression, cavitation is not observed. In fact, it is believed that the cavities, which might be produced at the tensile flow situation, can be removed during subsequent compressive flow. Due to this last belief, the metal-formers often do not care too much about the fine porosities and cavities, which may nucleate during the forming process, unless the coalescence of these cavities leads to total failure of the product.

In the situation of billets containing the large central burst cavities, two major factors should be highlighted. The first is the tensile stress/strain situation that may occur on the entry of the ingot to the rolling stand and which is assisted by the friction between the rolls and the oxidized surface of the ingot. The oxide scale on the investigated Cr-Mo steel does not easily strip-off as well as it limits the possibility of sliding between billet and rolls, thus contributing to high tensile forces. The second factor is the segregation of inclusions in particular this of AlN. For the purpose of de-oxidation and grain refinement the investigated steel contained almost 300 ppm of Al and 100 ppm of N. Concentration of these and other inclusion particles in certain regions of the ingot structure resulted in differences of hot deformability, in particular in capacity of material to flow with different rates at the same applied external stress. This phenomenon is consistent with the formation of the large central burst cavities having shear-type walls.

In this chapter the whole complexity of the problem was neither addressed nor solved. Our aim was mainly to describe the phenomena, which assist the formation of the cracks – the central burst cavities. It appeared that the differences in local ductility under dynamic deformation conditions could create the situation, which gets out of control. The hot deformation tests: SICO and hot tensile testing, gave results showing the differences in local material properties, while metallography revealed the effects: micro-structure and cracks, as well as indicated the distribution of particles which might assist the problem.

In result, measures were taken at steelworks to eliminate the problem, in particular by reducing the amount of AlN in the cast as well as introducing stirring to reduce segregation in the ingots.

## Conclusions

1. Large diameter continuously cast steel ingots have different hot formability/ductility properties in different zones of their cross-section.
2. When hot rolling such materials comprising regions of different ductility, internal cavities may form due to differences in local flow rates of the material.
3. High density of grain-refining particles and inclusions evidently strengthens the material and reduces its hot ductility.
4. Internal burst cavities initiate in the high strength, low ductility zones, in which due to segregation of strengthening particles high probability of voids formation around these particles occurs.
5. Dynamic tensile stresses of the hot rolling process, especially these created due to friction between the rolls and the work piece/billet, assist the formation of the central burst cavities.

## References


1. Avitzur B, in: "Metal Forming – the Application of Limit Analysis", Publ M Dekker Inc, New York, 1980, pp. 108–109.
2. Turucz S, "The effect of the Roll-Gap Shape Factor on Internal Defects in Rolling", J Mater Process Technol, Vol. 60, 1996. pp. 275–282.
3. Backofen WA, "Deformation Processing", Addison-Wesley Publ Co Inc, 1972.
4. Gladman T, Holmes B, and McIvor LD, "Effect of Second–Phase Particles on the Mechanical Properties of Steel", J Iron and Steel Institute, 1971, p. 78.
5. Ferguson HS; "Fundamentals of Physical Simulation", in Proc Intl Symposium on Physical Simulation, TU Delft 1992, pp. 1–21.
6. Metals Handbook 8th Edition, Lyman T (ed), Vol. 1, "Properties and Selection of Metals", ASM, Metals Park OH, 1961, p. 61.
7. Nicolaou PD, Semiatin SL and Lombard CM; "Simulation of the Hot-tension test under Cavitating Conditions", Metall Mater Trans A, Vol. 27, 1996, pp. 3112–3119.

# Part V
# Ductility-Dip Cracking

# Effect of Filler Metal La Additions on Micro-Cracking in Multi-Pass Laser Overlay Weld Metal of Alloy 690

K. Nishimoto[1], K. Saida[1], M. Sakamoto[1], W. Kono[2]

[1] Osaka University, Division of Materials & Manufacturing Science, Osaka, Japan
[2] Toshiba Co., Power & Industrial Systems R&D Centre, Japan

## Abstract

The effect of La addition on microcracking susceptibility in the multipass laser overlay welds of alloy 690 (Inconel 690) was investigated. Multipass laser overlay welding of alloy 690 was conducted on alloy 132 (Inconel 132) in air (in-air welding) and under water (underwater welding) by using a YAG laser with the power of 2 kW. Argon gas shielding was applied in both cases with the gas flow rate of 20–60 L/min. Five kinds of filler metals, which were varied in La content from 0 to 0.09 mass%, were used. Microcracks in both weld metals produced by in-air and underwater welding were categorised as ductility-dip cracks. Ductility-dip cracks occurring in the weld metal could be completely prevented by using 0.01–0.03 mass% La containing filler metals in both in-air and underwater welding. However, excessive La addition more than 0.04 mass% to the filler metal adversely led to the occurrence of weld metal solidification/liquation cracks also in both in-air and underwater welding. The improved ductility-dip cracking susceptibility was attributed to the desegregation of impurity elements of P and S to grain boundaries due to the scavenging effect of La. Excessive La addition to the weld metal caused the formation of a Ni-La intermetallic compound with low melting point, resulting in solidification/liquation cracking. Thermo elasto-plastic analysis indicated that the plastic strain-temperature curve slightly intersected the ductility-dip temperature range (DTR) of the La-free weld metal in the cooling stage of overlay welding, however, the plastic strain-temperature curves did not cross the DTRs of the La-added weld metals. The analysed results were fairly consistent with the experimental results that a few ductility-dip cracks occurred in the La-free weld metal, while no ductility-dip cracks occurred in the La-added weld metal of multipass laser overlay welds.

## Introduction

The Ni-base superalloy, alloy 600 (Inconel 600), has been commonly used in the reactor vessel, steam generator and pressuriser of nuclear power plants (pressurised-water nuclear reactors). However, serious damage such as the primary water stress corrosion cracking (PWSCC) has been found in alloy 600 components [1]. Another Ni-base superalloy, alloy 690 (Inconel 690), is the material of choice for pressurised-water nuclear reactor components because of its superior PWSCC resistance compared to alloy 600. It is planned that alloy 690 will replace the existing alloy 600 in ageing nuclear power plants [2, 3]. In particular, the cladding of alloy 690 onto the existing alloy 600 components and welds is one of the most practical countermeasures for the preventive maintenance and/or repair of PWSCC [4]. Recently, the laser cladding technique, especially when performed underwater, has been developed and applied in practice for overlay welding and seal welding of reactor components [5]. This technique possesses the advantages of low heat input and reliable welding, compact welding machine dimensions for welding in narrow spaces and a wide choice of weld materials.

On the other hand, it has been pointed out that alloy 690 is highly susceptible to hot cracking under heavy-restraint conditions such as the welding of thick components [6]. Microcracking (microfissuring) in the reheated weld metal during multipass welding can be of particular concern. According to previous studies [7, 8, 9, 10, 11], microcracking in alloy 690 welds has been characterised as ductility-dip cracking. The authors have reported that the primary cause of microcracking (ductility-dip cracking) in alloy 690 was the reduction of hot ductility during cooling in welding attributable to the grain boundary segregation of impurity elements such as P and S [12, 13]. Furthermore, we have also proposed an effective method for preventing microcracks in alloy 690 welds using La-containing alloy 690 filler metals, and confirmed that microcracks in multipass gas tungsten arc (GTA) welds of alloy 690 could be completely prevented using the filler metal with 0.01 mass% La added [14].

The objectives of the present study are to evaluate the microcracking susceptibility in the multipass laser overlay weld metal of alloy 690, and to clarify the effect of La addition to the filler metal with the objective of improving microcracking susceptibility of the weld metal. We have conducted microstructural analysis of the weld metal to clarify the mechanism by which La addition improves microcracking behaviour. Furthermore, in order to comprehend the occurrence of ductility-dip crack during laser overlay welding, we have made theoretical approaches, employing numerical analysis, to model the behaviours such as the grain boundary segregation of P, S and the development of thermal strain in laser overlay welds.

## Materials and Experimental Procedures

### Materials

The base metal used in this study to simulate alloy 600 weld metal was alloy 132 (Inconel 132, DNiCrFe-1J) overlay weld metal (shielded metal arc welding: SMAW, 4 layers), surface-ground to approx. 10 mm thick, on a 70 mm thick carbon steel, with the dimensions shown in Fig. 1. Five kinds of filler metals of alloy 690/alloy 52 (Inconel 52, ERNiCrFe-7) in which the La content varied between 0 and 0.07 mass% were used. The commercial filler metal NF0 (La-free) was used as a standard of comparison for microcracking susceptibility. The chemical compositions of the base metal and filler metals are shown in Tables 1 and 2, respectively. The diameters of the filler metals are 1.2 mm.

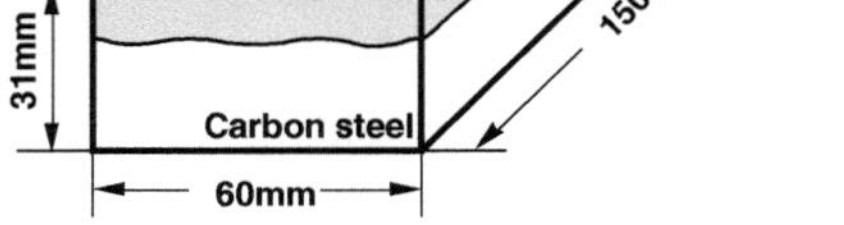


**Fig. 1.** Schematic illustration and dimensions of the base metal

**Table 1.** Chemical composition of base metal alloy 132 (DNiCrFe-1J) used (mass%)

| | C | Si | Mn | Ni | Cr | Cu | Fe | P* | S* | Nb+Ta |
|---|---|---|---|---|---|---|---|---|---|---|
| DNiCrFe-1J (Alloy 132) | 0.04 | 0.21 | 2.80 | Bal. | 14.43 | 0.02 | 8.88 | 40* | 30* | 1.86 |

* : ppm

**Table 2.** Chemical compositions of filler metals alloy 690/alloy 52 (ERNiCrFe-7) used (mass%)

| Filler metal | C | Si | Mn | Ni | Cr | Co | Mo | Ti | Al | Fe | P* | S* | B* | La |
|---|---|---|---|---|---|---|---|---|---|---|---|---|---|---|
| NF0 | 0.01 | 0.22 | 0.25 | Bal. | 29.74 | 0.02 | 0.02 | 0.11 | 0.10 | 10.31 | 50* | 10* | <10* | – |
| NF1 | 0.01 | 0.22 | 0.25 | Bal. | 29.9 | 0.02 | 0.02 | 0.10 | 0.10 | 10.30 | 60* | 10* | <10* | 0.01 |
| NF2 | 0.01 | 0.21 | 0.25 | Bal. | 29.6 | 0.02 | 0.02 | 0.10 | 0.09 | 10.23 | 60* | 10* | <10* | 0.03 |
| NF3 | 0.01 | 0.22 | 0.25 | Bal. | 29.6 | 0.02 | 0.02 | 0.10 | 0.10 | 10.27 | 60* | 10* | <10* | 0.04 |
| NF4 | 0.01 | 0.22 | 0.26 | Bal. | 29.5 | 0.21 | 0.02 | 0.11 | 0.09 | 10.13 | 70* | <10* | <10* | 0.07 |

* : ppm

## Experimental Procedures

Figure 2 illustrates the set-ups for multipass laser overlay welding (3 layers, 10-9-8 passes) of alloy 690 on alloy 132 was carried out in air and under water using a YAG laser, and the conditions are summarised in Table 3. The

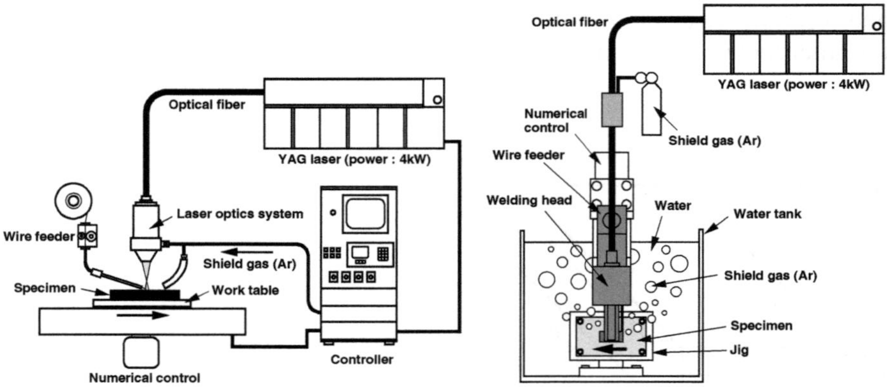


(a) In-air welding (b) Underwater welding

**Fig. 2.** Schematic illustration of laser overlay welding apparatus

**Table 3.** Laser overlay welding conditions used

| Cladding condition | A | B | C | D | E |
|---|---|---|---|---|---|
| Laser power (W) | 1000 | 1000 | 1100 | 1200 | 1200 |
| Laser travelling speed (mm/s) | 5.0 | 11.7 | 8.3 | 5.0 | 11.7 |
| Wire feeding speed (mm/s) | | | 10.0-18.3 | | |
| Shiled gas flow rate (Ar, L/min) | | | 20 (in–air)<br>60 (underwater) | | |

welding conditions were varied within the following ranges: laser power of 1.0–1.2 kW, laser travelling speed of 5.0–11.7 mm/s, wire feeding speed of 10.0–18.3 mm/s and shield gas flow rates (Ar) of 20 L/min for in-air welding and 60 L/min for underwater welding. The microcracking susceptibility was evaluated as the total crack length in the cross-section of the overlay weld metal. The weld metal was sectioned and prepared for optical microscope and scanning electron microscope (SEM) observation by grinding and then polishing with #400–#1500 grit emery paper. The final polishing was performed with 1 μm alumina paste. Electrolytic etching was performed with a 10% oxalic acid ethanol solution at 1.5–3.0 V for 25–30 s. The element distribution in the weld metal was analysed by EPMA.

## Microcracking Behaviour in Overlay Weld Metal

### Characterisation of Microcracks

The cross-sectional views of the laser overlay weld metal with varying La content, using Condition A in Table 3, are shown in Fig. 3. In the case of using the La-free filler metal (NF0), a few microcracks occur in the weld passes of the second and third weld layers where they were reheated by the subsequent weld beads (designated as "under-bead cracking"). Several large microcracks can be seen in the weld metal (NF4) containing the high proportion of La (0.07 mass%), mainly in the final weld layer. However, no microcracks can be observed in the weld metal in the intermediate case using 0.01 mass% La added to the filler metal (NF1). Figure 4 shows the microstructures of microcracks in the weld metals of NF0 and NF4. A microcrack in the La-free weld metal (NF0) occurs at a small distance from the fusion line and is propagated at the austenitic grain boundary (G.B.). From the fact that the crack surface indicates smooth and intergranular fracture without any trace of melting, the majority of microcracks in the La-free weld metal can be characterised as being ductility-dip cracks. On the other hand, a microcrack in the weld metal containing 0.07 mass% La (NF4) originates at the fusion line and is propagated at the solidification

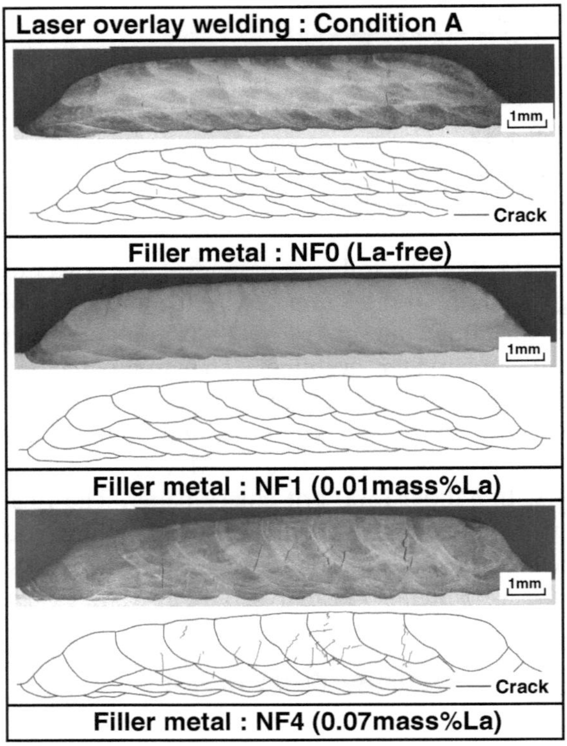


**Fig. 3.** Macrostructures of laser overlay weld metals with various La contents

boundary. The surface morphology of this crack indicates a dendritic structure with a trace of a liquid film. It follows that the majority of microcracks in the weld metal containing the larger amount of La can be characterised as being solidification/liquation cracks.

## Effect of La addition to Filler Metal on Microcracking Susceptibility

The effect of La addition to the filler metal on microcracking susceptibility in the overlay weld metal is shown in Fig. 5. The total crack length was evaluated for various combinations of the filler metal and welding condition. The types of microcracking were not distinguished in the present evaluation, however, the microcracking susceptibility in the La-free weld metal can be

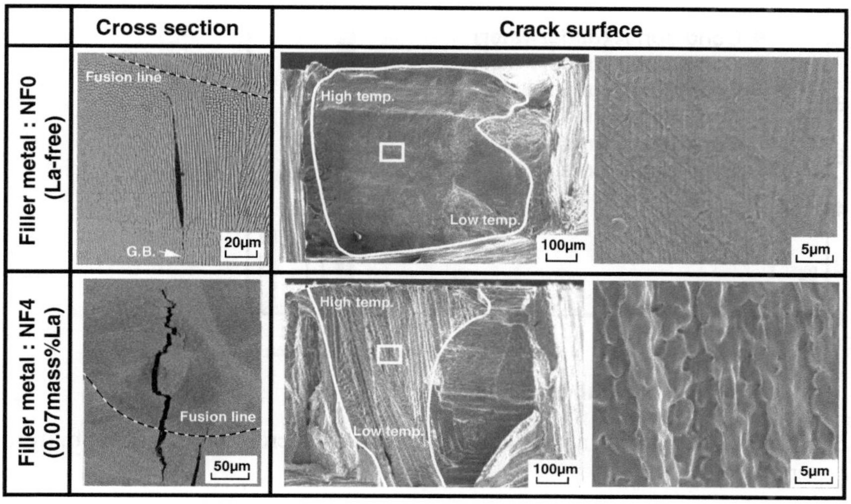


**Fig. 4.** Microstructures of cracks in laser overlay weld metals

approximately regarded as ductility-dip cracking susceptibility and that in La-added weld metal as solidification cracking susceptibility. Microcracking susceptibility in the La-free weld metal is not very high and microcracks occur for only a fraction of the welding conditions. No microcracks occur in the weld metal using 0.01–0.03 mass% La added to the filler metal (NF1 and NF2) at any welding condition including the weld conditions where microcracks occurred in the La-free weld metal. However, microcracking susceptibility increases again when the La content of the filler metal exceeds 0.04 mass%, except for Conditions D and E. In the present experimental welding conditions, microcracking susceptibilities at low laser power (Conditions A and B) tend to be higher than those at high laser power (Conditions D and E). Furthermore, there is not much difference in microcracking susceptibility between underwater overlay welding and in-air overlay welding.

## Microcracking Mechanism in Overlay Weld Metal

### Microstructural Analysis of Weld Metal with Added La

In order to elucidate the mechanism of change in microcracking susceptibility with La addition, microstructural analyses of the overlay weld metals were carried out. The preliminary SEM observation of the overlay

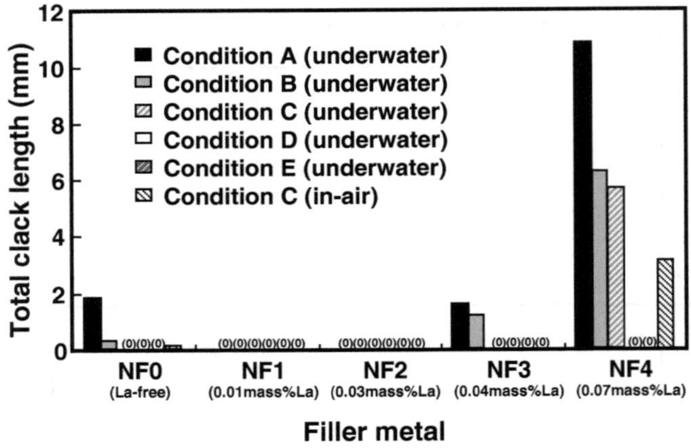


**Fig. 5.** Effects of La-addition and welding conditions on microcracking susceptibility in laser overlay weld metal

weld metals containing La revealed the presence of a cellular structure in all the weld metals; further, there was no significant difference in the morphology and evenness of grain boundaries between the overlay weld metals with different La contents [14]. The mapping of elements in the overlay weld metals NF2 and NF4 analysed by EPMA is shown in Figs. 6 and 7, respectively. P, S and O tend to be found at the same locations as La in the weld metal NF2 (arrows in Fig. 6), and La tends to be found also at different locations from P and S in the weld metal NF4 (arrow in Fig. 7). Taking these results together with the quite similar results reported in the previous study [14], we could identify the products in the La-added weld metals as La phosphide, sulphide and/or oxide in the La-added weld metal, and a Ni-La intermetallic compound might also be present in the overlay weld metal NF4.

## Mechanism of Improvement in Microcracking Susceptibility by La Addition

The improvement mechanism of microcracking in the alloy 690 overlay weld metal by the La addition to the filler metal is discussed. By analogy with the microcracking mechanism in the multipass GTA weld metal, we expect the ductility-dip cracking in the overlay weld metal would also be caused by the reduction of hot ductility attributed to grain boundary segregation of impurity elements such as P and S. From the fact that La phosphide and sulphide are formed in the La-added weld metals, microsegregation of P and S to a grain boundary would be depressed. In other words, La scavenges P and S in the weld metal [14]. Consequently, we deduce that ductility-dip cracking susceptibility would be improved by the amelioration of hot ductility of the weld metal [12-14]. On the other hand, excessive La-addition to the filler metal leads to the unwanted solidification and/or liquation cracking in the overlay weld metal. Excessive La-addition also leads to the formation of a Ni-La intermetallic compound in the weld metal which has a low eutectic point compared to the Ni matrix [15]. Therefore, cracks would occur in the weld metal for an excessive La content due to the enlargement of BTR (the brittle temperature range between solidus and liquidus) and/or the occurrence of local liquation of a Ni-La intermetallic compound during reheating [14].

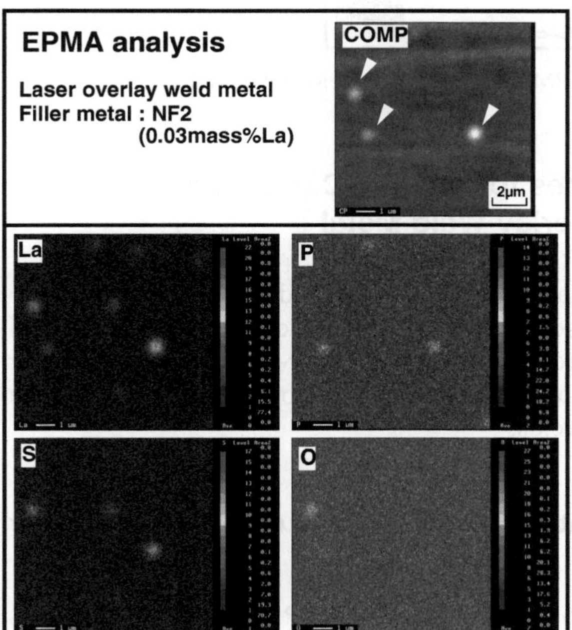


**Fig. 6.** Element analysis by EPMA for La-added weld metal NF2

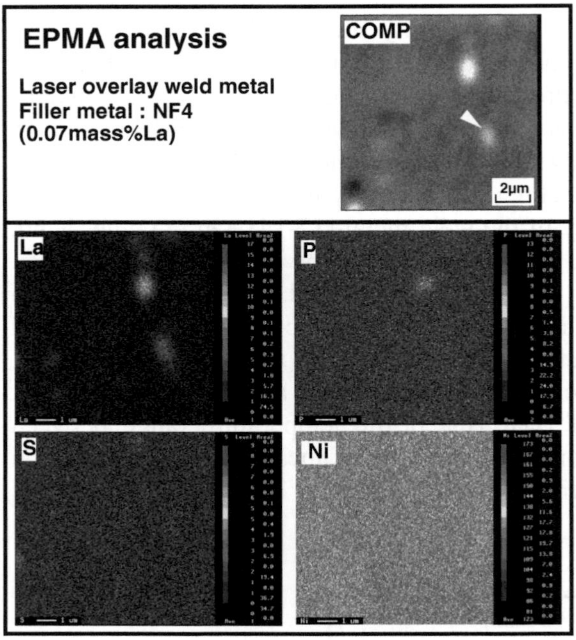


**Fig. 7.** Element analysis by EPMA for La-added weld metal NF4

## Theoretical Approaches to Ductility-Dip Cracking in Mutipass Laser Overlay Weld Metal

In order to better understand the conditions under which ductility-dip cracking occurs during laser overlay welding, we have used numerical analysis for theoretical approaches to such ductility-dip cracking behaviour as grain boundary segregation of P, S and the development of thermal strain in multipass laser overlay welds.

### Grain Boundary Segregation in Weld Metal during Multipass Laser Overlay Welding

In order to estimate grain boundary segregation of P and S during multipass laser overlay welding (reheating) process, the theoretical analysis was conducted under the condition that the overlay weld metal was subjected to welding thermal cycles generated by two successive overlay weld passes.

### *Theoretical Model of Microsegregation in Reheated Weld Metal*

As described above, ductility-dip cracks in the alloy 690 overlay weld metal would propagate preferentially along austenitic grain boundaries which were coincident with solidification boundaries (dendrite boundaries) [12]. As a result of this conclusion, the theoretical model for microsegregation used in the present simulation considered solidification segregation as well as grain boundary segregation. The calculation procedures for microsegregation are similar to those used previously for multipass GTA welding [13, 16, 17]. Specially, segregation is assumed to happen during the solidification stage of welding; subsequent segregation during the following cooling/reheating stage occurs under the initial condition of non-uniform distribution of segregating species formed during solidification segregation. In order to simplify the computation procedure, segregation behaviour of P and S to a grain boundary was calculated for pseudo-binary systems of (Ni-30Cr-10Fe)-P and (Ni-30Cr-10Fe)-S using the finite differential method (FDM) scheme. Furthermore, the co-segregation effect of P with S was not considered in the present simulation. The model analysed here uses one-dimensional diffusion in a rectangular triangle assuming that the morphology of a dendrite cell is basically a hexagonal prism. Further details of the theoretical models used in the present study can be found in Refs. [13, 16, 17]. The conditions used in the calculations are summarised in Table 4, i.e., the initial P and S contents in the weld metal are 50 ppm and 16 ppm, respectively, and the cooling and reheating rates were varied between four

**Table 4.** Calculation conditions for computer simulation of grain boundary segregation of P and S

| | | |
|---|---|---|
| Initial concentration (mass%) | P | 0.0050 |
| | S | 0.0016 |
| Cooling rate / reheating rate (K/s) | GTA welding | 30(20) / 50 |
| | Laser overlay welding-1 | 60(40) / 100 |
| | Laser overlay welding-2 | 150(100) / 250 |
| | Laser overlay welding-3 | 300(200) / 500 |

( ) : Cooling rate between liquidus and solidus

levels in order to simulate laser overlay welding and GTA welding. Material constants used for the numerical simulation are also given in Ref. [13].

### Microsegregation to Grain Boundary during Multipasss Welding

The computer simulation was carried out with the peak temperature in both the second and third cycles being 1500 K. Figure 8 shows the relation between the elapsed time in welding thermal cycles and the calculated P and S concentrations at a grain boundary for cooling/reheating rates of 150/250 K/s. In the solidification process, P and S concentrations in the rest liquid phase increase with the progress of solidification, resulting in remarkably high amount of P and S bring segregated to a grain boundary at the completion of solidification. However, P and S concentrations are drastically reduced to their equilibrium concentrations when the weld metal cools only by approx. 200 K. After that, they increase again with further cooling (drop in temperature) from approx. 1400 K to 700 K, and they saturate at the constant level below 700 K. In the reheating process, P and S concentrations at a grain boundary keep constant temporarily or increase slightly with a rise in temperature. However, they decline rapidly at temperatures over approx. 1100 K. They once more increase again during the cooling stage in the reheating process, and then change cyclically thereafter in synch with the thermal cycle. The effects of different cooling

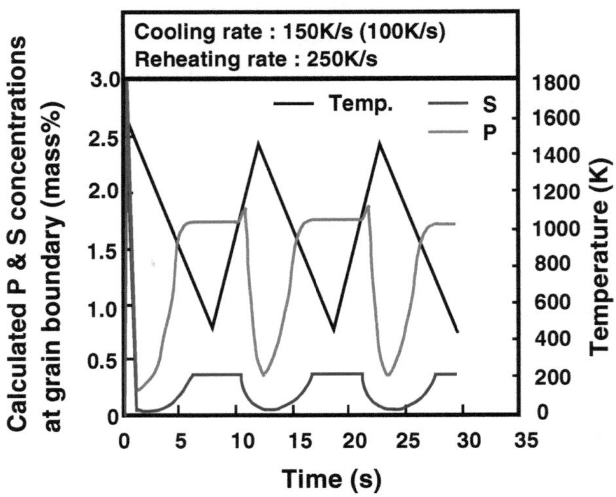


**Fig. 8.** Calculated P and S concentrations at grain boundary during multipass laser overlay welding thermal cycle

and reheating rates on the P and S concentrations at a grain boundary at 1500 K, which represents the typical temperature range of ductility-dip cracking, are shown in Fig. 9. The calculated P and S concentrations increase slightly with increases in the cooling and reheating rates as well as the number of weld passes. However, the values of the cooling and reheating rates do not much affect grain boundary segregation for the range of conditions used here. The approximate P and S concentrations at a grain boundary attained during welding were respectively about 70 and 25 times (approx. 3500 ppm and 390 ppm) as high as their initial concentrations. We conclude that microsegregation of P and S to a grain boundary affects microcracking susceptibility significantly, but grain boundary segregation tends to be reduced at higher temperatures during the reheating thermal cycle.

## Thermal Strain Analysis during Multipass Laser Overlay Welding

It has been generally understood that hot cracking problems, such as solidification and ductility-dip cracks occur when the strain-temperature

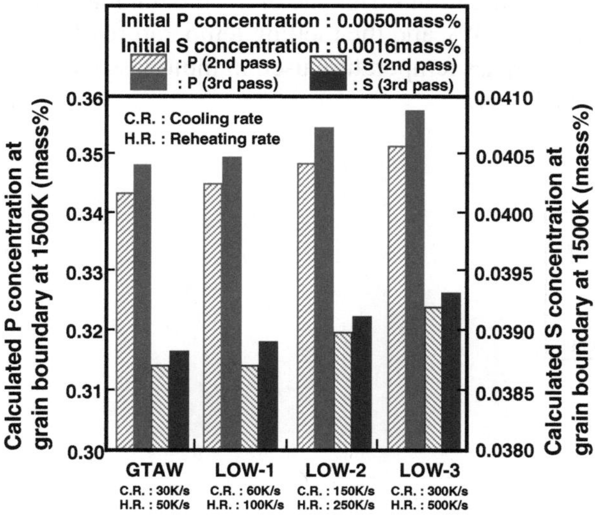


**Fig. 9.** Effect of welding condition (cooling and reheating rates) on P and S concentrations at grain boundary at 1500 K during multipass gas tungsten arc welding (GTAW) and laser overlay welding (LOW)

curve crosses the brittle temperature ranges (solidification brittle temperature range (BTR)/ductility-dip temperature range (DTR)) in the welding process. In the present study, we have used thermal stress/strain analysis to predict the occurrence of ductility-dip cracks in the multipass laser overlay weld metal.

### Model Details

The high temperature stress and strain in multipass welds were analysed by two-dimensional thermo elasto-plastic FEM (calculation code: Quick Therm). The model used for analysis of multipass laser overlay welds (3 layers, 6 passes) on alloy 690 is illustrated in Fig. 10. The deposited metal of each weld pass modelled in the present analysis has the rectangular shape with 1873 K (estimated temperature of the molten metal) dimensioned as 2 mm wide and 0.5 mm thick as illustrated in figures; the penetration of each weld pass is consequently determined by the heat conduction from this rectangular heat source. The position used for the analysis of stress/strain in the multipass weld metal is also indicated in the bottom figure; it is the reheated location in the first weld pass adjacent to the fusion boundaries of the subsequent weld passes. The temperature dependencies of the various material constants of the alloy 690 employed in the FEM analysis are summarised in Fig. 11. The solidification shrinkage of the molten metal was introduced as a volume change of 3%, and the melting temperature of alloy 690 was set at 1660 K. The interpass temperature used in multipass welding used for the FEM analysis is 473 K.

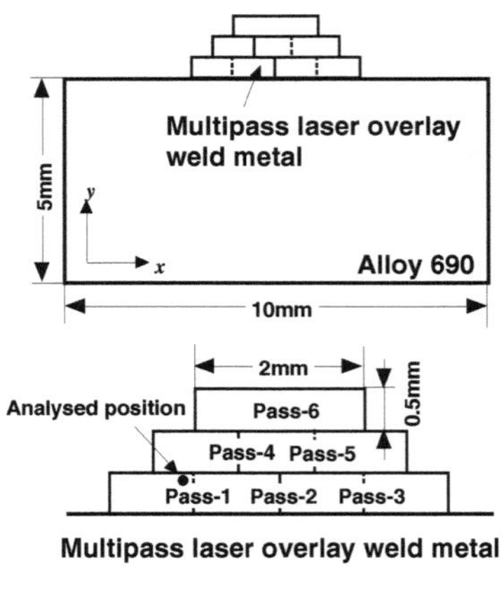


**Fig. 10.** Analysed models for stress and strain in multipass laser overlay welds

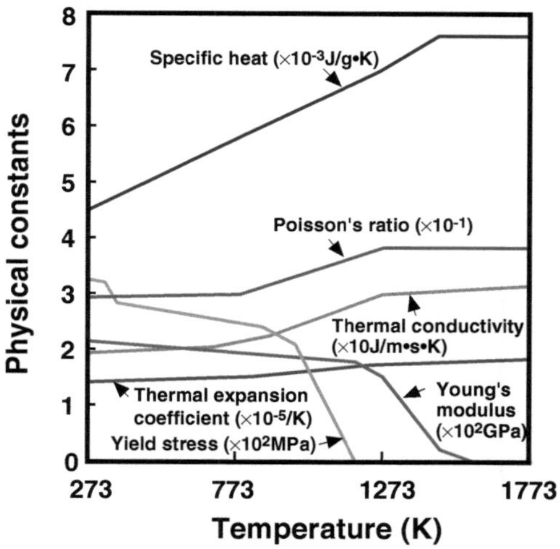


**Fig. 11.** Temperature dependencies of material constants of alloy 690 used in FEM analysis

### *Change in Temperature, Stress and Strain during Multipass Welding Process*

The calculated change in temperature, stress $\sigma_x$ (perpendicular to the weld line/crack) and total strain (thermal strain + elastic strain + plastic strain) at the analysed position during multipass laser overlay welding are shown in Fig. 12. The peak temperature of subsequent weld passes decreases gradually until the third weld pass, and then rises again before decreasing after the fourth weld pass. The peak temperature is located in the ductility-dip cracking temperature range (approx. 1630–1300 K) in the second and fourth weld passes. The crack opening stress ($\sigma_x$) changes cyclically in synch with the weld thermal cycle and increases with number of weld passes, i.e., it is compressive in the reheating stage, while rapidly changing to tension in the cooling stage of welding. The total strain also changes cyclically corresponding to the weld thermal cycle and tends to be compressive.

### *Prediction of Ductility-Dip Cracking*

The calculated results of the plastic strain at the analysed position in the first weld bead perpendicular to the weld line/crack during laser overlay welding

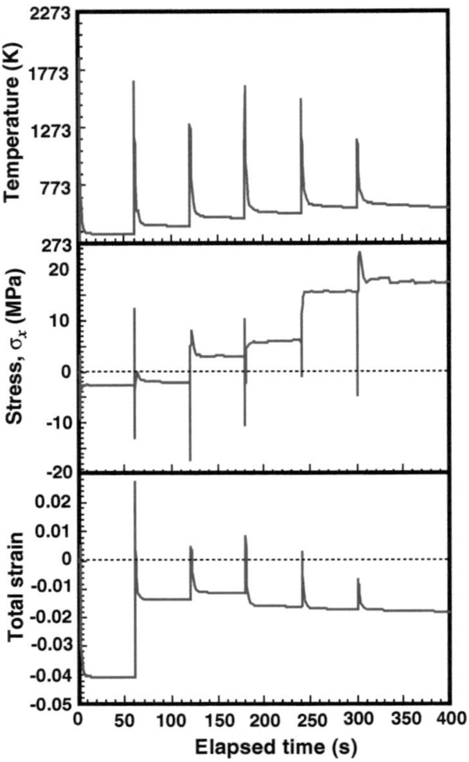


**Fig. 12.** Change in temperature, crack opening stress $\sigma_x$ and total strain during multipass laser overlay welding

is shown in Fig. 13. In the present study, the plastic strain was expressed as the increment in plastic strain [18] from the moment of solidification (approx. 1660 K) of the first weld pass. A solid line indicates the increment in plastic strain under the condition that the crack opening stress $\sigma_x$ is tensile (corresponding nearly to the cooling stage) and a dashed line indicates that the crack opening stress $\sigma_x$ is compressive (corresponding nearly to the reheating stage). The plastic strain increases monotonically with cooling in the first weld pass. However, in subsequent welds, once in the reheating stage, it decreases, and then increases again in the cooling stage. It follows that the plastic strain in the weld metal accumulates with increasing number of weld passes. The relationship between the plastic strain-temperature curve and the DTR, as a function of the La content in the weld metal are also shown in Fig. 13. In the present study, the DTRs of the laser overlay weld

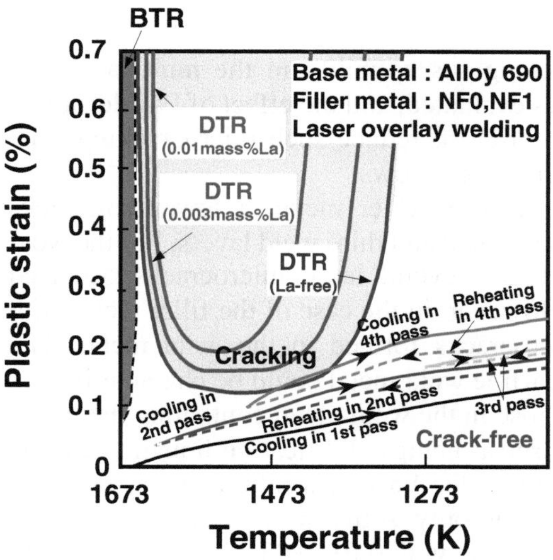


**Fig. 13.** Relation between plastic strain-temperature curve and ductility-dip temperature ranges (DTRs) in multipass laser overlay welding

metals with various La contents were assumed to be comparable to those of the GTA weld metals determined by the Varestraint test [19]. The plastic strain-temperature curve of the fourth weld pass slightly intersects the DTR of the La-free weld metal in the cooling stage of overlay welding, i.e., ductility-dip cracking probably occurs. However, no weld passes have the plastic strain-temperature curves that cross the DTRs of the weld metals with added La. Assuming that ductility-dip cracking will occur when the plastic stain-temperature curve intersects the DTR when the crack opening stress $\sigma_x$ is tensile, these calculated results are reasonably consistent with the experimental results that a few ductility-dip cracks occurred in the La-free weld metal, while no ductility-dip cracks occurred in the La-added weld metal of multipass laser overlay welds.

We conclude that ductility-dip cracking occurs during multipass laser overlay welding when the plastic strain-temperature curve intersects the DTR under tensile stress. The present prediction suggests that ductility-dip cracking can be prevented by reducing the increment in plastic strain during multipass welding.

## Conclusions

In the present study, microcracking susceptibility in the multipass laser overlay alloy 690 weld metal was evaluated, and the effect of La addition to the filler metal was clarified in order to reduce such microcracking. The results obtained may be summarised as follows:

(1) In the case of using the La-free filler metal, a few microcracks occurred in the weld passes of the second and third weld layers. For the weld metal with 0.07 mass% added La, several large microcracks occurred mainly in the final weld layer. However, in the case of the filler metal with 0.01 mass% La added, no microcracks formed in the weld metal. The majority of microcracks in the La-free weld metal could be characterised as being ductility-dip cracks, and those in the weld metal containing the larger amount of La could be characterised as being solidification/liquation cracks.

(2) Microcracking susceptibility in the La-free weld metal was not very high and microcracks occurred for only some welding conditions. No microcracks occurred in the weld metal with 0.01–0.03 mass% La added to the filler metal under any welding conditions. However, microcracking susceptibility increased again when the La content of the filler metal exceeded 0.04 mass%. Furthermore, there was not much difference in microcracking susceptibility between underwater overlay welding and in-air overlay welding.

(3) The EPMA analysis revealed that P, S and O tended to be distributed at the same locations as La in the weld metal NF2, and La also tended to be distributed at different locations from P and S in the weld metal NF4. The products in the La-added weld metals were identified as La phosphide, sulphide and/or oxide, and a Ni-La intermetallic compound was additionally identified in the overlay weld metal containing the larger amount of La.

(4) The ductility-dip cracking susceptibility was improved by the amelioration of hot ductility of the weld metal due to the desegregation of impurity elements of P and S to grain boundaries by the scavenging effect of La. Excessive La addition resulted in the formation of a Ni-La intermetallic compound, therefore, solidification and liquation cracks could occur in the weld metal because of the enlargement of the BTR and/or the occurrence of local liquation during the reheating process.

(5) Theoretical analysis of microsegregation of P and S to a grain boundary during the multipass weld thermal cycle revealed that grain boundary P and S concentrations in the ductility-dip cracking temperature range respectively attained values about 70 and 25 times of their initial concentrations. However, the cooling and reheating rates did not greatly

affect grain boundary segregation of P and S for the range of conditions used here.

(6) The occurrence of ductility-dip cracking in the multipass alloy 690 laser overlay weld metal was predicted by simulation using a mechanical approach. The plastic strain-temperature curve in the La-free weld metal did not cross the DTR in the first weld pass, however, it intersected slightly the DTR during the cooling portion of subsequent welds. On the other hand, for the weld metals containing La, none of the plastic strain-temperature curves of any weld passes crossed the DTR. The laser overlay welding test revealed that ductility-dip cracking would occur in the overlay weld metal when the plastic strain-temperature curve intersected the DTR.

## References

1. B.P. Miglin and G.J. Theus: "Stress Corrosion Cracking of Alloy 600 and 690 in All Volatile Treated Water at Elevated Temperature", EPRI Report NP-5761M, Electric Power Research Institute, Polo Alto, USA, (1988).
2. S.D. Strauss: "Inconel 690 is Alloy of Choice for Steam-Generator Tubing", Power, 140-2 (1996), p. 29–30.
3. J. Miyaguchi, K. Tabuchi and T. Yamamoto: "Towards Maintenance Service Supporting Secure Nuclear Energy", Mitsubishi Juko Giho, 40-2 (2003), p. 92–95 (in Japanese).
4. Y. Nakamura, K. Dambayashi, H. Sakamoto, Y. Fujiya and Y. Tsukamoto: "Field Application of the Cladding of PWR Reactor Vessel Outlet Nozzle", Mitsubishi Juko Giho, 43-1 (2006), p. 4–5 (in Japanese).
5. Y. Kanazawa and M. Tamura: "Underwater YAG Laser Welding Technique", Toshiba Review, 60-10 (2005), p. 36–39 (in Japanese).
6. W. Wu and C.H. Tsai: "Hot Cracking Susceptibility of Fillers 52 and 82 in Alloy 690 Welding", Metall. Mater. Trans., 30A-2, (1999), p. 417–426.
7. V.R. Davé, M.J. Cola, M. Kumar, A.J. Schwartz and G.N.A. Hussen: "Grain Boundary Character in Alloy 690 and Ductility-Dip Cracking Susceptibility", W.J., 83-1, (2004), p. 1s–5s.
8. N.E. Nissley and J.C. Lippold: "Development of the Strain-to-Fracture Test", W.J., 82-12, (2003), p. 355s–364s.
9. M.G. Collins and J.C. Lippold: "An Investigation of Ductility Dip Cracking in Nickel-Based Filler Materials - Part I", W.J., 82-10, (2003), p. 288s–295s.
10. M.G. Collins, A.J. Ramirez and J.C. Lippold: "An Investigation of Ductility Dip Cracking in Nickel-Based Filler Materials - Part II", W.J., 82-12, (2003), p. 348s–354s.
11. M.G. Collins, A.J. Ramirez and J.C. Lippold: "An Investigation of Ductility Dip Cracking in Nickel-Based Filler Materials - Part III", W.J., 83-2, (2004), p. 39s–49s.

12. K. Nishimoto, K. Saida and H. Okauchi: "Microcracking Susceptibility in Reheated Weld Metal, - Microcracking in Multipass Weld Metal of Alloy 690 (Part 1) -", Sci. Technol. Welding & Joining, Vol. 11 No. 4 (2006), p. 455–461.
13. K. Nishimoto, K. Saida, H. Okauchi and K. Ohta: "Microcracking Mechanism in Reheated Weld Metal, - Microcracking in Multipass Weld Metal of Alloy 690 (Part 2) -", Sci. Technol. Welding & Joining, Vol. 11 No. 4 (2006), p. 462–470.
14. K. Nishimoto, K. Saida, H. Okauchi and K. Ohta: "Prevention of Microcracking in Reheated Weld Metal by La Addition to Filler Metal, - Microcracking in Multipass Weld Metal of Alloy 690 (Part 3) -", Sci. Technol. Welding & Joining, Vol. 11 No. 4 (2006), p. 471–479.
15. K. Nishimoto, I. Woo and M. Shirai: "Underlying Mechanism in the Improvement of HAZ Cracking Susceptibility by Rare Earth Metal Addition, - Study on Weldability of Cast Alloy 718 (Report 8) -", Quarterly J. JWS, 20-1, (2002), p. 87–95 (in Japanese).
16. K. Nishimoto, H. Mori and H. Hirata: "Effect of Grain Boundary Segregation of Sulphur on Reheat Cracking Susceptibility in Multipass Weld Metal of Fe-36%Ni Alloy", Proc. of the 7th Int. Symp. JWS, Kobe (2001), p. 827–838.
17. K. Nishimoto, H. Mori, K. Esaki, S. Hongoh and M. Shirai: "Effect of Sulphur and Thermal Cycles on Reheat Cracking Susceptibility in Multipass Weld Metal of Fe-36%Ni Alloy", IIW Doc. IX-1934-99 (1999).
18. M. Shibahara, S. Sone, H. Serizawa, H. Murakawa, K. Nishimoto and K. Saida: "FEM Analysis for Ductility-Dip Cracking Under Multi-Pass Welding", J. Kansai Soc. Naval Architects, Japan, 241 (2004), p. 177-186 (in Japanese).
19. K. Saida, M. Sakamoto and K. Nishimoto: "Mechanical Approach for Prediction of Microcracking in Multipass Weld Metal of Ni-Base Alloy 690", International Welding/Joining Conference-Korea 2007, Seoul, Korea (2007), p. 24–25.

# Ductility-Dip Cracking in High Chromium, Ni-Base Filler Metals

J.C. Lippold[1], N.E. Nissley[2]

[1]The Ohio State University Columbus, Ohio, USA
[2]Exxon Mobil Upstream Research Houston, Texas, USA

## Abstract

Ductility-dip cracking (DDC) is an elevated temperature, solid-state cracking phenomenon that is observed in austenitic weld metals. In this study, the DDC susceptibility of several high-chromium, nickel-base filler metals was evaluated using the strain-to-fracture (STF) test technique. These filler metals were of the Ni-30Cr type and included INCONEL® Filler Metals 52 and 52M supplied by Special Metals Welding Products Company, and Sanicro 68HP® and Sanicro 69HP® supplied by Sandvik AB. In addition, two experimental Ni-30Cr filler metals were evaluated which contained variations in other alloy additions, including niobium additions up to 2.5 wt% and molybdenum additions of 4 wt%. A wide range in DDC susceptibility was observed with these filler metals, including large heat-to-heat variations in filler metals with similar compositions. The experimental filler metals containing Nb and Mo were found to be remarkably resistant to DDC. Cracking susceptibility is primarily associated with the type and form of precipitate that forms along the weld metal migrated grain boundaries. The formation of Nb-rich, M(C,N) at the end of solidification has the most profound effect on DDC, since these precipitates are most effective in pinning the boundaries. The formation of $M_{23}C_6$ carbides during weld cooling or subsequent reheating can also affect DDC susceptibility in these filler metals.

## Introduction

Ductility-dip cracking (DDC) is an intermediate temperature, solid-state grain boundary embrittlement phenomenon that has been observed in Ni-base weld metals, particularly those with high-Cr levels conforming to AWS A5.14, ERNiCrFe-7 and ERNiCrFe-7A. The precise mechanism for

DDC is unclear, although large grain size and long, straight grain "migrated" grain boundaries in the weld metal have been shown to increase susceptibility. The effect of filler metal composition and the related nature of grain boundary precipitation phenomena are key elements in the understanding of DDC, as recently described by Ramirez and Lippold [1, 2]. The strain-to-fracture (STF) test has proved to be an efficient and effective method for quantifying susceptibility to DDC [3] and has been used here to compare a number of high-Cr, Ni-base filler metals. This chapters is a continuation of the research that has been reported previously at IIW annual assemblies [4, 5]. Readers are referred to those chapter for the results of preliminary studies of the DDC susceptibility of FM52 and FM52M. This chapter provides additional STF data on other high-Cr, Ni-base filler metals and compares this data to the FM52 and FM52M results reported previously.

Filler metals conforming to AWS A5.14, ERNiCrFe-7 (FM52, Sanicro 68HP) and ERNiCrFe-7A (FM52M) are frequently used in nuclear power plant applications, often for welding Ni-base Alloy 690. These filler metals have good resistance to stress-corrosion cracking and are used as a replacement for lower-Cr, FM82 in environments where stress corrosion cracking is a concern, such as in high-purity, primary water in nuclear power plants. DDC has been observed when using these filler metals, particularly in multipass welds made under medium to high restraint conditions.

## Experimental Procedures

Weld deposits of eight (8) different high-Cr filler metals were tested using the strain-to-fracture (STF) test. This test was developed at The Ohio State University for the evaluation of elevated temperature grain boundary cracking in austenitic materials. Details of the test development and procedures can be found in Ref. [3]. The compositions of the filler metals tested in this investigation are provided in Table 1.

The FM52 and Sanicro 68HP filler metals conform to AWS ERNiCrFe-7. The three FM52M filler metals were produced to conform to ERNiCrFe-7A with Nb in the range from 0.5–1.0 wt%. Two Ni-30Cr variants containing elemental variations and Nb and Mo (FM52X-D and FM52X-H), and Sanicro 69HP with nominally 1.8 wt% Nb were also tested.

The testing procedure used for the eight filler metals is essentially identical to that used previously for FM52 and FM52M [5]. These represent the FM52 and FM52M-A filler metals in Table 1. STF samples were machined from multipass welds. An autogenous, gas tungsten arc spot weld was made in the gage section of these samples under controlled conditions to produce a radial pattern of grain boundaries in the spot weld.

Full temperature-strain plots were produced for the FM52, 52M, and 52X-H filler metals, and the Sanicro 68HP and 69HP filler metals. Filler metals described as B, C, and D were only tested at 950°C. This temperature was selected since it is normally the temperature that represents the minimum threshold strain for cracking across the entire DDC range and where the highest degree of cracking (number of cracks) occurs as strain is increased in the range from 0–10%.

**Table 1.** Chemical composition (wt %) of filler metals

| Element | 52 | 52M A | 52M B | 52M C | 52X D | 52X H | Sanicro 68HP | Sanicro 69HP |
|---|---|---|---|---|---|---|---|---|
| Ni | 60.12 | 60.37 | 59.54 | 60.38 | 55.5 | 53.9 | 58.81 | 57.0 |
| Cr | 29.09 | 30.04 | 30.06 | 29.5 | 29.6 | 30.2 | 29.94 | 30.0 |
| Fe | 8.88 | 8.42 | 8.22 | 8.04 | 8.04 | 8.2 | 8.80 | 9.15 |
| C | 0.026 | 0.020 | 0.02 | 0.03 | 0.02 | 0.02 | 0.029 | 0.005 |
| Nb | 0.02 | 0.85 | 0.83 | 0.82 | 1.0 | 2.5 | <0.01 | 1.83 |
| Mn | 0.25 | 0.81 | 0.80 | 0.77 | 0.72 | 0.8 | 0.90 | 1.21 |
| Ti | 0.50 | 0.21 | 0.22 | 0.19 | 0.2 | 0.2 | 0.68 | 0.39 |
| Al | 0.71 | 0.10 | 0.11 | 0.12 | 0.05 | 0.04 | 0.75 | 0.20 |
| Si | 0.17 | 0.03 | 0.09 | 0.12 | 0.03 | 0.2 | 0.13 | 0.11 |
| Cu | 0.011 | 0.02 | 0.02 | 0.01 | <0.05 | – | <0.01 | <0.01 |
| Mo | 0.05 | 0.02 | – | – | 4.0 | 4.0 | <0.01 | <0.01 |
| P | 0.0044 | 0.004 | 0.003 | 0.004 | 0.015 | 0.004 | <0.003 | 0.004 |
| S | 0.003 | 0.001 | 0.001 | 0.001 | 0.001 | 0.001 | 0.001 | 0.001 |
| Co | – | 0.007 | – | – | – | – | <0.005 | <0.010 |
| Zr | – | 0.015 | 0.01 | 0.008 | 0.006 | 0.007 | – | – |
| B | – | 0.004 | 0.003 | 0.001 | 0.004 | 0.001 | 0.0008 | 0.0014 |

The degree of DDC in a given sample is determined by counting the number of cracks in the gage section of the sample using a stereo microscope at magnifications up to 30X. Cracks that are only resolvable above 30X are not included in the crack count. The number of cracks reported for a given test sample represents the number of cracks per spot weld. In samples where the spot welds on either side of the sample were of the same composition, the total number of cracks in both welds was divided by two.

Samples were prepared for metallographic analysis by electrolytic etching with 10% chromic acid at 1.5-2 V for 30 s. Scanning electron microscopy (SEM) and EDS analysis was performed on a Sirion SEM/FEG and Phillips XL-30 ESEM/FEG at 15 kV.

## Results and Discussion

### Temperature-Strain Curves

STF test results for FM52 and FM52M that have been published previously [5] are shown in Fig. 1. Each symbol represents a single test with an "X" for a strain-temperature condition where cracking was not observed and a circular symbol for samples where cracking occurred. The number adjacent to each point represents the number of cracks that were observed on the surface of the sample under a binocular microscope using the procedure described previously. A curve separates samples that cracked from those that did not and represents the minimum strain for cracking as a function of temperature. The threshold strain (minimum strain for cracking in the ductility-dip temperature range) was similar in both materials, approximately 2% at 950°C. The temperature range where the threshold strain occurred (900–1000°C) is consistent with other STF testing, and the threshold strain is similar to results on the same heat of IN52 previously reported by Collins et al. [6]. As a reference, the STF curve previously developed by Collins et al. for Filler Metal 82 is included.

Results of STF tests for Sanicro 68HP and 69HP are provided in Fig. 2. These filler metals exhibit a slightly higher threshold strain (2.5%) than FM52 and 52M and have less severe cracking at higher strains at 950°C. No increase in the minimum strain for cracking could be identified at the high temperature end of the DDC range (>1150°C), possibly due to the overlap of the DDC and liquation cracking range.

STF results for the experimental 52X-H filler metal are presented in Fig. 3. The threshold strain for cracking for this filler metal is considerably higher than for the filler metals shown in Figs. 1 and 2. In the temperature range from 750 to 950°C, 52X-H is extremely resistant to DDC as indicated by both a high threshold strain for cracking and the small number of cracks that occur above threshold.

In Fig. 4, The STF results for 52X-H are compared to STF data previously generated for Type 304L and Filler Metal 82 weld metals [6, 7, 8]. This shows that the DDC resistance of 52X-H is superior to FM-82 over most of the DDC temperature range. FM-82 is relatively resistant to DDC, except under very high restraint conditions. Austenitic stainless steel weld metal that contains ferrite is known to be essentially immune to DDC. Thus, the curve for Type 304L with Ferrite Number (FN) 6 represents STF behavior characteristic of an extremely resistant weld metal. Based on these comparative curves, it is anticipated that 52X-H weld metal will be very resistant to DDC under actual fabrication conditions.

### Comparison of DDC Susceptibility at 950°C

Development of a full STF curve for a single filler metal requires 30–40 samples and may not be appropriate for preliminary screening of filler metal compositions. In this investigation, only filler metals 52, 52M-A, 52X-H, and Sanicro 68HP and 69HP were tested over the full DDC temperature range. For filler metals 52M-B, 52M-C, and 52X-D, testing was only conducted at 950°C. This temperature was selected because the minimum in DDC behavior usually occurs at, or near, 950°C and that cracking response at this temperature correlates well with actual fabrication behavior.

A comparison of the eight filler metals at 950°C is provided in Fig. 5. The horizontal bar in the column above each filler metal represents the threshold strain for cracking for that material while the numbers represent the number of cracks that were present in the sample as a function of strain. All the FM52 and FM52M weld metals exhibit a threshold strain in the range from 1–2%, while the Sanicro weld metals are at approximately 2.5%. There is a considerable difference in cracking susceptibility, based on number of cracks at applied strains above 2%, for these filler metals.

As reported previously [5], FM52 and FM52M-A exhibit extensive cracking above the threshold strain. FM52M-C exhibits similar behavior, but FM52M-B does not show extensive cracking until the strain exceeds 6%. The Sanicro 68HP and 69HP weld metals do not show extensive cracking until strains reach almost 8%.

The experimental weld metals containing Nb levels above 1 wt% and 4 wt% Mo exhibit both the highest threshold strains and the best resistance to cracking above the threshold. FM52X-D has a threshold strain of 3%, while that for FM52X-H is approximately 6%. It should be noted that all the other filler metals showed extensive cracking above 6% strain, but the experimental filler metals were very resistant to DDC at these strain levels.

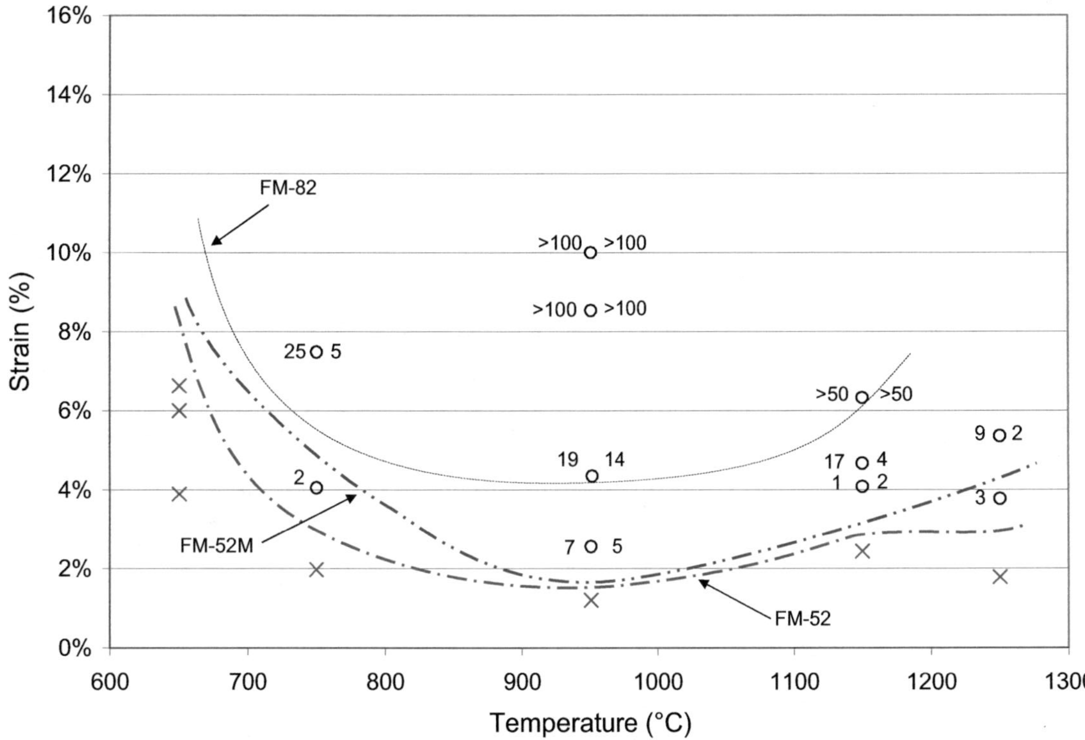


**Fig. 1.** STF results for FM52 and FM52M. Numbers to the left of the open symbols are the number of cracks for the FM52 weld metal, while the numbers to the right are for FM52M. A curve for FM-82 is included for comparison

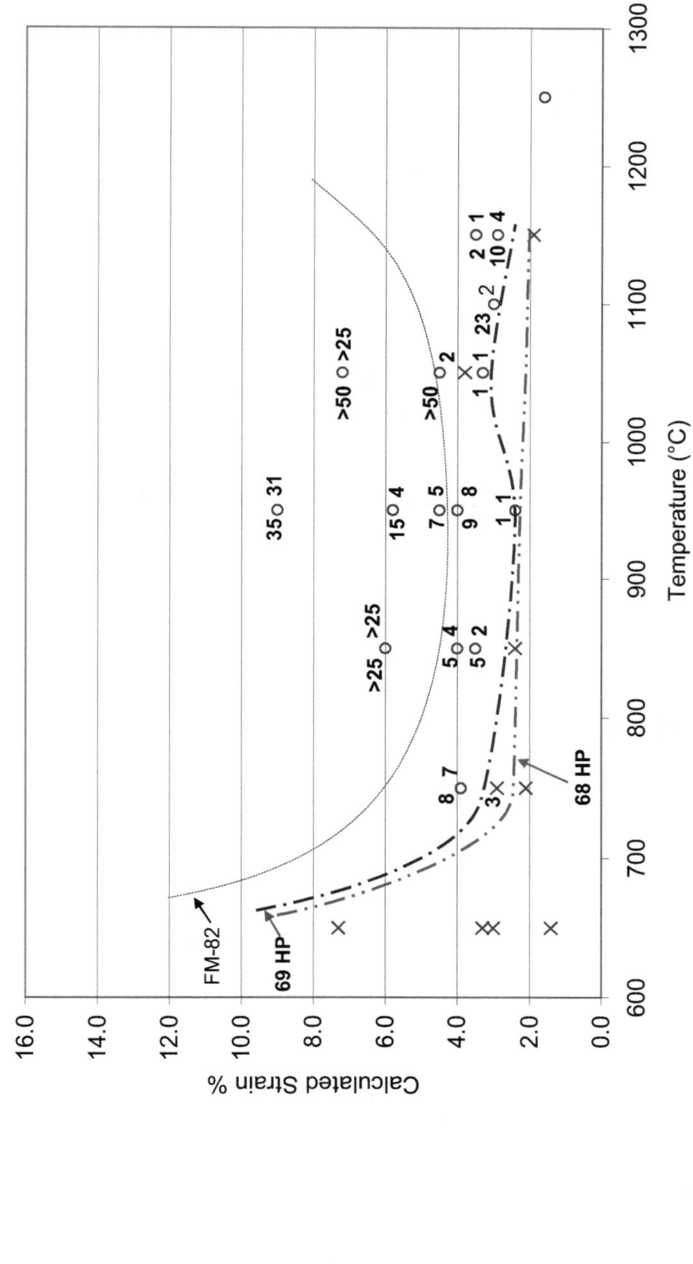


**Fig. 2.** STF results for Sanicro 68HP and 69HP filler metals. Numbers to the left of the open symbols are the number of cracks for the 68HP weld metal, while the numbers to the right are for 69HP

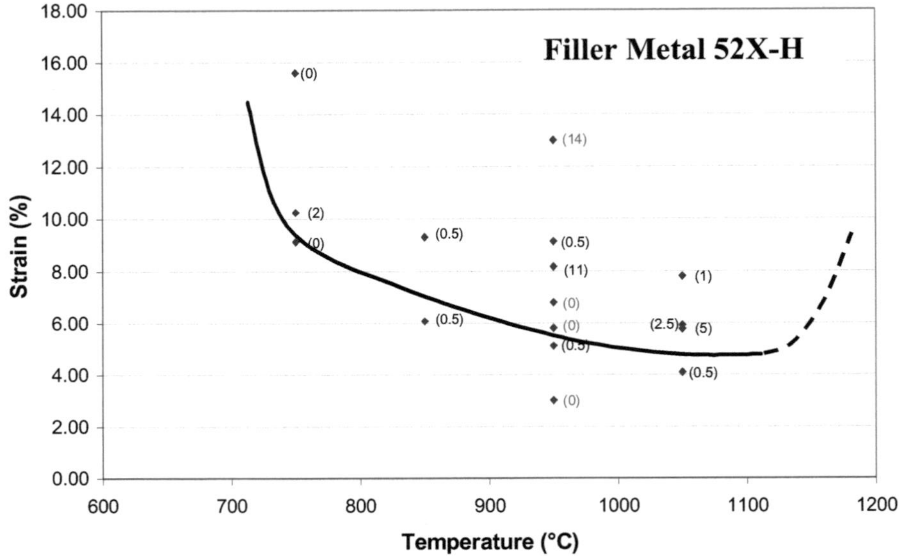


**Fig. 3.** STF results for Filler Metal 52X-H. No data was collected above 1050°C, so a hypothetical extension of the curve is indicated by the dashed line

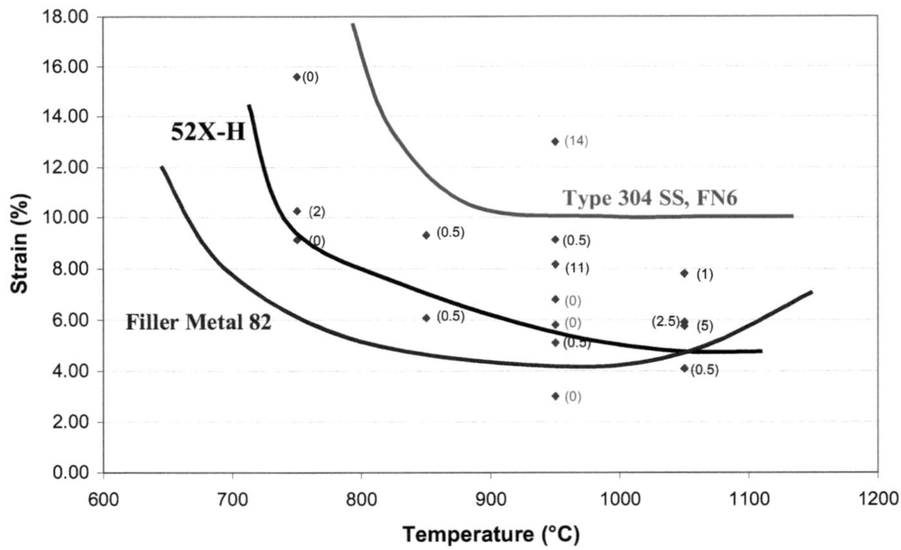


**Fig. 4.** Comparison of STF results for Filler Metal 52X-H with those of Type 304L (Ferrite Number 6) and Filler Metal 82 weld metal

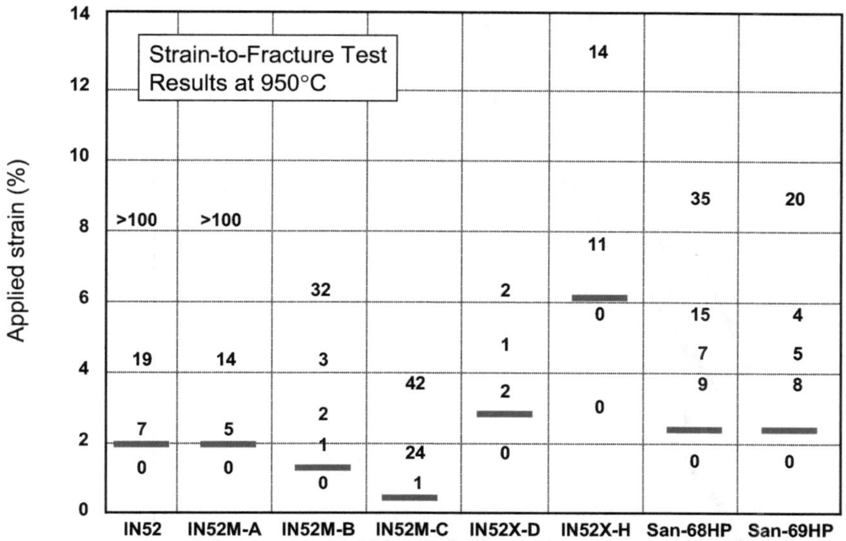


**Fig. 5.** Strain-to-fracture test results for high-Cr, Ni-base weld metals tested at 950°C. Horizontal bars represent the threshold strain for cracking. The numbers represent the individual cracks counted on the sample surface

## Effect of Composition and Weld Metal Microstructure on DDC

Metallographic examination was conducted on STF samples for FM52, FM52M-A, B, and C and FM52X-H. As has been reported previously [5], the FM52 filler metal exhibited large grain size and relatively long straight migrated grain boundaries. A representative microstructure of both FM52 and FM52M weld metals containing DDC along migrated grain boundaries is shown in Fig. 6. As can be seen, the grain size is large and the boundaries are quite straight. When examined using electron microscopy techniques (SEM and TEM), the migrated grain boundaries were found to be decorated with small, Cr-rich carbides of the $M_{23}C_6$ type. An SEM micrograph of a typical boundary in FM52 is shown in Fig. 7 and a higher magnification TEM image in Fig. 8. These carbides form in the solid-state as the weld metal cools to room temperature (or during reheating) and seem to have little or no effect on boundary pinning in the FM52 and FM52M filler metals. The microstructure and morphology of the grain boundaries and carbides in FM52 and are consistent with weld metals that have a high susceptibility to DDC, i.e. virtually no pinning of the grain boundaries.

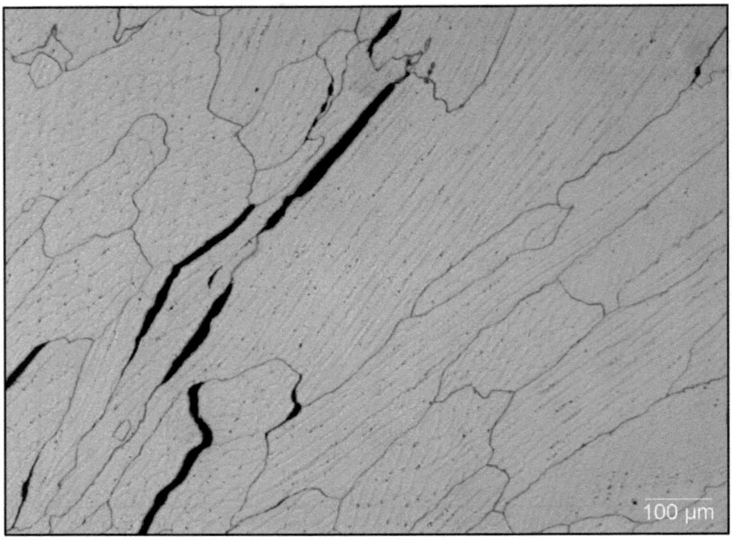


**Fig. 6.** Weld metal microstructure in a FM52 STF sample tested at 950°C and 8.5% strain. Ductility dip cracks are evident along the weld metal migrated grain boundaries

**Fig. 7.** SEM secondary electron image of a migrated grain boundary in FM52 showing small (<0.5 µm) $M_{23}C_6$ carbides along the boundary

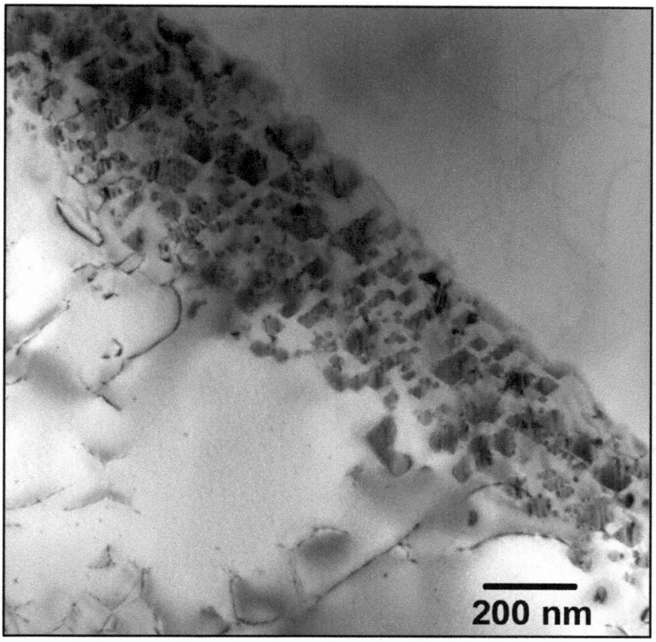


**Fig. 8.** TEM micrograph of a migrated grain boundary in FM52. The precipitates were identified as $M_{23}C_6$ and exhibited a cube-on-cube orientation relationship with the austenite matrix (Courtesy of Dr. Antonio Ramirez)

An SEM image of the FM52X-H weld metal microstructure is shown in Fig. 9. The fraction of Nb-rich M(C,N) precipitates is much higher in this weld metal compared to the FM52M weld metals. These precipitates form at the end of solidification and are effective at pinning the migrated grain boundaries, resulting in more tortuous grain boundaries in this weld metal.

Further evidence of grain boundary tortuosity can be obtained by using orientation imaging microscopy (OIM™) in the SEM. Using this technique which analyzes electron backscattered diffraction (EBSD) patterns, the nature of the migrated grain boundaries can be clearly differentiated. Patterns for FM52M-C and 52X-H are shown in Fig. 10. Note that the FM52M-C boundaries are generally straight, while the FM52X-H boundaries are extremely jagged and tortuous. This clearly demonstrates the effect of Nb-rich, M(C,N) precipitates that form at the end of solidification on increasing the tortuosity of the weld metal migrated grain boundaries.

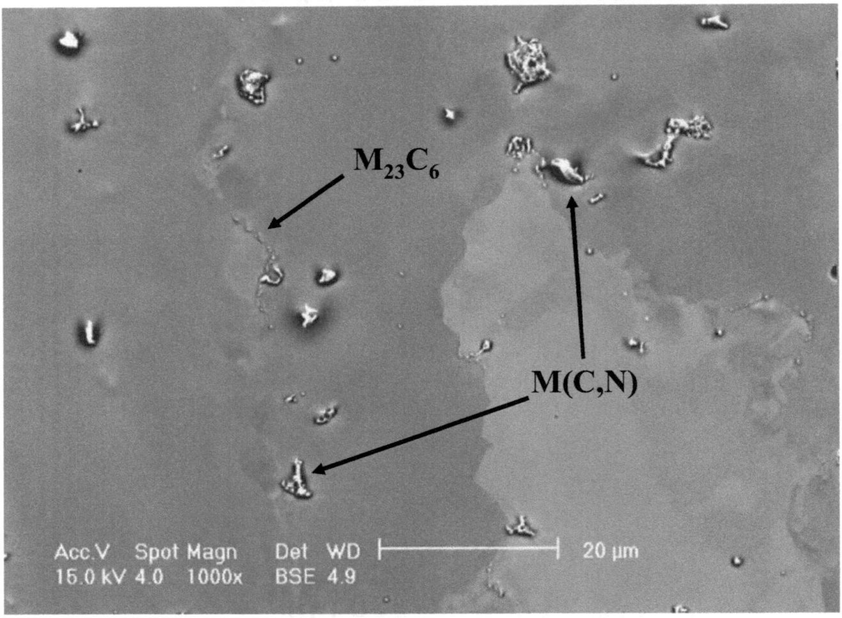


**Fig. 9.** Weld metal microstructure of FM52X-H, SEM secondary electron image. M(C,N) precipitates are visible in the interdendritic regions and $M_{23}C_6$ carbides were found along the migrated grain boundaries. Grain boundary pinning by the large M(C,N) precipitates is evident

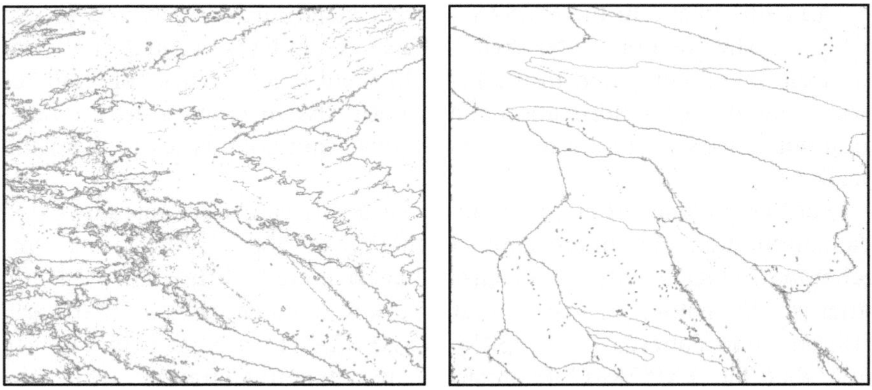

**Fig. 10.** Electron backscattered diffraction (EBSD) patterns showing the nature of the weld metal migrated grain boundaries in FM52X-H (*left*) and FM52M-C (*right*)

## Mechanism of Ductility-Dip Cracking

The behavior of the high-Cr, Ni-base filler metals again shows the importance of the nature and morphology of precipitates on resistance to DDC. Nb additions above the level prescribed in ERNiCrFe-7A (0.5–1.0 wt%) lead to the formation of a higher fraction of Nb-rich M(C,N) at the end of weld solidification. Because these precipitates are present at the end of solidification they are very effective in pinning the migrated grain boundaries that form from the original solidification grain boundaries. The fraction of these solidification- related precipitates is apparently quite important, since when the fraction is low, such as with FM52M-A and FM52M-C, grain boundary pinning is minimal and DDC susceptibility is high. Increasing Nb content above 1 wt% has a profound effect on DDC resistance as evidenced by the behavior of FM52X-D, FM52X-H, and Sancro 69HP reported here and previous data for FM82 [6]. This behavior is consistent with the model proposed by Ramirez and Lippold [2] and shown in Fig. 11.

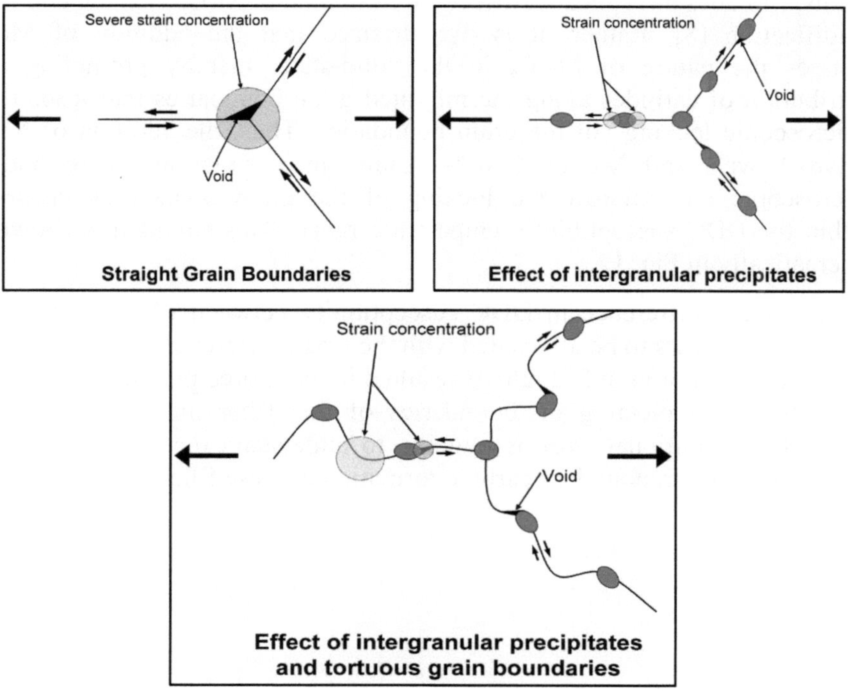


**Fig. 11.** Schematic showing grain boundary strain accumulation as a function of boundary geometry and precipitation behavior. (From Ramirez and Lippold [2])

Precipitate formation and grain boundary pinning from Nb-rich particles forming at the end of solidification does not completely explain the DDC susceptibility that is summarized in Fig. 11. For example, FM52M-A and FM52M-B have essentially identical compositions (Table 1) yet their susceptibility to DDC is markedly different. FM52M-C shows a very high susceptibility to DDC, yet (with the exception of boron) its composition is nearly identical to A and B. Sanicro 68P which is compositionally equivalent to FM52 (with the exception of higher Mn) has better DDC resistance at 950°C. This suggests that grain boundary segregation and precipitation that occur in the solid-state must also have some influence on DDC susceptibility. It has previously been shown by Collins et al. [6] that hydrogen can significantly reduce DDC resistance, and Ramirez and Lippold [2] have speculated that other interstitial elements (such as oxygen) may also be detrimental.

The addition of molybdenum to the experimental filler metals 52X-D and 52X-H appears to provide additional DDC resistance. Based on Scheil-Gulliver calculations of the solidification behavior of these filler metals, Mo does not appear to affect the formation of M(C,N) at the end of solidification [8]. Rather, it is hypothesized that the addition of Mo changes the nature of $M_{23}C_6$ in the solid-state, thereby promoting a distribution of carbides along the migrated grain boundaries that leads to "microscopic locking" of the grain boundaries. Thus, the addition of Nb above 1 wt% and Mo at 4 wt% results in a situation where both macroscopic and microscopic locking of the grain boundaries occurs within the DDC susceptibility temperature range. This situation is shown schematically in Fig. 12.

The large difference in DDC susceptibility between FM52M-B and FM52M-C appears to be associated with the small difference in boron. The higher level of boron in FM52M-B resulted in more precipitation of $M_{23}C_6$ along the weld metal grain boundaries of this filler metal relative to FM52M-C. Additional work is ongoing to understand the effect of both Mo and B on grain boundary carbide formation in these filler metals.

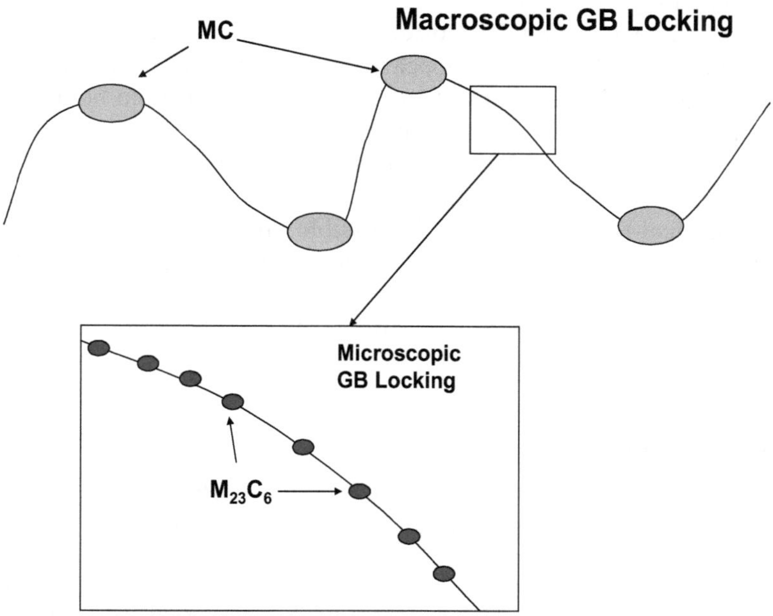


**Fig. 12.** Schematic showing the combination of grain boundary tortuosity that provides macroscopic grain boundary locking (resistance to sliding) and microscopic locking by the formation of grain boundary precipitates. This combination appears to provide the best resistance to ductility dip cracking

## Conclusions

1. Using the strain-to-fracture test, a wide variation in DDC susceptibility was observed among Ni-base filler metals conforming to ERNiCrFe-7 and ERNiCrFe-7A.
2. Two experimental filler metals with Nb content exceeding 1 wt% and Mo additions of 4 wt% were the most resistant to cracking, based on STF test results.
3. DDC resistance is associated with macroscopic and microscopic locking of the weld metal grain boundaries that prevents grain boundary sliding at elevated temperatures.
4. Macroscopic boundary pinning by Nb-rich M(C,N) at the end of solidification increases the boundary "tortuosity". It appears that Nb levels above 1 wt% are required to promote sufficient tortuosity to

prevent cracking by a boundary locking mechanism. When Nb is increased to 2.5 wt%, boundary tortuosity due to pinning is significant.

5. Microscopic boundary locking appears to be promoted by additions of Mo and B. Both of these elements affect the nature of grain boundary $M_{23}C_6$ precipitation. Since only one level of Mo (4 wt%) was investigated, it was not possible to determine if lower levels of Mo may also be effective in reducing cracking.
6. Significant heat-to-heat variation is possible among filler metals that meet the same specification. Based on our current understanding of the DDC mechanism, there was no way to predict the susceptibility to DDC in ERNiCrFe-7 and -7A filler metals based solely on composition.

## Acknowledgements

The authors wish to thank the Special Metals Welding Products Company for providing the FM52, FM52M, and FM52X welded samples, and Sandvik AB for providing the Sanicro 68HP and 69HP consumables. Our thanks to Kenny Izor, an undergraduate research assistant at OSU, for performing some of the STF testing. We also wish to thank Mr. Sam Kiser of Specials Metals and Mr. Claes-Ove Pettersson from Sandvik Materials Technology, R&D for useful discussions and insight.

## References

1. A.J. Ramirez and J.C. Lippold, 2004. High Temperature Behavior of Ni-base Weld Metal, Part I, Ductility and microstructural characterization, Materials Science and Engineering A, 380, p. 259–271.
2. A.J Ramirez and J.C. Lippold, 2004. High Temperature Behavior of Ni-base Weld Metal. Part II, Insight into the mechanism for ductility dip cracking, Materials Science and Engineering A, 380, p. 245–258.
3. N.E. Nissley and J.C. Lippold, 2003. Development of the strain-to-fracture test for evaluating ductility-dip cracking in austenitic alloys, Welding Journal, 82(12):355s–364s.
4. J.C. Lippold, N.E. Nissley, and A.J. Ramirez, 2005. Recent advances in the understanding of ductility-dip cracking, IIW-IX-2161-05, Prague, July 11–14, 2005.
5. J.C. Lippold and N.E. Nissley, 2006. Further investigations of ductility-dip cracking in high chromium, Ni-base Filler Metals, IIW-IIC-326-06, Quebec City, August 27–31.

6. M.G. Collins and J.C. Lippold. 2003. An investigation of ductility-dip cracking in Ni-base filler metals-Part 1, Welding Journal, 82(10):288s–295s.
7. J.C. Lippold and A.J. Ramirez, 2004. Insight into the Mechanism of Ductility Dip Cracking, IIW-IX-2123–04, July 2004, Osaka, Japan.
8. N. Nissley, 2006. Intermediate temperature grain boundary embrittlement in Ni-base weld metal. PhD Dissertation, The Ohio State University.

# Thermodynamic and Kinetic Approach to Ductility-Dip Cracking Resistance Improvement of Ni-base Alloy ERNiCrFe-7: Effect of Ti and Nb Additions

A.J. Ramirez[1], C.M. Garzón[1-2]

[1] Brazilian Synchrotron Light Laboratory, Campinas-SP, Brazil
[2] National University of Colombia, Physics Department, Bogotá, Colombia

## Abstract

Previous research has suggested that significant improvement in ductility-dip cracking resistance of ERNiCrFe-7 weld metal can be obtained if second phase precipitation during welding is optimized with some carbide and nitride forming element additions. Therefore, a theoretical and experimental work has been conducted to address the effect of Nb and Ti additions to the precipitation of second-phase particles during welding of a ERNiCrFe-7-like alloy. The critical precipitation temperatures, phase fractions and atomic partitioning in the microstructure were among the factors analyzed in the light of DDC resistance improvement.

## Introduction

Ni-base filler metal AWS A5.14 ERNiCrFe-7 (NiCr30Fe9), also known as filler metal 52 (FM-52), which is a nearly matching filler metal for alloy 690, is widely used for joining solid solution strengthened Ni-base alloys, particularly alloys 600 and 690, and for dissimilar welding between several Ni-base alloys, stainless and carbon steels. ERNiCrFe-7 exhibits high resistance to corrosion and to stress corrosion cracking at moderate temperatures. However, when highly restrained welded joints are performed, this alloy is susceptible to ductility dip cracking (DDC), which is a solid state failure characterized by intergranular cracking occurring at homologous temperatures ranging from approximately 0.5 to 0.7. This type of failure plagues different alloy systems that present a FCC structure in these intermediate temperatures, as Ni-alloys, Cu-alloys, steels and stainless steels and has been the object of research for quite a long time [1, 2, 3, 4, 5]. Among the factors that have been reported to have influence on the DDC phenomenon are: alloy chemical composition (with special influence of impurities and interstitials), segregation to grain boundaries, precipitation behavior, grain boundary migration and pinning, grain boundary orientation relative to the applied strain and recrystallization [6-12]. Depending on the system, some of these factors become more important and eventually control the cracking mechanism. Therefore, as a result of this apparently changing mechanism the technical and scientific literature on DDC may become quite confusing and controversy involving these different controlling factors has appeared. For example, the dominant factor on the DDC of Ni-base alloys with moderate to high levels of impurities has been proposed to be impurity segregation to the gain boundaries [11]. On the other hand, the DDC of very "clean alloys', as most commercial Ni-base alloys, can not be directly linked to intergranular segregation of impurities [10]. However, S segregation has also been observed in these "clean" alloys [13].

Based on microstructural analysis up to the nanoscale and the exhaustive study of DDC behavior of as-welded Ni-base filler metals AWS A5.14 ERNiCrFe-7 (FM-52) and ERNiCr-3 (FM-82), it was verified that the difference on DDC resistance between these two alloys is due to the large amount of primary intergranular precipitates found in ERNiCr-3 (NiCr20Mn3Nb), which is a near-matching filler metal for alloy 600 [8-10]. These primary precipitates in as-welded ERNiCr-3 are (NbTi)(CN) cabonitrides and their effect on DDC resistance is based on the capability of such eutectic-like precipitates to preclude gain boundary (GB) migration at the very end of solidification. The GB pinning results in tortuous GBs that would make GB sliding more difficult and strain concentration at triple points less likely to happen, resulting in less severe void formation, which improves DDC resistance [6, 8-10]. Figure 1 presents schematically the proposed effect of GB tortuosity on GB sliding, strain concentration at triple points and intergranular precipitates, and void formation.

In order to verify the Nb rich carbonitrides role on the DDC resistance of ERNiCr-3, welds were prepared using the DDC susceptible filler metal ERNiCrFe-7 with additions of Nb and Ti to induce a larger amount of primary precipitation. The remarkable DDC resistance improvement of these modified ERNiCrFe-7 welds was verified using the strain-to-fracture test

[14-15]. Figure 2 presents the comparison of strain-to-fracture results obtained for the unmodified ERNiCrFe-7 alloy (circles) and the Nb and Ti additions modified alloys (inverted triangles). It is evident in this figure the reduction in the number of cracks for the modified alloys.

Other researchers and welding consumable producers have been working on the development and testing of DDC resistant alloys based on ERNiCrFe-7. A patented alloy with 2,5% Nb and 4% Mo additions has shown a remarkable performance with cracking starting at 10% strain, compared to 2% strain in the original ERNiCrFe-7 [16]. La additions to alloy 690 have also been proposed with promising results controlling the damaging effect of impurities segregation to GBs [17]. However, some of the proposed modifications can solve the DDC problems at the cost of compromising the high temperature resistance, the creep properties, the microstrutural stability at high temperatures, and the corrosion resistance of the alloy 690. Therefore, exhaustive and expensive testing may be ahead before some of these improved alloys can be used in actual fabrication and/or repair.

The advances on the formulation of DDC resistant Ni-based alloys reveal that some understanding has been obtained to date regarding the DDC phenomena. However, there is still a long way to fully understand this phenomenon and how to control it not just for Ni-based alloys, but other important systems as steels and stainless steels, where it causes problems during continuous casting and hot-rolling. Therefore, the authors have engaged on a project to understand some of the factors controlling DDC in FFC alloys. The present report concentrates on the thermodynamic and kinetic approach to optimize second phase precipitation on Ni-base alloy AWS A5.14 ERNiCrFe-7 in order to improve DDC resistance[18-19]. The precipitation of primary nitrides and carbonitrides is optimized through the proposed addition of Ti and Nb. Secondary precipitation of $M_{23}C_6$ and $\gamma'$ were also analyzed. The addition of Ti and Nb up to 3%-wt has been evaluated using CALPHAD-based numerical simulation of microstructural evolution during weld metal solidification and cooling.

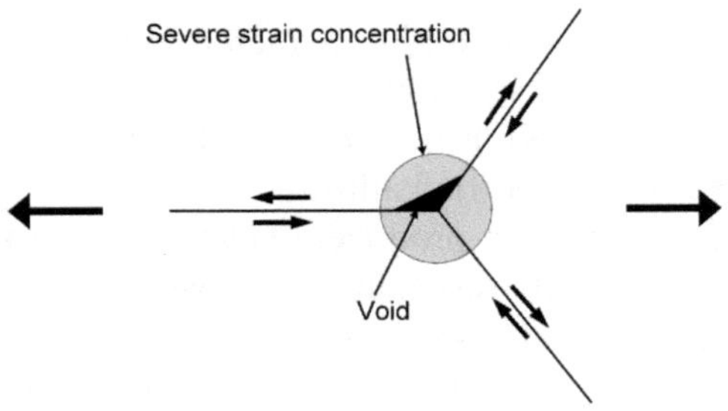


a. **Straight Grain Boundaries**

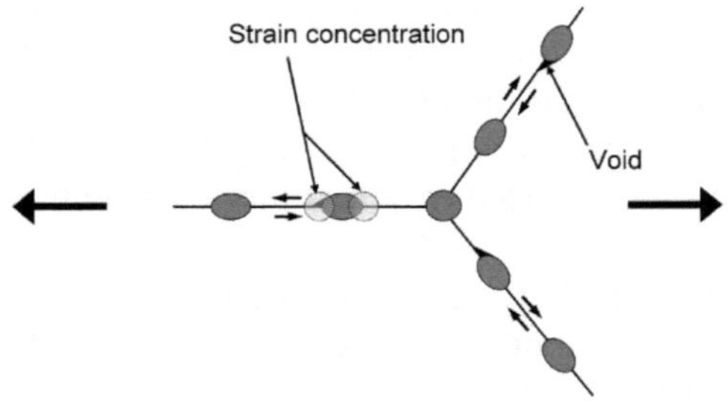


b. **Effect of intergranular precipitates**

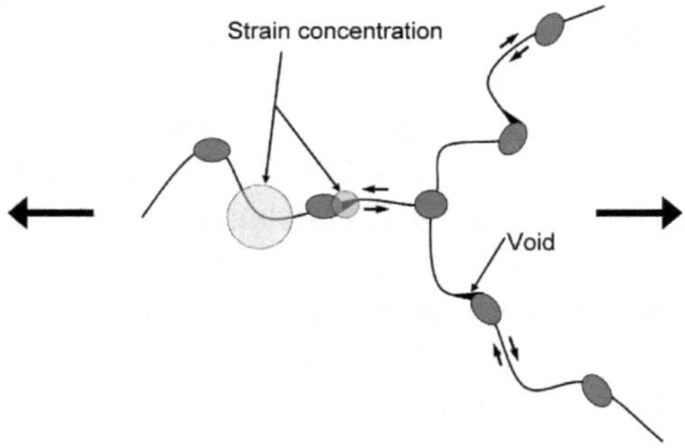


c. **Effect of intergranular precipitates and tortuous grain boundaries**

**Fig. 1.** Influence of intergranular precipitates on GB sliding, strain concentration and void formation. (**a**) Straight GBs result in severe GB sliding and strain concentration at the triple points; (**b**) intergranular precipitates lock the GBs reducing GB sliding; (**c**) tortuous GBs and intergranular precipitates lock efficiently the GBs reducing GB sliding, causing further strain distribution, and reducing the likeness of fracture initiation [10]

Actual weld metal samples have also been characterized and compared with the simulations. Among the obtained results can be highlighted the kinetic phase diagrams where the microstructure stability during cooling after welding is related to Ti- and Nb-content. Atomic partitioning between the austenitic matrix and the precipitates was also evaluated.

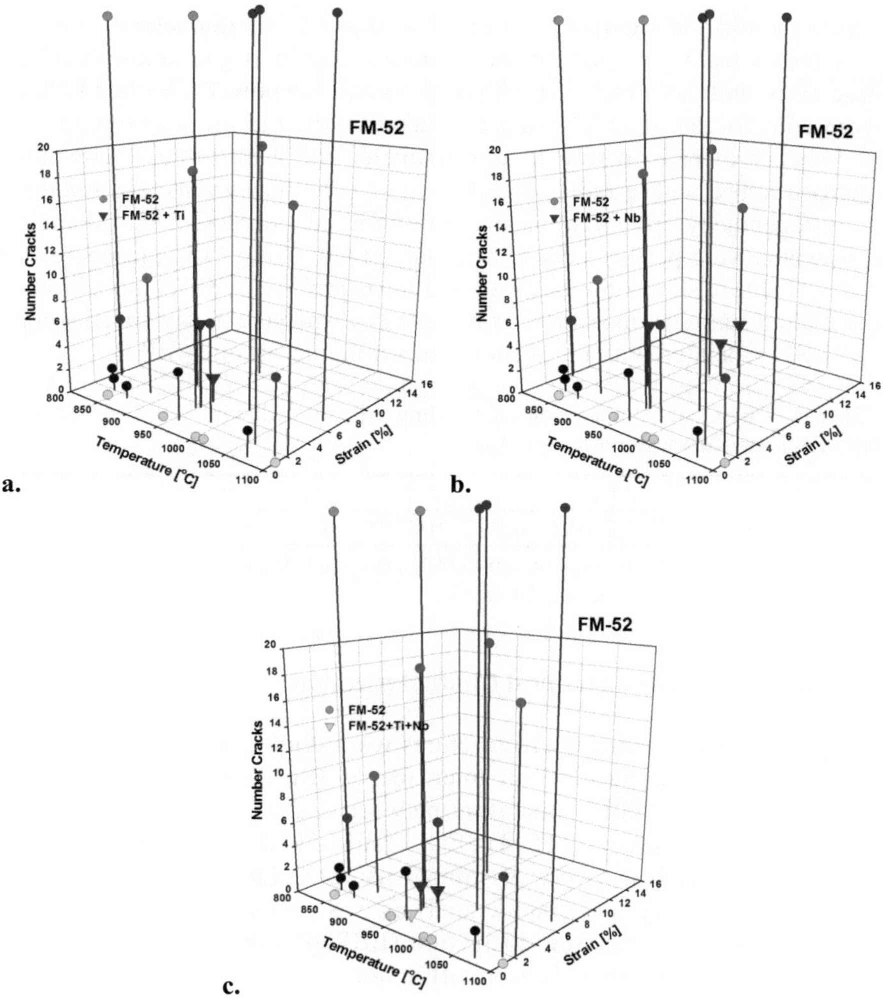


**Fig. 2.** Comparison of Strain-to-fracture results on unmodified (circles) and modified (inverted triangles) ERNiCrFe-7 alloy with small (**a**) Ti, (**b**) Nb, and (**c**) Nb+Ti additions [15]

## Materials and Experimental Procedures

### Materials

The nominal chemical composition of the studied alloy AWS A5.14 ER-NiCrFe-7 used as a master alloy is presented in Table 1. The base alloy was in the form of welding filler metal wire of 1.2 mm diameter.

Samples were obtained by arc melting using inert gas atmosphere the base alloy with additions of high purity Ti and Nb powder. Melted buttons were cold forged at 25°C using a uniaxial hydraulic press to obtain flat-like shapes of proximately 15 mm diameter and 3 mm thick. As-welded microstructure with a radial distribution of columnar grains was obtained by autogenously spot welding on top of forged samples. 5 mm diameter spot welds were performed with manual GTAW using 115 A (ampere) for 3–5 s. Table 2 shows the targeted and measured chemical composition of the melted samples, measured using energy-dispersive X-ray spectroscopy (XEDS) associated with the scanning electron microscope (SEM).

**Table 1.** Chemical composition of welding filler metal wire AWS A5.14 ER-NiCrFe-7 used as master alloy (wt.%)

| Fe | Cr | Al | Ti | Mn | Si | C | N | S | P | Nb+Ta | Ni |
|---|---|---|---|---|---|---|---|---|---|---|---|
| 10.1 | 29.1 | 0.71 | 0.51 | 0.25 | 0.13 | 0.027* | 0.015* | <0.001 | 0.003 | <0.01 | Bal. |

Notes: * N-content was between 0.008 and 0.0170 and carbon content was between 0.025 and 0.029 wt.%.

### Chemical and Microstructural Characterization

The produced as-welded microstructures were studied using optical microscope (OM), scanning electron microscopes (SEM) JEOL® JSM-6330F and JEOL® JSM-5900LV operating between 5 and 20 kV, and transmission electron microscope (TEM) JEOL® JEM-3010 operating at 300 kV. The chemical analyses were performed using XEDS associated with the SEM and TEM. Crystallographic analysis was performed on the SEM using a backscattered electron diffraction (EBSD) system by *HKL*® and on the TEM by selected area electron diffraction.

Conventional grinding and polishing practices were used for OM and SEM sample preparation. Final polishing for EBSD analyses was achieved using colloidal silica on a vibrating polishing system. Electrolytic etching for OM and SEM analyses was performed using a 40%-vol $HNO_3$ aqueous solution at 25°C and applying between 1.3 and 1.7 V during 60-120 s.

TEM sample preparation was performed using carbon replica extraction of second phases.

**Table 2.** Chemical analysis of Ti and Nb content of studied samples (wt.%)

| Sample Designation | Ti %-wt | Nb %-wt |
|---|---|---|
| ERNiCrFe-7 | 0.52 | -* |
| 0.7Ti/1Ti/1.5Ti/3Ti | 0.74/0.95/1.45/2.7 respectively | -* |
| 0.5Nb/1Nb/1.5Nb/3Nb | 0.52 | 0.6/1.2/1.6/2.9 respectively |
| 1Nb-1Ti | 0.95 | 1.1 |

Notes: *: Measured content lower than the SEM-XEDS detection limit for the used procedure.
For a confidence interval of 67%: Ti-content, between ± 0.05 and ± 0.15, and Nb-content, between ± 0.1 and ± 0.2, increasing standard deviations corresponding to increasing alloying contents in the table.

## Simulation Methodology

Calphad-based numerical work consisted of simulating the following phenomena:

- Simultaneous growth during weld pool solidification, against a liquid region, for both austenite dendrite and primary MX precipitate;
- Dendritic micro-segregation of alloying elements;
- Solid state precipitation of second phases during cooling inside the austenitic dendrites.

The geometry of the modeled system, which is schematically shown in Fig. 3, with a size of 5 μm, was defined based on the experimentally observed half the interdendritic spacing. Simulations started from a fully liquid system, which was cooled at the reasonable welding cooling rate of $100°C \cdot s^{-1}$ down to 327°C (600 K). As usual on some of these simplified models was assumed a planar and sharp phase interfaces, as well as the establishment of a local equilibrium condition at these interfaces. Diffusion in each region was computed by solving the multicomponent non-steady state atomic diffusion Equation 1 [20]. This equation was solved with Dictra® software [21-22], which uses the finite-difference method and assesses thermodynamic and kinetic parameters from specific databases. The commercially available databases Ni-data® and Mob2® were used for the calculations performed. The migration rate for each phase interface was computed by assuming that interface migration is controlled by the mass balance obtained from fluxes of diffusing elements across the interface (Eqs. 2 and 3).

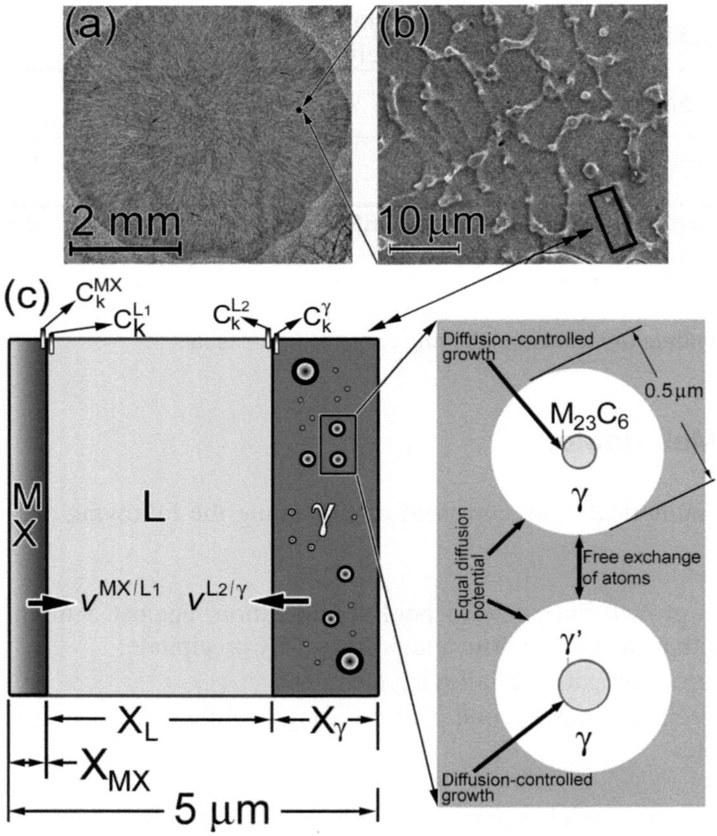


**Fig. 3.** Schematic representation of the simulated system. (**a**) OM top image of the welding spot; (**b**) SEM micrograph showing alloying micro-segregation at interdendritic regions; (**c**) scheme of the one dimensional system used for numerical simulations and its relevant dimensions

$$J_K = -\sum_{i=1}^{n} L'_{ki} \frac{\partial \mu_i}{\partial z} \tag{1}$$

where $J_K$ is the diffusional flux of the atomic alloy component k, $L'_{ki}$ is a proportionality factor that depends on the mobility of the atomic species, $\mu_i$ is the species chemical potential in the alloy, and z is the spatial coordinate.

$$v^{MX/L1}(c_k^{MX} - c_k^{L1}) = J_k^{MX} - J_k^{L1}, \quad k = 1,2, \ldots, n \tag{2}$$

$$v^{L2/\gamma}(c_k^{L2} - c_k^{\gamma}) = J_k^{L2} - J_k^{\gamma}, \quad k = 1,2, \ldots, n \tag{3}$$

where $v^{MX/L1}$ and $v^{L2/\gamma}$ are the precipitate–liquid and liquid-austenite phase interface migration rates, respectively, $c_k^{MX}$, $c_k^{L1}$, $c_k^{L2}$ and $c_k^{\gamma}$ are the concentrations of the atomic component k close the interface in the precipitate, liquid near to the precipitate, liquid near to the austenite, and austenite near to the precipitate, respectively, and $J_k^{MX}$, $J_k^{L1}$, $J_k^{L2}$ and $J_k^{\gamma}$ are the corresponding diffusional fluxes.

Within the dendrite, occurrence of diffusion-controlled precipitation of second phases was accounted by allowing the growth of spherical precipitates at different distances from the dendrite surface. Thus, the dendrite was divided in several sub-systems; each one consisting of a 0.5 μm diameter spherical cell (1/20 of measured dendrite width), which contains a spherical precipitate at the center and matrix phase in the outer case. During the cooling the precipitation of both $M_{23}C_6$ and γ' phases was arbitrarily allowed to start within these cells when driving force was greater than $10^{-3}$. Further complication with the simulation of the nucleation stage of these precipitation reactions was avoided by assuming the prior existence of 1 nm diameter second-phase nucleus. The chemical composition of this nucleus is such that leads to equal chemical potential at both sides of the precipitate–matrix phase interface. Diffusion inside γ' was neglected due to lack of diffusion data for this phase in the Mob2® database. The average chemical composition of each cell was set equal to the previously computed average chemical composition at this point of the dendrite in the system containing primary MX precipitates and dendritic austenite.

## Results and Discussion

### Effect of Ti and Nd Additions on the Thermodynamics and Kinetics

Figures 4 and 5 show pseudo-binary phase diagrams relating phase domains with Ti- and Nb-content for temperatures between 327 and 1727°C for the alloy ERNiCrFe-7 (10.1Fe–29.1Cr–0.71Al–0.25Mn–0.13Si–0.027C–0.015N wt-%). The liquidus temperature is strongly increased when Ti is added to ERNiCrFe-7 alloy but not when Nb is added, whereas the solidus temperature slightly decreases with both Ti and Nb additions. Therefore, solidification cracking problems are expected when large amounts of these two elements are added to ERNiCrFe-7, as experimentally verified [15]. Precipitation of primary MX carbonitrides is expected in this alloy. In the unmodified ERNiCrFe-7 alloy, primary precipitates are expected to be Ti(N,C). In Nb-modified alloys is expected the substitution of metallic specie Ti for Nb in MX precipitates, whereas in both Ti- and Nb-modified alloys substitution of non-metallic specie N for C is also expected. For example, at 977 °C, the content of Nb in MX precipitates changes from 2.4 to 50.0 Nb wt.%, with additions between 0.5 and 3 Nb wt.% to the alloy, respectively. While, the ratio between N-content wt-% and C-content wt-% in MX precipitates changes from 1.7 in the unmodified ERNiCrFe-7 alloy to (i) 0.6 in an alloy containing 3.0 Ti wt-% and (ii) 0.6 in an alloy containing 3.0 Nb wt.%. The liquidus temperature increase with the additions of Ti is due to the precipitation within the liquid of MX precipitates, which at high temperatures are predominately TiN.

At 700°C the equilibrium presence of secondary phases $M_{23}C_6$, gamma-prime ($\gamma'$), and Cr-rich BCC is expected in the unmodified alloy. On the other hand, the addition of Ti or Nb induces the stabilization of eta ($\eta$) and delta ($\delta$) phases, respectively. No significant chemical composition changes of secondary precipitates as a consequence of alloying additions are expected.

Computed kinetic phase diagrams that relate phase domains with the Ti- and Nb-content for temperatures between 727°C and 1727°C are presented in Fig. 6. The strong influence of Ti and Nb additions on critical phase transformation temperatures during cooling following welding can be seen. Solidification temperature is strongly reduced by both Ti and Nb additions. However, the Nb effect becomes more pronounced for small additions. For example, according to the calculations 3%-wt of Ti in this alloy will reduce the kinetic solidification temperature to ~1150°C, while 3%-wt of Nb in this alloy could bring the solidification temperature down to ~900°C. Thus, Nb additions will have a more severe effect on solidification cracking than Ti additions.

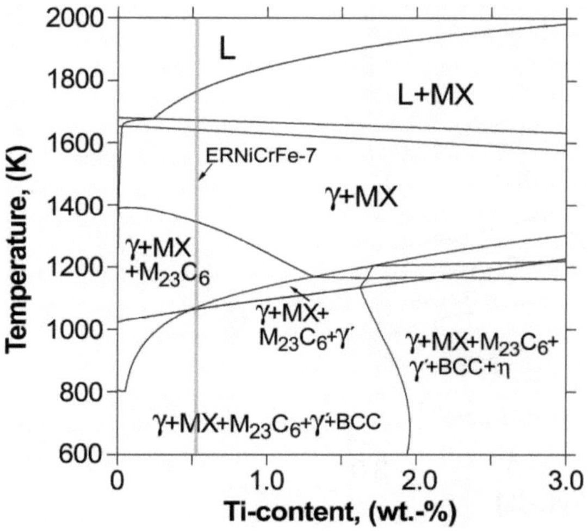


**Fig. 4.** Equilibrium phase domains as a function of temperature and addition of Ti to the Ni-base alloy ERNiCrFe-7

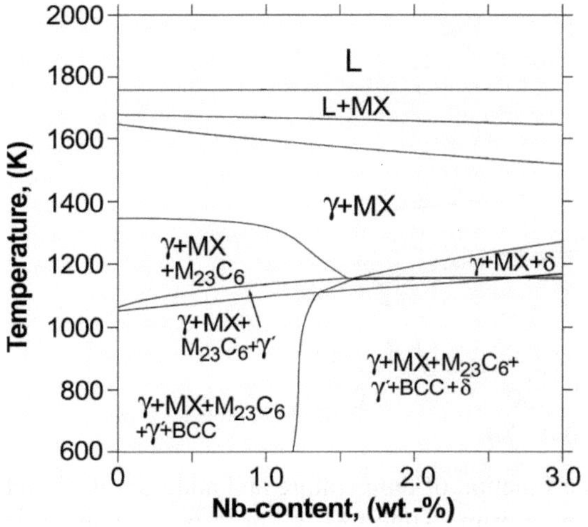


**Fig. 5.** Equilibrium phase domains as a function of temperature and addition of Nb to the Ni-base alloy ERNiCrFe-7

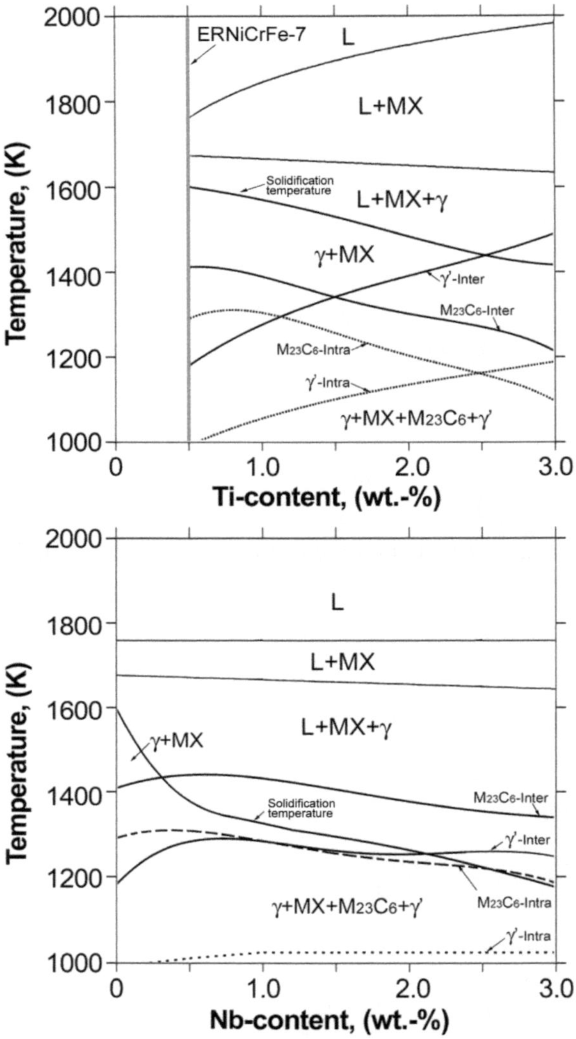


**Fig. 6.** Kinetic phase domains as a function of temperature and addition of Ti and Nb. L: liquid, γ: austenite, γ': gamma-prime, -Inter: $M_{23}C_6$ or γ' precipitated near the dendrite surface, -Intra: $M_{23}C_6$ or γ' precipitated at dendrite core

Alloying additions and the resultant segregation also induce a decrease on $M_{23}C_6$ precipitation temperature and strong increase on γ' precipitation temperature, with a stronger effect caused by Ti. As can be seen in the predicted kinetic diagrams the model accounts for intra-dendritic and inter-dendritic second phase precipitation.

## Micro-Segregation During Solidification

Figure 7 shows SEM micrographs of three representative samples, namely, master alloy ERNiCrFe-7, 3Ti and 3Nb. The low magnification images reveal the cellular solidification pattern. Also some details from the inter-dendritic (or intercellular) regions and large primary precipitates are presented. It should be highlighted in these images that not all the contrast observed at the inter-dendritic regions on the low magnification images is due to MX precipitates. In fact part of this contrast is due to segregation. In fact, EBSD analysis (not shown) revealed that these regions have the same crystallographic structure and orientation as the surrounding matrix. However, these highly segregated inter-dendritic regions probably have large amounts of nanometer-sized secondary precipitates, as will be presented later. Stereological analysis showed that the distance between dendrite (or cell) triple intersections follows a normal distribution, with average inter-dendrite spacing of ~7–12 μm.

Figure 8 shows computed average alloying content in the region near the dendrite surface, this region being defined as the 0.5 μm tick outmost layer of the dendrite (5% of the dendrite width). It is evident the accentuated tendency of Ti and Nb species to segregate towards the inter-dendritic regions, as well as to induce strong and moderate negative inter-dendritic segregation of Fe and Cr, respectively. On the other hand, C and N segregation towards the austenite inter-dendritic regions increases when ERNiCrFe-7 is modified with minor additions of Nb and/or Ti (up to ~1 wt.%), meanwhile this segregation decreases when Nb and/or Ti additions are intensified. Ti additions are more efficient inducing N depletion from austenite inter-dendritic regions than Nb additions. However, both Ti and Nb additions display a similar effect with regard to C depletion from austenite interdendritic regions. Notice that the C and N depletion from austenite inter-dendritic regions is a consequence of primary carbonitrides precipitation at this region.

This segregation has a strong effect on the inter- and intra-dendritic precipitation of MX carbonitrides, $M_{23}C_6$ and $\gamma'$ and these precipitates will have an effect on the DDC resistance of the microstructure and on the alloy performance. Therefore, a good understanding of the microstructure, chemistry, and kinetics of these alloys will provide important insights on the DDC phenomenon. The following sections describe the second phase precipitation based on the simulation and characterization work.

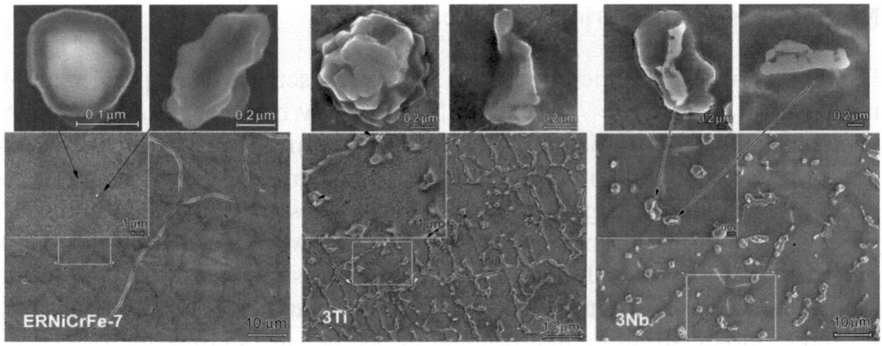


**Fig. 7.** SEM micrographs of three representative samples ERNiCrFe-7, 3 Ti, and 3 Nb. Detail of inter-dendritic regions

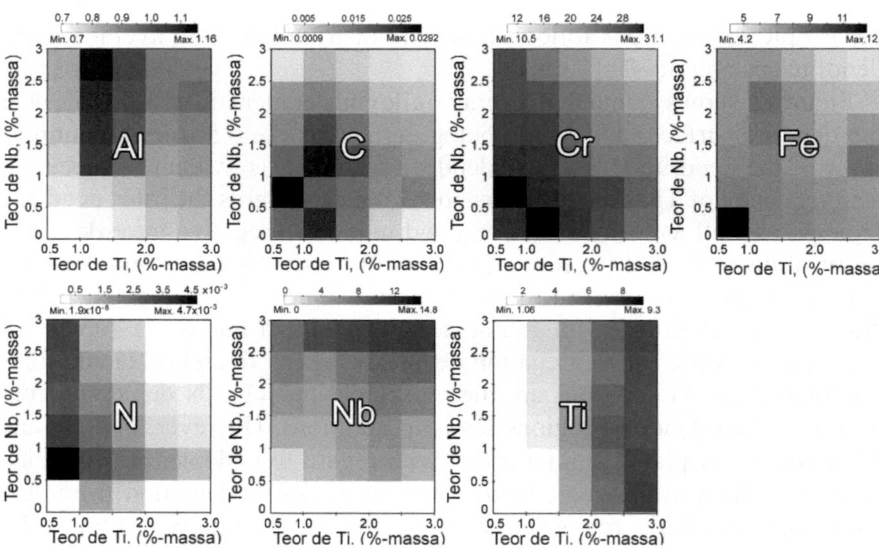


**Fig. 8.** Computed average alloying content at the 0.5 µm tick outmost layer of the austenite dendrite as a function of Ti and Nb additions

## Primary Carbonitride Precipitation

Figure 9 shows TEM micrographs and XEDS spectra of a MX precipitate observed in a carbon replica of unmodified ERNiCrFe-7 alloy. Figure 9d shows precipitate morphology; Fig. 9c shows the selected area diffraction (SAD) pattern at the axis zone $[011]_{MX}$ from the region arrowed in Fig. 9d; Fig. 4b shows a lattice resolution image of region arrowed in Fig. 9d;

Fig. 9a shows an XEDS spectrum colleted focusing a 10 nm diameter electron probe at the middle region of the shown MX precipitate. The SAD pattern reveals the FCC crystallographic structure with ~0.424 nm lattice parameter of the observed precipitate, which corresponds to the previously reported NaCl type-structure of MX carbonitride [23]. The Cu peak observed in the XEDS spectrum is due to the TEM Cu mesh upon which the carbon replica is supported.

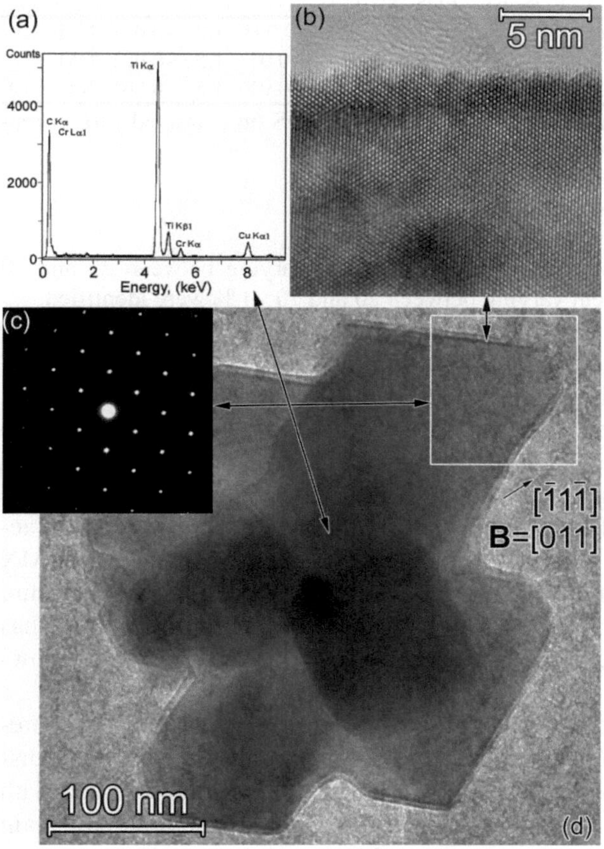


**Fig. 9.** MX precipitate observed in carbon replica of unmodified ERNiCrFe-7 sample. (a) XEDS spectrum colleted with a 10 nm diameter electron probe at the middle region of the particle; (b) lattice resolution image detail; (c) SAD pattern; (d) general view of precipitate morphology

Table 3 presents XEDS measurements and numerical simulations of average metallic fraction chemical composition of MX particles. The analysis of this chemical composition data reveals a minimal solubility of Ni, Fe,

Cr, and Al in the MX precipitates. The measured high Cr and Ni content on the MX precipitates was verified to come from a spurious X-ray signal generated in the surrounding carbon film containing un-dissolved matrix.

**Table 3.** TEM-XEDS Chemical analysis and numerical calculations (NC) of metallic elements fractions in MX primary precipitates (wt.%)

| Phase | Sample | Ti | | Nb | | Cr | | Ni | | Fe | | Al | |
|---|---|---|---|---|---|---|---|---|---|---|---|---|---|
| | | EDS | NC | EDS | NC | EDS | NC | EDS | NC | EDS | NC | EDS | NC |
| MX | MA/1Ti /3Ti | 90.0 | 99.8 | - - | - - | 4.5 | 0.2 | 4.5 | <0.001 | 1.0 | <0.001 | NM | $<10^{-8}$ |
| | 1Nb | 25 | 14.8 | 65 | 85 | 5.0 | 0.2 | 3.5 | <0.01 | 1.5 | <0.001 | NM | $<10^{-8}$ |
| | 1Nb-1Ti | 45* | 38 | 42* | 62 | 6 | 0.2 | 5 | <0.01 | 1.5 | <0.001 | NM | $<10^{-8}$ |

Notes: EDS: Experimentally measured by TEM-XEDS on extracted carbon replica;
NC: Numerical calculation;
MA: Master Alloy ERNiCrFe-7;
NM: Non measurable value;
* : MX particles with average Nb-content varying between 20 and 70 wt.% and Ti-content varying between 20 and 70 wt.% were identified.

An heterogeneous chemical composition pattern inside the MX precipitates is predicted by Scheil and diffusion calculations; a core consisting of titanium nitride TiN and a shell consisting of titanium and/or niobium carbide (Ti, Nb)C. This gradual core-shell chemical structure does not mean a change in the crystallographic structure since both phases have the same crystalline structure and very similar lattice parameter. Therefore, what actually happens is the gradual exchange of metallic specie Ti in MX stoichiometry for Nb and of non-metallic specie N for C as temperature decreases during cooling. Similar behavior on the MX precipitation has been previously reported on Ni-base alloys [24] and high strength low-alloy steels [25].

Table 4 summarizes the main stereological features of primary MX precipitates observed on the SEM. Primary MX precipitates were evenly dispersed within the microstructure, although minor presence of clusters with up to four precipitates was also observed (not shown). Preference for grain boundary precipitation was not observed. MX precipitates with near equiaxial, elongated and ramified morphologies were observed (Fig. 7). Nb additions induced a prevalence of both near equiaxial and elongated primary MX morphologies and Ti additions induced a prevalence of elongated and ramified morphologies. In general terms, in the samples alloyed with Nb the MX precipitates displayed irregular holes or imperfections (Fig. 7). The effect of alloying additions on observed MX morphologies can to a certain extent be explained by the relationship between the MX transformed fraction and temperature. Figure 10 presents the measured and calculated total volumetric fraction of primary MX precipitates present in the microstructure for different alloying additions.

**Table 4.** SEM-measured size and volumetric fraction of MX precipitates. Numerically computed volumetric fraction of MX particles are also shown for comparison

| Sample | | Average size [μm$^2$] | Vol.-fraction [vol.-%] | | Morphologic Character Distribution | | |
|---|---|---|---|---|---|---|---|
| | | | SEM | NC | Near equiaxial | Elongated | Ramified |
| ERNiCrFe-7 | | 0.035 ± 0.01 | 0.08 ± 0.04 | 0.10 | High | Low | Low |
| Ti | 0.7Ti | 0.05 ± 0.02 | 0.1 ± 0.05 | 0.12 | High | Low | Low |
| | 1Ti | 0.05 ± 0.02 | 0.1 ± 0.05 | 0.13 | High | Low | Low |
| | 1.5Ti | 0.07 ± 0.03 | 0.18 ± 0.07 | 0.19 | High | Moderate | Low |
| | 3Ti | 0.11 ± 0.05 | 0.33 ± 0.10 | 0.24 | Moderate | Moderate | Moderate |
| Nb | 0.5Nb | 0.08 ± 0.03 | 0.16 ± 0.07 | 0.10 | High | Moderate | Low |
| | 1Nb | 0.11 ± 0.05 | 0.22 ± 0.08 | 0.15 | High | Moderate | Low |
| | 1.5Nb | 0.14 ± 0.06 | 0.28 ± 0.11 | 0.19 | Moderate | Moderate | Low |
| | 3Nb | 0.18 ± 0.06 | 0.43 ± 0.14 | 0.24 | Low | High | High |

Notes: NC: Numerically computed.

As previously discussed, according to phase diagrams in Figs. 4 and 5, growth of MX during cooling takes place initially directly as precipitation inside the liquid phase, i.e. above γ solvus temperature, then cooperative growth of both MX and γ take place from the liquid phase. The primary MX formed above the γ solvus temperature will be called pro-eutectic MX and the primary MX formed below γ solvus will be called eutectic MX. Figure 11 shows the calculated volumetric faction ratio of pro-eutectic MX to total MX as a function of alloying. On the other hand, Figures 12 and 13 show the computed MX growth kinetics as a function of Nb and Ti content. The fraction of pro-eutectic MX with regard to the total MX smoothly increases with Ti additions and steeply decreases with Nb additions. On the other hand, the higher the alloying content the slower the proportional kinetics of MX growth. Thus, for high additions of Ti and Nb a significant fraction of the total MX grows at very low temperatures (~1100–1200°C). However, the total fraction of MX also increases with alloying content and consequently the reduction on the proportional kinetics should be carefully evaluated. Nevertheless, the proportional precipitation kinetics during the weld cooling is definitely slowed down by the higher Nb and Ti additions. Nb additions displayed a more accentuated effect than Ti additions on decreasing the fraction of pro-eutectic MX and retarding the proportional MX growth kinetics. The Ti addition effect on promoting higher pro-eutectic MX fractions can be understood in light of the predominant phase formed, which in this case is Ti nitride TiN and/or Ti carbonitride Ti(CN), different from the Nb additions that predominantly promote Nb Carbide NbC and/or Nb carbonitride Nb(CN). Therefore, TiN being a phase that becomes stable at higher temperatures than NbC, it is expected that higher fractions of pro-eutectic MX are formed when Ti instead of Nb is added to ERNiCrFe-7 alloy. However, Ti stabilizes γ′, which in principle is a undesired phase in this type of solid solution strengthen Ni-base alloy.

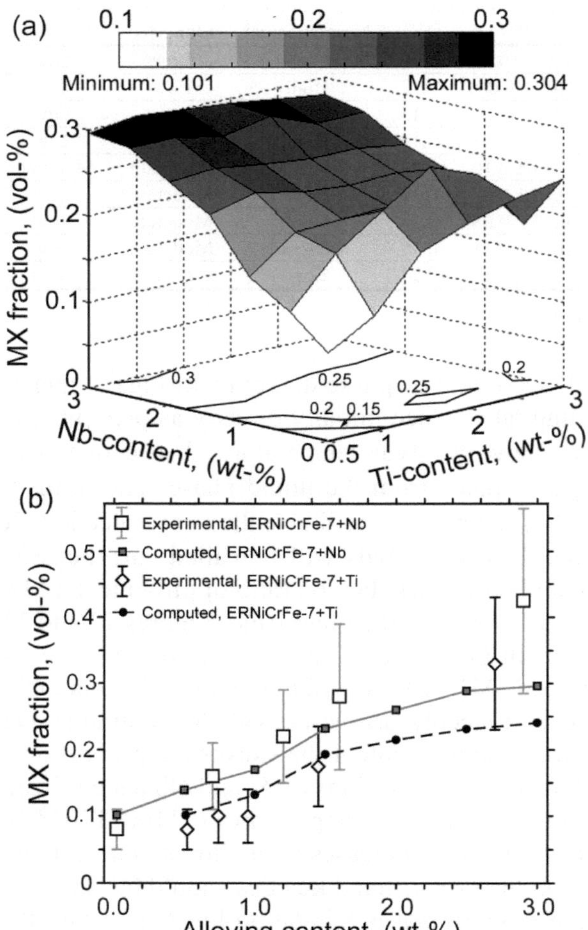


**Fig. 10.** (**a**) Computed MX fraction as a function of Ti and Nb additions; (**b**) computed and SEM-measured MX fraction as a function of Ti or Nb additions

In general terms, both the MX precipitate size and volumetric fraction were increased when the alloying addition was larger (Table 4). However, the increase in volumetric fraction was higher than the increase in average size. This is due to solute segregation at interdendritic regions that simultaneously increases driving force for MX precipitation and the number of cell-triple intersections, the density of nucleation sites and the concomitant number of particles per unit area being consequently increased. For modified alloys with high alloying contents (3Nb or 3Ti wt-%), the experimentally verified intensification of MX precipitation was significantly higher than the one numerically predicted, which most probably arises from errors associated with neglecting nucleation events in the numerical algorithm as well as with simplifying the analyzed 3D micro-segregation phenomena to a 1D system. However, an actually wider solubility range of either carbon or nitrogen inside the MX phase with regard to the one described by the Ni-data® database can also be a factor affecting the relationship predicted between alloying additions and primary carbonitride precipitation, i.e. the actual carbon and/or nitrogen content of MX precipitates can be lower than the one numerically predicted.

It is a reasonable approximation to relate the MX precipitates displaying irregular morphologies with eutectic particles precipitated at lower temperatures, where diffusion distances are shorter due to limited diffusivity of substitutional atomic species in austenite. On the other hand, it is a good approximation to relate the MX particles displaying regular, almost polygonal or equiaxial morphologies with pro-eutectic particles precipitated at high temperatures, where minimization of interfacial energy prevails over minimization of diffusion distances. Thus, it can be inferred that Nb and Ti additions induce an increase in MX particles with irregular morphologies due to MX particles in modified alloys having more eutectic character and they are precipitated at lower temperatures when compared to MX particles in the unmodified alloy. Irregular primary precipitates are expected to produce less cracking resistance microstructures due to the facilitated nucleation of cavities at such intergranular particles when compared to smooth round-like precipitates. Therefore, Nb and/or Ti additions should be optimized in order to have the right particle temperature formation, the right number and distribution of particles within the macrostructure and finally the right particle morphology. From the point of view of morphology, the lower the addition of these elements would produce needed round-like primary MX precipitates.

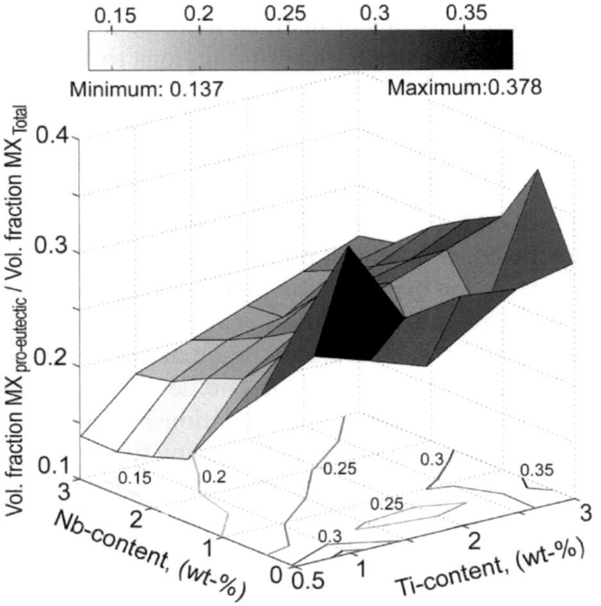


**Fig. 11.** Computed ratio between vol.-fraction of MX precipitated above γ-solvus temperature ($MX_{Pro\text{-}Eutectic}$) and total vol.-fraction of primary MX ($MX_{Total}$).

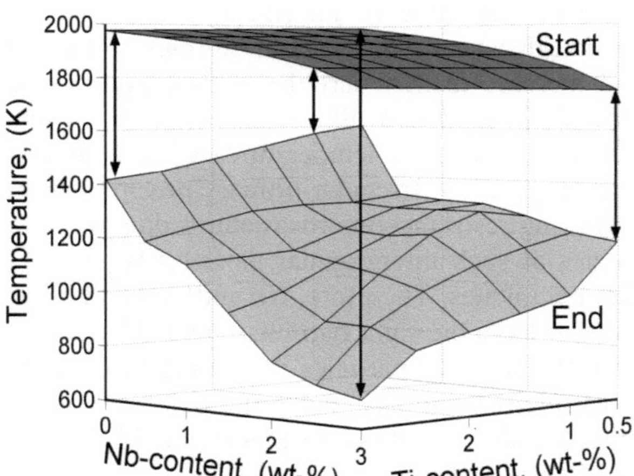


**Fig. 12.** Computed growth kinetics of primary MX. The start and end of MX precipitation are shown as a function of Nb and Ti content of the alloy and temperature. Notice the reverse order of alloying content axis.

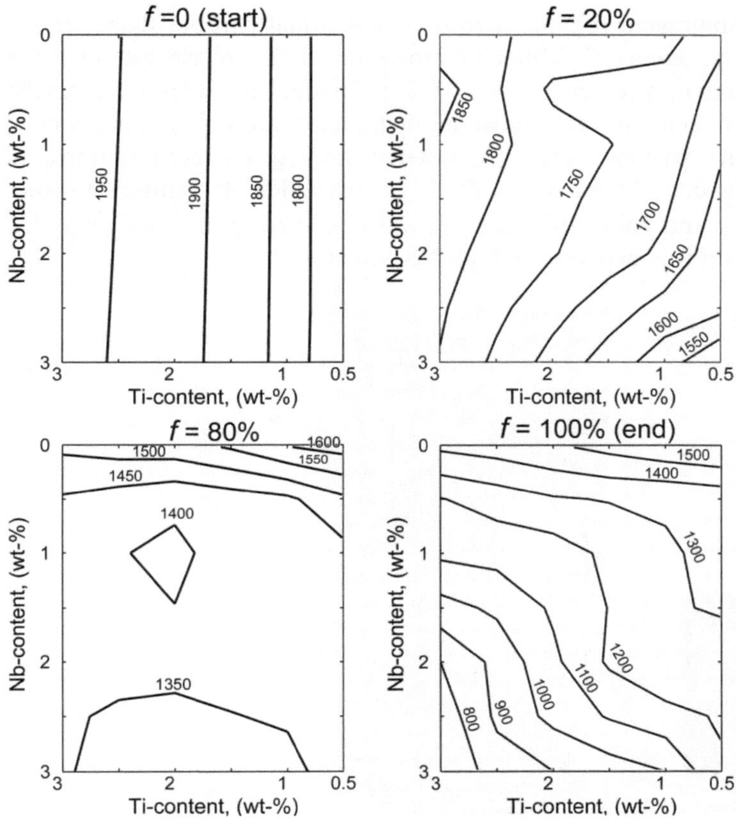


**Fig. 13.** Computed growth kinetics of primary MX. The 0%, 20%, 80% and 100% transformed fractions are shown as a function of temperature, and Nb and Ti content of the alloy. Notice the reverse order of alloying content axis

## Secondary Precipitation of $M_{23}C_6$ and γ'

As an example, Figs. 14 and 15 show some of the results from the TEM analyses of $M_{23}C_6$ and γ' precipitates observed in carbon replicas extracted from the master alloy ERNiCrFe-7 and the sample 1Nb-1Ti. There are shown in these figures the overall precipitate morphology, enlarged detail with lattice resolution image, SAD patterns of the region shown, and XEDS spectra colleted focusing an 5 nm diameter electron beam at the centre of the correspondingly enlarged second-phase particles.

There were not detected significant changes in $M_{23}C_6$ or γ' morphology as a consequence of alloying additions. $M_{23}C_6$ carbides precipitated in densely agglomerated regions with their individual particle shape slightly deferring from equiaxial. Minor precipitation of this phase was observed with a vol. fraction lower than ~0.01%. Individual particles were smaller than ~30 nm diameter and most of them were around 5 nm diameter. Gamma-prime particles also precipitated in agglomerated regions, although not as densely packed as the $M_{23}C_6$ carbides. A significant dispersion in $\gamma'$ size and volumetric fraction was observed. $\gamma''$ has not been identified; however it is expected in high Nb alloys.

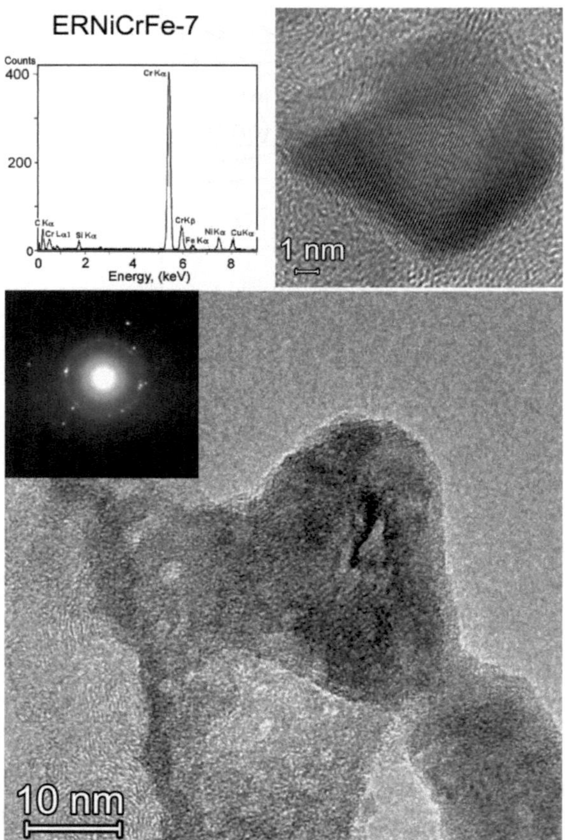


**Fig. 14.** $M_{23}C_6$ precipitates observed in carbon replicas from master alloy ERNiCrFe-7. XEDS spectra of precipitate colleted using 5 nm diameter electron beam at the center of precipitate shown on the lattice resolution image detail; SAD pattern

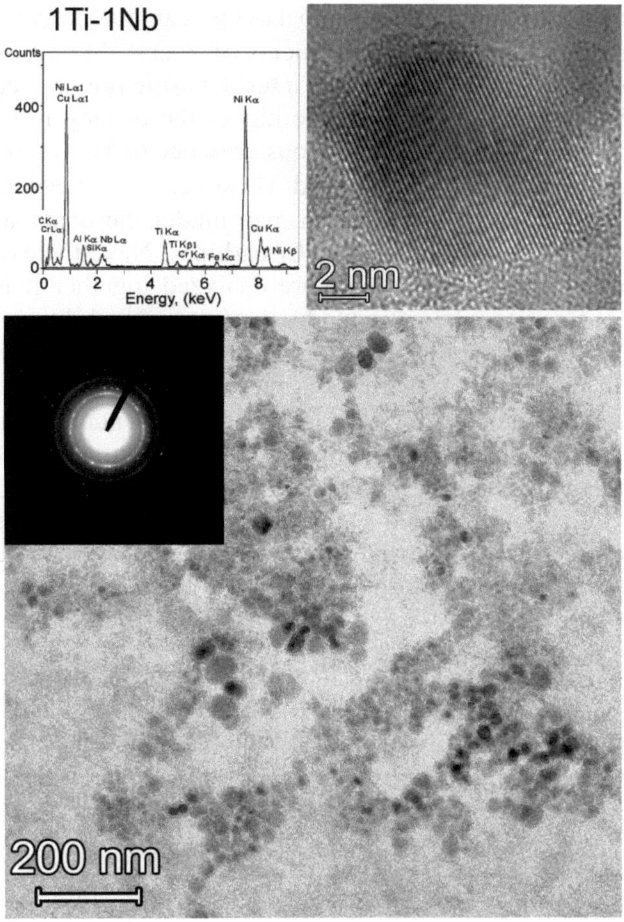


**Fig. 15.** γ' or γ" precipitates observed in carbon replicas from alloy 1Ti-1Nb. XEDS spectra of precipitate colleted using 5 nm diameter electron beam at the center of precipitate shown on the lattice resolution image detail; SAD pattern

Table 5 presents a summary of the semi-quantitative description of $M_{23}C_6$ and γ' particles analyzed in five samples. It is also presented the computed driving force for precipitation of secondary phases averaged at the region near the dendrite shell, i.e. the outmost layer equivalent to 5% of dendrite width. Care should be taken because the database used apparently does not differentiate between γ' and γ"; therefore, both phases may be mixed and will be called here γ'. Work is ongoing in this matter. Driving force for γ' precipitation steadily increases with alloying additions, whereas the driving force for $M_{23}C_6$ precipitation is minimum when high or non alloying additions are made. These two behaviors can qualitatively be explained by the average Ti- or Nb- (constituents of γ' and γ'') and C-content (constituent of $M_{23}C_6$) segregated to the inter-dendritic region (see Fig. 8). The higher the alloying segregation, the higher the driving force for secondary precipitation. Although, simultaneous presence of $M_{23}C_6$ and γ'was numerically predicted, it was not observed. However, it is expected that a microstructure containing γ' precipitation may hinder the observation and identification of small fractions of $M_{23}C_6$ carbides. Nevertheless, when the numerical and experimental results are analyzed together it is found that $M_{23}C_6$ precipitation can be inhibited by the faster γ' stability increase compared to the $M_{23}C_6$ stability rise (samples 1Nb, and 1Ti1Nb). This is probably due to the competition between these two phases for the nucleation sites, with $M_{23}C_6$ nucleation being overrun by γ' nucleation. The prevalence of γ′ nucleation is due to the lower volumetric atomic diffusion required and the lower energy interface formed when compared with $M_{23}C_6$ nucleation. It is not clear to the date if intergranular $M_{23}C_6$ are damaging for the DDC resistance. However, is believed that when these precipitates appear in the microstructure the ductility dip cracks have already had the chance to form.

**Table 5.** Semi-quantitative description of $M_{23}C_6$ and γ' particles experimentally observed and driving force for their precipitation near the dendrite surface at 1027°C and 727°C. Driving force, ΔG/RT, is reported as a dimensionless magnitude

| Sample | TEM Characterization | Driving Force, ΔG/RT | | | |
|---|---|---|---|---|---|
| | | $M_{23}C_6$ | | γ' | |
| | | 1027°C | 727°C | 1027°C | 727°C |
| FM52 | Minor precipitation of $M_{23}C_6$ | 0.13 | 0.96 | -0.12 | 0.23 |
| 1Ti | Minor precipitation of $M_{23}C_6$ | 0.18 | 1.01 | -0.12 | 0.24 |
| 1Nb | Minor precipitation of γ' | 0.22 | 1.12 | -0.04 | 0.30 |
| 1Nb-1Ti | Moderate precipitation of γ' | 0.37 | 1.32 | -0.004 | 0.33 |
| 3Ti | Profuse precipitation of γ' | -0.23 | 0.79 | 0.09 | 0.45 |

Note: Minor: aprox. between 0.001 and 0.01 vol.%.
Moderate: aprox. between 0.05 and 0.5 vol.%.
Profuse: aprox. between 1 and 10 vol.%.

XEDS spectra (Figs. 14 and 15) showed the occurrence of minimal solubility of Ni, Fe, Al, Nb and Ti in $M_{23}C_6$ and Fe and Cr in γ'. In a similar way as presented above for XEDS analysis of MX particles, it is expected that Ni and Fe content of $M_{23}C_6$ particles and Fe and Cr content of γ' particles are lower than revealed by the XEDS analyses, which are degraded by spurious X-ray signal coming form regions away from the analyzed particles. Minor changes of chemical composition inside second-phase precipitates were predicted by kinetic simulations, and therefore these results are not shown. These numerical results show that atomic element stoichiometry in second-phase precipitates can be described as $Cr_{26}C_6$ and $Ni_3$(Ti,Nb,Al). XEDS analyses and numerical simulations of average chemical composition of $M_{23}C_6$ and γ' revealed that Ti additions did not caused significant composition changes of $M_{23}C_6$ or γ'. In contrast, in Nb-modified alloys, the here so-called γ' particles displayed substitution of Ti and Al for Nb. Therefore, the new formed phase may be γ''.

## Summary

Based on the exposed results and previously reported data on the weldability and performance of alloy ERNiCrFe-7, it is possible to establish appropriate composition ranges for the modification of such alloy with Nb and Ti additions. Thus, the kinetic phase diagrams presented indicate that alloy ERNiCrFe-7 modified with Ti-content up to ~1.0 wt-% and Nb-content between ~0.5 and ~1.0 wt-% will present optimized as-welded microstructure with regard to DDC resistance and in some degree to corrosion and SCC resistances because of (i) $M_{23}C_6$ precipitation can be slightly inhibited with regard unmodified ERNiCrFe-7, (ii) although some γ′ precipitation is induced, still low driving force for γ' precipitation is established, and (iii) profuse MX precipitation above the ductility-dip temperature range (i.e. ~0.2–0.25 vol.% of MX above ~1150 °C) takes place. Intense precipitation of primary carbonitrides above the ductility-dip temperature range and minor precipitation of secondary chromium carbides during the whole welding cooling cycle are desired because [1-5] primary carbonitride precipitation results on increased DDC resistance due to the formed tortuous grain boundaries, whereas chromium rich carbide precipitation at low temperature induces chromium depletion of the matrix phase near grain boundaries and, consequently, may reduce corrosion resistance as well as SCC resistance of the alloy.

It was observed that Ti or Nb additions are useful to induce increasing precipitation of primary carbonitrides and simultaneously to inhibit secondary precipitation of chromium rich carbides. This is due to alloying additions increasing the fraction of C partitioned in the (Ti,Nb)C phase with regard to C partitioned in the γ phase. However, these Ti or Nb additions lead to strong atomic segregation near the interdendritic regions and, as a result, to both intense decreasing of solidification temperature and precipitation of N- and C-free second phase, namely γ', at the interdendritic regions in as-welded microstructures. However, high Ti additions compromise the thermal stability of this alloy due to γ'-induced precipitation hardening during following heat treatments and service.

## Acknowledgements

The authors would like to acknowledge the Brazilian Synchrotron Light Laboratory-LNLS, CNPq (contract 150523/2005-0), and FAPESP (contract 04/04526-8) for funding. Arc melting of the alloys was performed at State Univ. of Campinas with the invaluable help of Prof. S. Gama and Dr. A. Coelho. Special thanks to Sam D. Kiser from Special Metals Welding Products for providing the Ni-base alloys.

## References

1. Rhines FN, Wray PJ (1961) Investigation of the Intermediate Temperature Ductility Minimum in Metals. Trans. ASM 54: 117–128.
2. Yeniscavich W (1966) A Correlation of Ni-Cr-Fe Alloy Weld Metal Fissuring with Hot Ductility Behavior. Weld. J. 45(8): 344s–355s.
3. Ouchi C, Matsumoto K (1982) Hot ductility in Nb-bearing High-strength Low Alloy Steels. Trans. Iron Steel Inst. Japan 22(3): 181–189.
4. Mintz B, Yue S, Jonas JJ (1991) Hot Ductility of Steels and its Relationship to the Problem of Transverse Cracking During Continuous Casting. Inter. Mater. Rev. 36(5): 187–217.
5. Duvall DS, Owczarski WA (1967) Further Heat-Affected-Zone Studies in Heat-Resistant Nickel Alloys. Weld. J. 46(9): 423s–432s.
6. Collins MG, Lippold JC (2003) An Investigation of ductility Dip Cracking in Nickel-Based Filler Materials – Part I. Weld. J. 82(10): 288s–295s.
7. Collins MG, Ramirez AJ, Lippold JC (2003) An Investigation of Ductility-Dip Cracking in Nickel Based Weld Metals – Part II. Weld. J. 82(12): 348s–354s.
8. Collins MG, Ramirez AJ, Lippold JC (2004) An Investigation of Ductility-Dip Cracking in Nickel Based Weld Metals – Part III, Weld. J. 83(2): 39s–49s.
9. Ramirez AJ, Lippold JC (2004) High Temperature Behavior of Ni-base Weld Metal – Part I. Ductility and Microestructural Characterization. Mater. Sci. Eng. A. 380: 259–271.

10. Ramirez AJ, Lippold JC (2004) High Temperature Behavior of Ni-base Weld Metal – Part II. Insight Into the Mechanism for Ductility Dip Cracking. Mater Sci. Eng. A 380: 245–258.
11. Nishimoto K, Saida K, Okauchi H (2006) Microcracking in Multipass Weld Metal of Alloy 690 Part 1 – Microcraking Susceptibility in Reheated Weld Metal. Sci. Tech. Weld. Joining 11(4): 455–461.
12. Nishimoto K, Saida K, Okauchi H, Ohta K (2006) Microcracking in Multipass Weld Metal of Alloy 690 Part 1 – Microcraking Mechanism in Reheated Weld Metal. Sci. Tech. Weld. Joining 11(4): 462–470.
13. Capobianco TE, Hanson ME (2005) Auger Spectroscopy Results from Ductility Dip Cracks Opened under Ultra-High Vacuum. In: Proc. 7th Int. Conf. on 'Trends in Welding Research', Pine Mountain, GA, USA, May 2005, ASM International: 767–772.
14. Nissley NE, Lippold JC (2003) Development of the Strain-to-Fracture Test. Weld. J. 82(12): 355s–364s.
15. Ramirez AJ, Sowards JW, Lippold JC (2006) Improving the Ductility-Dip Cracking Resistance of Ni-Base Alloys. J. Mat. Proc. Tech. 179: 212–218.
16. Lippold JC, Nissley NE (2006) Further Investigations of Ductility-Dip Cracking in High Chromium, Ni-base Filler Metals. Document IIW-IIC-326-06: 1–10.
17. Nishimoto K, Saida K, Okauchi H, Otha K (2006) Microcracking in Multipass Weld Metal of Alloy 690 Part 1 – Prevention of Microcraking in Reheated Weld Metal by Addition of La to Filler Metal. Sci. Tech. Weld. Joining 11(4): 471–479.
18. Garzon CM, Ramirez AJ (2007) Second Phase Precipitation in As-Welded Microstructures of ERNiCrFe-7 Ni Alloy Modified with Ti and Nb Additions. Submitted to Mater. Sci. Eng. A.
19. Garzon CM, Ramirez AJ (2007) Kinetic phase diagrams for assessing second-phase precipitation during welding a Ni-30Cr-10Fe alloy modified with Ti and Nb additions. Submitted to Scripta Materialia.
20. Andersson JO, Ågren J (1992) Models for numerical treatment of. multicomponent diffusion in simple phases. J. Appl. Phys. 72: 1350–1355.
21. Borgenstam A, Engström A, Hoglund L, Ågren J (2000) J. Phase Equilibria 21: 269–280
22. Andersson JO, Helander T, Höglund L, Shi P, Sundman B (2002) Thermo-Calc & DICTRA, Computational Tools for Materials Science. Calphad 26: 273–312
23. Christensen AN (1978) Acta Chem. Scand. Ser. A 32(1): 89–90.
24. Huang X, Zhang Y, Liu Y, Hu Z (1997) Effect of Small Amount of Nitroten on Carbide Characteristics in unidirectional Ni-Base Superalloy, Metall. Mater. Trans. A 28A(10): 2143–2147.
25. Craven AJ, He K, Garvie LAJ, Baker TN (2000) Complex Heterogeneous Precipitation in Titanium–Niobium Microalloyed Al-killed HSLA Steels—1. (Ti,Nb)(C,N) Particles, Acta Mater. 48: 3857–3868.

# Index

MIX
Papier aus verantwortungsvollen Quellen
Paper from responsible sources
FSC® C105338

If you have any concerns about our products,
you can contact us on
**ProductSafety@springernature.com**

In case Publisher is established outside the EU,
the EU authorized representative is:
**Springer Nature Customer Service Center GmbH**
**Europaplatz 3, 69115 Heidelberg, Germany**

Printed by Libri Plureos GmbH
in Hamburg, Germany